Petroleum Contaminated Soils

Volume 3

Paul T. Kostecki
Edward J. Calabrese

Technical Editor
Charles E. Bell

School of Public Health
Environmental Health Sciences Program
University of Massachusetts, Amherst

Library of Congress Cataloging-in-Publication Data

(Revised for vol. 3)

Petroleum contaminated soils.

On spine: Contaminated soils.
"The proceedings of the Second National Conference on the Environmental and Public Health Effects of Soils Contaminated with Petroleum Products, which took place at the University of Massachusetts, Amherst, September 28, 29, and 30, 1987"—V. 1, pref.
"The proceedings of the Third National Conference on Petroleum Contaminated Soils held at the University of Massachusetts, Amherst, September 19–21, 1988"—V. 2, pref.
Vol. 3 contains the proceedings of a conference on hydrocarbon contaminated soils, held Sept. 1989, Amherst, Mass.
Includes bibliographies and index.
Contents: v. 1. Remediation techniques, environmental fate, risk assessment—v. 2. Remediation techniques, environmental fate, risk assessment, analytical methodologies—v. 3. Remediation techniques, environmental fate, risk assessment, analytical methodologies.
1. Soil pollution—Environmental aspects—Congresses. 2. Petroleum products—Environmental aspects—Congresses. 3. Risk assessment—Congresses. I. Kostecki, Paul T. II. Calabrese, Edward J., 1946- . III. National Conference on the Environmental and Public Health Effects of Soils Contaminated with Petroleum Products (2nd: 1987: University of Massachusetts) IV. National Conference on Petroleum Contaminated Soils (3rd: 1988: University of Massachusetts) V. Contaminated soils.
TD878.P47 1989 628.5′5 88-13987
ISBN 0-87371-135-1 (v. 1)

LEWIS PUBLISHERS, INC.
121 South Main Street, Chelsea, Michigan 48118

PRINTED IN THE UNITED STATES OF AMERICA

Preface

Hydrocarbon contamination of soils has been a major focus of regulatory agencies and the research community for the past decade. While the research base upon which to provide public health and regulatory guidance to state and federal agencies was extremely limited in the early to mid portion of the 1980s, the past five years have witnessed a rapid maturation in the field, especially with respect to analysis, fate modeling, public health risk assessment, and remediation. The field has also seen the creation and development of an international advisory organization (CHESS—The Council for Health and Environmental Safety of Soils) comprised of representatives of the federal and state governments, the private sector, and academia. CHESS is designed to provide technical information exchange to affected parties of soil contamination and has as its ultimate goal the development of a clear and adaptable decision framework methodology for assessing and remediating soil contaminated sites.

This book, the proceedings of the fourth national conference on hydrocarbon contaminated soils (September 1989, Amherst, MA), is designed to provide a benchmark for the 1990s, since it captures and assesses the progress of the 1980s while helping to shape the directions of regulatory and research initiatives of the 1990s. The conference assesses where the field is with respect to the identification of critical current issues as well as emerging concerns (Chapter 1–6), analytical methods and environmental fate (Chapters 7–12), remedial techniques (Chapters 13–19) risk assessment (Chapters 20–25), and regulatory considerations (Chapters 26–28).

The book, therefore, represents a carefully interwoven synthesis of this maturing field, offering not only a historical perspective but also a perspective on future developments, all within the context of practically oriented solutions. Thus, this book will assist those facing the challenge of effectively solving the problem of petroleum contaminated soils, including the federal and state regulatory community, public health officials and affected industries, as well as the environmental consultant community. It is these groups who are deeply involved with developing creative, cost-effective assessments and solutions able to withstand the demands of regulatory requirements. In addition, this book gives students a broad perspective on the problem by integrating the various dimensions into an intelligible whole so that a proper, balanced appreciation of the problem will be discernible.

Acknowledgments

The Conference was a tremendous success because of the hard work, dedication, and support of many people and organizations. Special thanks to the following friends of the conference:

Dir. Barry Johnson and Cynthia Harris
Agency for Toxic Substances and Disease Registry
1600 Clifton Road, Northeast
Atlanta, GA 30333

Bruce Bauman and Members of the Groundwater Task Force
American Petroleum Institute
1220 L Street, Northwest
Washington, DC 20001

Chris Barkan
Association of American Railroads
50 F Street, Northwest
Washington, DC 20001

Vice Pres. Thomas Veratti
CON-TEST
39 Spruce Street
East Longmeadow, MA 01028

James Dragun
The Dragun Corporation
3240 Coolidge
Berkley, MI 48072

Vice Pres. Willard Potter and Ken Goldbach
Dunn Geoscience Corporation
Lincoln Center
299 Cherry Hill Road
Parsippany, NJ 07054

Vice Pres. Steven Jinks
EA Engineering, Science, & Technology, Inc.
Maple Building
3 Washington Center
Newburgh, NY 12550

Chr. Joseph Shefcek, John Novak, and Gayle Harreld
Utility Solid Waste Activities Group
Edison Electric Institute
1111 19th Street, Northwest
Washington, DC 20036

Mary McLearn and Gordon Newell
Electric Power Research Institute
3412 Hillview Avenue
Palo Alto, CA 94303

Ronald Brand and Vinay Kumar
EPA/Office of Underground Storage Tanks
401 M Street, Southwest
Washington, DC 20460

Pres. John Frawley
International Society of Regulatory Toxicology & Pharmacology
6546 Belleview Drive
Columbia, MD 21046

Dep. Commr. Ken Hagg, James Coleman, and Susan Fessenden
MA Department of Environmental Protection
1 Winter Street
Boston, MA 02108

Vice President Thomas Swoyer and Kate Strauser
Roy F. Weston Inc.
Weston Way
West Chester, PA 19380

Irwin Baumel
U.S. Army Biomedical Research and Development Lab
Fort Detrick
Frederick, MD 21701

Paul T. Kostecki, a Senior Research Associate/Adjunct Faculty in the Environmental Health Sciences Program at the University of Massachusetts, Amherst, received his PhD from the School of Natural Resources at the University of Michigan in 1980. He has been involved with risk assessment and risk management research for contaminated soils for the last five years, and is coauthor of *Remedial Technologies for Leaking Underground Storage Tanks* and coeditor of *Soils Contaminated by Petroleum Products* and *Petroleum Contaminated Soils,* Vols. I and II. Dr. Kostecki's yearly conference on petroleum contaminated soils draws hundreds of researchers and regulatory scientists to present and discuss state-of-the-art solutions to the multidisciplinary problems surrounding this issue. Dr. Kostecki also serves as Managing Director for the International Society of Regulatory Toxicology and Pharmacology's Council for Health and Environmental Safety of Soils (CHESS).

Edward J. Calabrese is a board certified toxicologist who is professor of toxicology at the University of Massachusetts School of Public Health, Amherst. Dr. Calabrese has researched extensively in the area of host factors affecting susceptibility to pollutants, and is the author of more than 240 papers in scholarly journals, as well as 10 books, including *Principles of Animal Extrapolation, Nutrition and Environmental Health,* Vols. I and II, *Ecogenetics, Safe Drinking Water Act: Amendments, Regulations and Standards, Petroleum Contaminated Soils,* Vols. I and II, and *Ozone Risk Communication and Management.* He has been a member of the U.S. National Academy of Sciences and NATO Countries Safe Drinking Water committees, and most recently has been appointed to the Board of Scientific Counselors for the Agency for Toxic Substances and Disease Registry (ATSDR). Dr. Calabrese also serves as Chairman of the International Society of Regulatory Toxicology and Pharmacology's Council for Health and Environmental Safety of Soils (CHESS).

Contents

PART I
ISSUES AND PERSPECTIVES

PART II
STATE OF RESEARCH

PART III
ANALYTICAL AND ENVIRONMENTAL FATE

PART IV
REMEDIAL TECHNOLOGIES

PART V
RISK ASSESSMENT

PART VI
REGULATORY CONSIDERATIONS

Petroleum Contaminated Soils

Volume 3

PART I

Issues and Perspectives

CHAPTER 1

Evaluating Migration of Petroleum Products in Soil to Determine Public Health Implications: The Health Assessment Process

Chebryll J. Carter, Agency for Toxic Substances and Disease Registry, Public Health Service, Atlanta, Georgia

The release of hazardous substances into the environment is becoming an increasingly widespread problem. The Agency for Toxic Substances and Disease Registry (ATSDR) was created to implement activities that will help to prevent or mitigate the adverse human health effects and diminished quality of life resulting from exposure to hazardous substances in the environment. One of the principal mechanisms used to achieve this goal is the health assessment process. The process is used as a tool to determine how environmental contamination can reach human populations and cause potential human health effects.

Spills, leaks, wastewater discharge, vapor/gas emissions, and similar releases related to petroleum refining and other petrochemical processes are contributing to the environmental contamination. Soil is one of the major environmental media that is being contaminated by such releases. Petroleum products have been identified as soil contaminants at many hazardous waste sites.

In this chapter, I will give the reader a public health perspective on evaluating the fate and transport of petroleum products in the soil as part of the Health Assessment process. I will also attempt to explain how this evaluation ultimately translates into assessing the public health risk associated with exposure to these types of contaminants.

THE ATSDR HEALTH ASSESSMENT

An ATSDR Health Assessment is the evaluation of data and information on the release of hazardous substances into the environment in order to assess any current or future impacts on public health, develop health advisories or other recommendations, and identify studies or actions needed to evaluate and mitigate or prevent human health effects. This statement is the formal definition of an ATSDR Health Assessment. A simpler definition is that a Health Assessment is an evaluation designed to answer the following questions, which are most frequently asked by people living around hazardous waste sites:

1. What contaminants are in the onsite and offsite environmental media?
2. What concentrations of contaminants have been detected?
3. Where are these contaminants located with respect to human populations?
4. Can the contaminants get to me?
5. If so, will my family get sick?
6. What is being done about the contamination?

To answer these questions, the health assessor must consult a variety of information sources to gain knowledge of the characteristics of the site. Information about the site's history, future, and past land uses in the area, community-specific databases for health outcomes, and other site-related factors (e.g., site access, ground cover) must be collected to evaluate the public health implications of the site.

CONTAMINANTS OF CONCERN (ONSITE AND OFFSITE) AND THEIR CONCENTRATIONS

What contaminants are in the onsite and offsite environmental media? What concentrations of contaminants have been detected?

Once the site characterization information has been evaluated, the health assessor must determine which contaminants present at the site are most likely to pose a public health concern. Such chemicals and compounds as benzene, xylene, polyaromatic hydrocarbons (PAHs), toluene, and metals are potential contaminants of concern when petroleum products have been released into the environment. Identifying the contaminants of concern is an iterative process, during which the following elements are examined: the range of contaminant concentrations present at the site, the number of times a contaminant was detected at levels of public health concern during a particular sampling event, the quality of environmental sampling data, and the potential for human exposure to those contaminants present.

The health assessor must also evaluate the physical and chemical properties of the contaminants to determine the potential the chemicals have to migrate and contaminate offsite environmental media. Water solubility, vapor pressure, biological concentration factors, soil/water partition coefficients, volatility, and biodegradation factors are just some of the relevant factors that greatly influence not only the time it takes for a contaminant to move in an environmental medium, but also influence what quantity of the contaminant will potentially be present in a particular environmental medium.

In addition, background levels are used to determine if the concentrations of contaminants detected at the site are within the range of typical values for the state, region, or nation. Contaminants that exceed maximum reported background levels should be listed as contaminants of concern. At some sites, chemical concentrations that are considered background may be at levels that are of potential concern to public health. In these instances, both the background and site-related contaminant concentrations are considered in determining the contaminants of concern.

PATHWAYS TO HUMAN EXPOSURE

Where are these contaminants located with respect to human populations? Can the contaminants get to me?

Identifying the environmental pathways of concern is the next step in the health assessment evaluation process. The term environmental pathways is used to describe the transformation and transport processes that could affect the contaminants of concern. These pathways are significant because they determine how individuals may come into contact with site-related contaminants.

Current and potential contamination of surface water, groundwater, soil, air, consumable plants, consumable animals, and animal products are all considered in the evaluation. Environmental pathways that should be considered involve the transport of gases, liquids, and particulate solids within a given medium, across the interfaces between air, water, and soil. For the purposes of this chapter, soil is the principal environmental medium of concern, and the specific environmental pathways that are considered of concern are contaminant migration from surface soil to air, surface soil to subsurface soil, and subsurface soil to groundwater.

The fate and persistence of chemicals in soil is a complex function of physical, chemical, and biological factors—the main ones being the specific chemical and its formulation, soil class, organic composition, specific physicochemical properties, type of cover vegetation, degree of soil cultivation, and nature of the microbial population.[1] For instance, the soil class (i.e., dominant grain size or grade) may help the health assessor determine permeability, or how persistent contaminants will be in the soil. The organic composition of the soil may influence how tightly some contaminants will bind to the soil. Porosity, absorption, and cation exchange capacity are also used to determine migration potential in the soil.

Hazardous materials will react differently in different soil-solvent systems. For example, the heavy metals found in petroleum products will tend to sorb to soil particles, making the metals virtually immobile. On the other hand, the volatile organic compounds, such as benzene and toluene, will tend to evaporate and volatilize from the surface soil into the air or migrate from subsurface soils into the groundwater. The soil environment represents a significant transport pathway for many volatile hazardous materials.

As previously mentioned, both the air and groundwater can be contaminated by chemicals migrating from the soil. In addition to these environmental media, the assessor may also consider surface water and sediment as potential routes of contaminant transport. Runoff from the site may transport surface soils off-site and potentially contaminate both sediment and surface water. Furthermore, food-chain contamination may occur through uptake of contaminants by plant roots and bioaccumulation in plants and in animals that consume these plants.

If the soil is found to have elevated concentrations of petroleum products and if there are completed environmental pathways, the assessor then determines if there are completed human exposure pathways. For soil, one would be concerned with ingestion (whether it be inadvertent or pica), inhalation of volatilized contaminants and contaminated reentrained dust particles, and direct contact.

PUBLIC HEALTH IMPLICATIONS

Will my family get sick?

The next step in the health assessment process is to evaluate the public health implications associated with the site. This process includes evaluating the toxicity of contaminants at the concentrations found in the soil; determining the presence of completed human exposure pathways; and, then, assessing possible health effects associated with exposure to the contaminants of concern at the levels detected.

Most people living in the vicinity of hazardous waste sites are exposed to chemical contaminants at low concentrations for long periods of time. The responses to exposure to this type of low-level, long-term exposure, if any, are likely to be nonspecific, i.e., an increase in the frequency of chronic diseases that are also present in nonexposed populations.[1] Low-level chemical exposures may play contributory, rather than primary, roles in the causation of an increased disease incidence, or they may not express their effects without the coaction of other factors.[1]

An integral part of this phase of the health assessment process is reviewing and evaluating health-related databases to assess past and current health-outcome parameters associated with populations living near the site. Evaluating health-outcome parameters is a community-specific, interactive process that involves the health assessors, data source generators, and the targeted community. Health-outcome databases to be considered include medical records, morbidity and mortality data, tumor and disease registries, and surveillance data.

FOLLOW-UP ACTIVITIES

What is being done about the contamination?

Both health- and environmental-related follow-up actions can be initiated as a result of the health assessment evaluation process. These follow-up activities are included in the recommendations and conclusions sections of the Health Assessment.

Health-related follow-up actions may include such measures as (1) writing Public Health Advisories to reflect concerns related to human exposure to site-related contaminants; (2) initiating emergency actions such as the evacuation of residences; (3) fencing the site to limit access to contaminated areas; and (4) determining whether health studies or other follow-up activities (i.e., health surveillance, exposure assessments, exposure and disease registries, epidemiological investigations) need to be undertaken. Environmental-related follow-up actions may include (1) requests for additional sampling to better characterize the nature and extent of contamination in a given environmental medium; (2) emergency soil removal or soil remediation; and (3) recommendations to the Environmental Protection Agency (EPA) or state designed to help determine which remedial alternatives will protect human health. ATSDR does not decide which of the alternatives will be implemented at the sites; EPA is responsible for choosing the remedial alternatives.

SUMMARY

In conclusion, the ATSDR Health Assessment is an analysis of the public health implications of the facility or release under consideration. It is based on professional judgment and weight of evidence analysis, and is part of the public record. Ultimately, the Health Assessment is conducted to help provide answers to the questions that people living in the area of the hazardous waste site will ask, and to prevent the diminished quality of life that can be caused by hazardous waste in the environment.

REFERENCES

1. Lippmann, M., and R.B. Schlesinger. *Chemical Contaminants in the Human Environment* (New York: Oxford University Press, 1979), pp. 170–175.

CHAPTER 2

EIL Insurance for Underground Storage Tanks: The Changing Market

William P. Gulledge, Front Royal Group, Inc., McLean, Virginia

Three major changes have occurred in the environmental impairment liability (EIL) insurance market over the past year. While EIL coverage for industrial exposures is still quite limited, more insurance carriers are now offering coverage for underground storage tank (UST) exposures. The second major change that has occurred focuses on differing underwriting requirements that UST owners and operators are subjected to in order to obtain an EIL policy. Finally, the growth of state funds to cover all or portions of the federal financial responsibility requirements has impacted the insurance marketplace in many states and presented more options to owners and operators.

THE INSURANCE MARKET

Approximately one year ago, there were only three to six major carriers offering EIL coverage for USTs. With the implementation of the January and October 1989 financial responsibility deadlines, more carriers have entered the UST insurance market. Today, approximately 15 companies offer full or partial coverage to meet the Environmental Protection Agency's (EPA's) financial responsibility requirements. The major companies are listed in Table 1. Any new insurance product, such as EIL insurance for USTs, goes through a shakedown period in

Table 1. UST Insurance Market Contacts

Agricultural Excess and Surplus Insurance Co. 515 Main Street/P.O. Box 2575 Cincinnati, OH 45201 513-369-5880
American Home Assurance Company (AIG) 70 Pine Street New York, NY 10270 202-770-5398
Environmental Protection Insurance Co. (Risk Retention Group) 111 Canal Street Chicago, IL 60606 312-715-0800
Federated Mutual Group 121 East Park Square Owatonna, MN 55060 800-533-0472
Front Royal Group, Inc. 7900 Westpark Drive, Suite A300 McLean, VA 22102 703-893-0900
James Group Services 230 West Monroe Street, 9th Floor Chicago, IL 60606 312-236-0220
Oilmen's Insurance 350 Fifth Avenue, Suite 6805 New York, NY 10118 212-629-4290
Petroleum Marketers Mutual Insurance Co. (PETROMARK) (Risk Retention Group) c/o The Planning Corporation 11347 Sunset Hills Road Reston, VA 22090 703-481-0200
Shand Morahan & Co. Shand Morahan Plaza Evanston, IL 60201 312-866-0716

which some carriers quickly enter and leave the market, and other carriers intend to offer the coverage for the long term.

The most notable exit from UST insurers during the past year was the Pollution Liability Insurance Association (PLIA). For a variety of reasons, PLIA chose not to renew existing policies or offer coverage to new owners and operators.

Other carriers who have been writing EIL coverage for several years are in the process of changing their EIL policy conditions to meet EPA UST insurance requirements. Requirements currently being addressed include coverage for onsite contamination, defense costs outside the primary limits of liability, and provisions for an extended claims reporting period. The EPA is also expected to revise the regulations on providing notice to the insured in the event a policy is cancelled for nonpayment of premium.

The number of insurance carriers offering UST EIL insurance is expected to stabilize at the current number of carriers. The final two financial responsibility deadlines, April and October, 1990, will likely generate additional interest among larger multiline insurance companies, but it is doubtful many of these larger carriers will enter the market. The vast majority of policies will continue to be written by specialty carriers.

COPING WITH DIFFERING UNDERWRITING REQUIREMENTS

Concurrently with the growth in insurance providers for UST exposures comes the growth in underwriting requirements. During the last year, underwriting criteria or technical standards used to define acceptable and preferred risks have not changed dramatically. Newer and upgraded UST components are still considered better environmental risks, and the owners of these facilities generally pay lower premiums for UST coverage. Premiums still range from a low of approximately $1,500 per retail petroleum site to a high of $10,000 for a higher risk site. The biggest change has come in the amount of up-front environmental data that is required to review a site.

More carriers are requiring technical data that goes beyond the EIL application itself. These data may include precision tests of tanks and piping, soil gas readings from the pump island and tank excavation areas, site assessments focusing on all potential environmental exposures from site operations and the surrounding area, and computerized reconciliation of product inventory records. Requiring this information has generated controversy with UST owners and operators. Requirements may differ from one insurance provider to the next, and most owners do not wish to duplicate several types of environmental analyses for each site desiring insurance coverage.

It is clear that the environmental data are useful and necessary to evaluate a site for EIL underwriting. The costs are being passed on to the owners of the facilities, and they ultimately will pass the costs on to the consumer. Environmental analyses are yielding some very positive benefits. While directly serving as both an insurance underwriting tool and a basis for the site owner to develop an upgrade plan, indirectly the site assessments and environmental monitoring techniques are uncovering incidents of past contamination. (Past contamination incidents may or may not be covered within an EIL policy.) Some carriers can expect additional claims for cleanup of past contamination, and premiums will rise accordingly.

GROWTH IN STATE FUNDS TO PROVIDE FINANCIAL RESPONSIBILITY COVERAGE

Encouraged by state-based petroleum marketing associations, many states during the past year passed legislation creating state supported pollution cleanup funds and financial responsibility programs. In 1988, only 5 or 6 states had such programs. By the end of 1989, 37 states had a cleanup fund or an insurance-like program in effect, or were in the process of creating a program. Table 2 provides a summary of existing and possible future state programs.

There is no standard characteristic state program. Many were developed from legislation that resulted from a perception that insurance to fulfill financial responsibility requirements was not available. As mentioned earlier in this discussion, this perception is not accurate. Many state programs have also been developed with mixed objectives and, sometimes, goals that are contradictory. Some states have created cleanup funds that are designed to pay for the cleanup of prior contamination incidents. The state may hire a cleanup contractor directly, or the site owner may hire a contractor and then be reimbursed by the state for his expense. In either case, the state must administer the cleanup fund and track all expenses.

Several states with cleanup funds have mixed program objectives by also including financial responsibility coverage. The goal of financial responsibility is to provide coverage for future contamination incidents. The states that have cleanup funds to address old contamination mixed with financial responsibility coverage have confused objectives. Several states have exacerbated the problem by including amnesty provisions or state cleanup for any petroleum-based pollution event, regardless of when it occurred or was discovered.

Each element of state financial responsibility programs differs from one state to the next. Many states provide primary insurance coverage with deductibles of $5,000 or $10,000. One state hopes to act as a reinsurer to state approved commercial primary carriers. The limits of liability and coverage can also differ dramatically from state to state. All programs cover cleanup costs, but many states do not include third party liability coverage as required under the federal regulations. Some states have separate limits for cleanup and third party liability, while others combine these limits. Most programs provide the full $1,000,000 limit for cleanup, but several states do not.

State funds are confusing to both owners who maintain UST facilities in multiple states and insurers attempting to compete or wrap their insurance program around a state fund. Two big items concern both parties. First, how well is the state program capitalized to adequately handle all pollution claims? Many state programs are funded by a wholesale tax on gasoline distribution. While presently maintaining a surplus of funds, these states are experiencing a steady increase in claims against the cleanup fund or financial responsibility program. Claims management has been poorly done on the state level, so reserving, as a commercial insurance entity would do, is not being practiced in the state program. With this trend continuing, one would expect many funds to be under-capitalized over a period of three to five years.

Table 2. State Financial Assurance Funds (As of November 17, 1989)

State	Facilities	S.I.R. Cleanup/Third Party	Fund Coverage: Cleanup	Fund Coverage: Third-Party	Comments
AL	N/A	$5K/$5K	$1,000,000	Included	All-in
AK	N/A	N/A	N/A	N/A	Fund proposed in '90
AR	N/A	$50K/50K + >550K	$1,050,000	$50K to $550K	All-in
AZ	N/A	N/A	N/A	N/A	Fund proposed in '90
CA	N/A	N/A	N/A	N/A	Fund proposed
CO	N/A	$10K/$25K	$1,000,000	Included	< = 100 USTS = $1 mil. agg. > 100 USTS = $2 mil. agg.
CT	N/A	$10K	$1,000,000	Included	All-in
DE	N/A	$100K/$300K	$1,000,000	$5,000,000	Optional participation
DC	N/A	N/A	N/A	N/A	No Fund
FL	N/A	$500/$500	$1,000,000	$1,000,000	Third party required to get cleanup coverage
GA	N/A	$10K/$10K	$1,000,000	Included	Optional participation $1–$2 mil. yearly agg.
HI	N/A	N/A	N/A	N/A	Fund being considered
ID	N/A	N/A	N/A	N/A	No fund
IL	N/A	$10K	$1,000,000	Included	All-in
IN	N/A	$100K	$1,000,000	N/A	No third party, all-in
IA	N/A	Greater of $5K or 25% of cleanup variable for third	No individual limit, $6 million limit on all cleanups	Up to federal requirements	Third party optional cleanup exp. 10-26-90
KS	1–12 USTs	$10K	$1,000,000	N/A	No third party
	13–99 USTs	$20K	$1,000,000		$1–$2 mil. yearly agg.
	100 + USTs	$60K	$1,000,000		
KY	N/A	N/A	N/A	N/A	Fund proposed in '90
LA	N/A	$10K/$10K	Federal limit	Federal limit	All-in
ME	N/A	N/A	N/A	N/A	No fund
MD	N/A	N/A	N/A	N/A	Fund proposal veto in '89

Table 2. Continued

		S.I.R.	Fund Coverage		
State	Facilities	Cleanup/Third Party	Cleanup	Third-Party	Comments
MA	N/A	N/A	N/A	N/A	Fund proposed
MI	1–100 USTs	$10K	$1,000,000	Included	$1–$2 mil. yearly agg.
	101 + USTs	$10K	$1,000,000	Included	All-in
MN	N/A	10% of first $250K, all over $250 K	90% of first $250K of costs	Included	Fund covers 90% of cleanup and third <$250K
MO	N/A	$25K/$100K 50% btwn $25K & $50K 25% btwn $50K & $100K	$1,000,000	$1,000,000 but not bodily injury	Third party covers only property damage, optional participation
MS	N/A	$100K/$300K	$1,000,000	$1,000,000	All-in, amnesty exp. 5-18-90
MT	N/A	50% of <$35K	$1,000,000	Included	All-in
NE	N/A	First $10K, 25% of rest*	$1,000,000	N/A	No third party, *max. S.I.R. = $25K
NV	N/A	$25K/$25K	$1,000,000	$1,000,000	$2 mil./$2 mil. annual agg. for cleanup/third all-in, ASTs optional
NH	1	$5K/$5K	$1,000,000	Included	All-in
	2–19	$20K/$20K	$1,000,000	Included	
	20+	$30K/$30K	$1,000,000	Included	
NJ	N/A	N/A	N/A	N/A	No fund
NM	N/A	$25K, 50% over $200K, max. of $425K S.I.R.	Limit set by Div. of Environment Improv.	All bodily injury and some property damage not covered	All-in
NY	N/A	N/A	N/A	N/A	Fund proposed
NC	N/A	$50K/100K	$1,000,000	Included	All-in
ND	N/A	$7.5K, 10% of $ <$100K, all over $100K	90% of cleanup costs btwn $7.5K–$100K	N/A	No third party, all-in
OH	N/A	$50K combined	$1,000,000	Included	Annual aggs. based on 100s of USTs, start $1 mil. up to $4 mil.
OK	N/A	$10K	$1,000,000	Included	All-in

Table 2. Continued

State	Facilities	S.I.R. Cleanup/Third Party	Fund Coverage Cleanup	Fund Coverage Third-Party	Comments
OR	N/A	N/A	N/A	N/A	No fund
PA	N/A	$75K/$150K	$1,000,000	Included	All-in
RI	N/A	N/A	N/A	N/A	Fund proposed in '90
SC	N/A	$100K/$300K	$1,000,000	Included	All-in, amnesty exp. 12-31-89
SD	N/A	$10K over $100K	Cleanup costs btwn 10K & 100K	N/A	No third party, All-in
TN	N/A	$50/$150K	$1,000,000	$1,000,000	All-in
TX	N/A	$10K	$1,000,000	N/A	No third party
UT	N/A	$25K/over $300K	$1,000,000	First $300K	All-in
VT	N/A	$10K/$0	$1,000,000	$1,000,000	Optional participation
VA	N/A	$50K/$150K	$1,000,000	$1,000,000	All-in
WA	N/A	N/A	N/A	N/A	Special reinsurance program
WV	N/A	N/A	N/A	N/A	Fund being considered
WI	N/A	$5K, 100% over $200K, total award not to exceed $195,000K	$200,000	Included	Covers 100% cleanup btwn $5K and $200K
WY	1–5 USTs	$10K/30K[a]	No limit	$1,000,000	All-in, annual agg. for third party = $970,000
	6–14 USTs	$15K/30K[a]	No limit	$1,000,000	
	15 + USTs	$50K/30K[a]	No limit	$1,000,000	

[a]Corrective actions in Wyoming are paid for by the Department of Environmental Quality from monies specifically allocated to its budget for such purpose. Corrective action monies are not drawn from the Financial Responsibility Account which is used to pay third-party claims.

The second item of concern is when, how, and who will respond when a pollution incident occurs. No insurer wants to give up the right to intervene on an insured site to contain and remediate contamination. Similarly, neither the insurer nor site owner wish to wait until the state selects and assigns a cleanup contractor to administer a claim against the state fund. Can either party completely trust the state to reimburse for expenses in a timely manner?

These issues and others not yet identified will continue to place state funds in controversy. It is likely, over the long term, that legislators will see that some state programs are under-capitalized and inefficient in operation. The commercial insurance market will replace some of these state-based financial responsibility programs.

SUMMARY

Each owner or operator of USTs should carefully compare and evaluate state funds and commercial insurance providers. Table 3 lists nine critical questions to ask. The UST insurance market will continue to evolve in the next year.

Table 3. How To Evaluate UST Insurance Providers

1. Does the policy comply with EPA regulations (corrective action, third-party liability, defense costs, first dollar coverage, multistate admittance, extended reporting period)?
2. If no, is the policy being revised?
3. Will the provider issue EPA Certificate of Insurance with the policy or equivalent documentation?
4. Does the policy specifically exclude certain environmental exposures (preexisting contamination, non-UST related)?
5. What are the application procedures and time period to receive a quote?
6. Are there additional requirements for underwriting (site assessments, tank test, environmental monitoring, inventory records review)?
7. What deductible or self-insured retention limits are available?
8. What risk management and claims response services are provided with the policy?
9. How is the premium determined (risk based, per site or per tank)?

PART II

State of Research

CHAPTER 3

Petroleum Release Decision Framework (PRDF)

Phillip J. Ludvigsen, Automated Compliance Systems, Inc., Bridgewater, New Jersey
David H. Chen, American Petroleum Institute, Washington D.C.
Curtis C. Stanley, Shell Oil Company, Houston, Texas
David Draney, Chevron Corporation, San Francisco, California

The magnitude and seriousness of environmental problems associated with petroleum releases from underground storage tanks warrant the need for powerful decisionmaking tools. The U.S. General Accounting Office has estimated the costs resulting from the most recent underground storage tank (UST) federal regulations (40 CFR Parts 280 and 281) to be in excess of $48 billion over the next 30 years. Without a structured approach to gathering and analyzing complex data as well as communicating results, the UST investigation and cleanup process could be hampered as a result of inefficiencies, delays, and cost overruns.

The Petroleum Release Decision Framework (PRDF) offers an integrated approach to facilitate rational decisionmaking for those involved in addressing hydrocarbon releases primarily from USTs. The framework software presents a logical methodology to collecting and archiving field information; characterizing the site; focusing on key decision parameters; and, if necessary, developing and evaluating a corrective action plan. The framework emphasizes, above all, an immediate response designed to protect the surrounding environment and public health by controlling both further hydrocarbon releases and the spread of potential contamination.

One technique for presenting a logical approach to problem solving is the use of hypermedia. Hypermedia is the interactive linking of text, graphics, databases, citations, models, spreadsheets, and any supporting information into a computer-based graphical framework. The strict theoretical vision of hypermedia proposes that every piece of information/knowledge be treated as a linkable element to other related elements. The power of hypermedia is not in its ability to decide where links should be placed, but its ability to establish a complex stream of links between nearly any information element.

For graphical decision trees, the hypermedia user selects the decision node of interest and immediately "zooms" to a higher resolution/level of detail. This means that hypermedia documents allow users to view only the information that is pertinent to making a given decision. The user also has instant access to documentation that is germane to the decision node being viewed. The PRDF software is structured to aid the user in comprehending the total solution (the forest) among numerous subtasks and associated decisions (the trees) while focusing on specific decisions and required information (the path). Figure 1 shows a conceptualization of this tiered design.

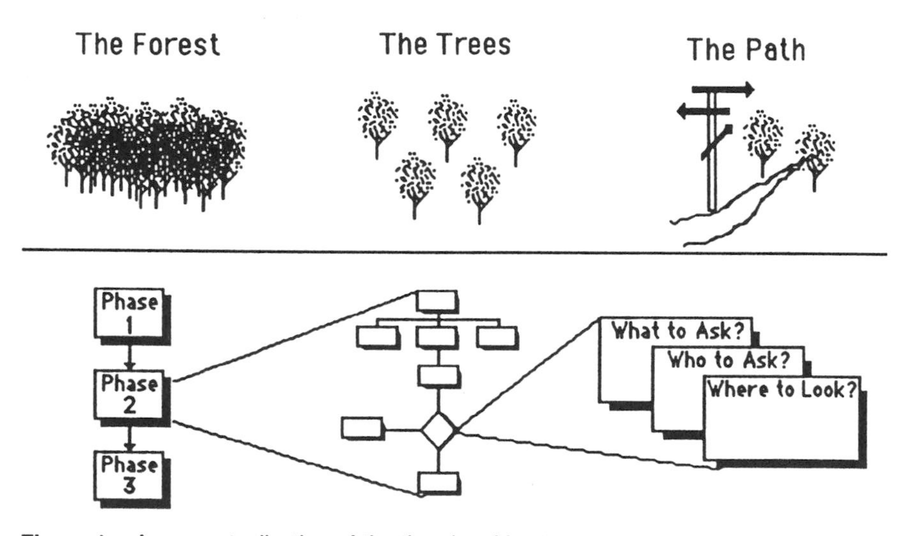

Figure 1. A conceptualization of the tiered architecture.

THE PDRF SOFTWARE

The objective of building the decision framework is to develop an integrated PC-based approach to facilitate rational decisionmaking for investigation and cleanup of petroleum releases to soil and groundwater. To meet this objective, the PDRF software is designed to serve seven critical roles:

1. *A decision aid* that facilitates rational decisionmaking for initial response and abatement, site assessments, and if necessary, site remediation;

2. *A training aid* that assists novice field personnel in learning a technically sound approach that can be customized by environmental managers to a given corporate or agency perspective;

3. *An electronic checklist and verification tool* of steps and decisions that environmental managers can establish for field personnel or hired contractors that would encourage consistency and thoroughness of information gathering, collating, and evaluation;

4. *A site-specific information management system* that allows key technical and regulatory information to be stored or retrieved in free format via system "scratch pads" and automatically generated "notebook" reports;

5. *An archive* of site-specific UST studies containing information that is specific to initial response and abatement, site assessment, and if necessary, site remediation; this archive is designed to facilitate the overall UST management process;

6. *Communication support* allowing for user-specific examples, explanations, and framework additions to be incorporated and then disseminated within a corporation or organization; and

7. *Future expansion* that will allow other decision support tools (mathematical models, simulation software, expert system advisors, object oriented databases, etc.) to be attached to the framework.

Considering a site where a petroleum release from an UST is suspected, the issue demands a logical way of confirming the suspicion. By using the PRDF tool, one is able to go through the complete flow of tasks and decisions in a consistent and logical manner.

For example, a novice user selects ("clicks on") the decision icon (triangle) for the first phase—"Initial Response and Abatement" (Figure 2); a decision tree appears; and the user addresses the first decision node—"Hydrocarbon Release Suspected or Confirmed?" (Figure 3). Being uncertain of what is meant by "suspected" vs "confirmed" from a regulatory standpoint, the user clicks on the "regulations" icon; a menu of pertinent federal regulatory categories appears (Figure 4). The user would most likely select the "Suspected Releases" option. A description of what is an acceptable response to a suspected release appears (Figure 5). The user can then click on the highlighted phrase "suspected release" and a regulatory definition appears (Figure 6).

By continuing through the decision tree in this manner, the PRDF software leads the user to discover what actions are warranted.

As a training aid, the PRDF software helps define a consistent and suitable approach to site-specific technical and regulatory data acquisition. Consider a novice environmental professional who has just been introduced to guidelines for investigations and cleanup of a petroleum contaminated site from a UST release. It usually requires a lot of effort to flip through paper manuals/documentation in search of appropriate tasks to perform. In contrast, by using the PRDF software, this novice user is led through all the appropriate tasks, as he/she works his/her way through the framework.

To facilitate the compilation, storage, and retrieval of key site-specific information, the PRDF software can be used as an information management system. In situations when, for instance, field personnel are collecting and compiling data, the software offers electronic "scratch pads" at every level within the framework. If the user wishes, each "scratch pad" can be automatically compiled into

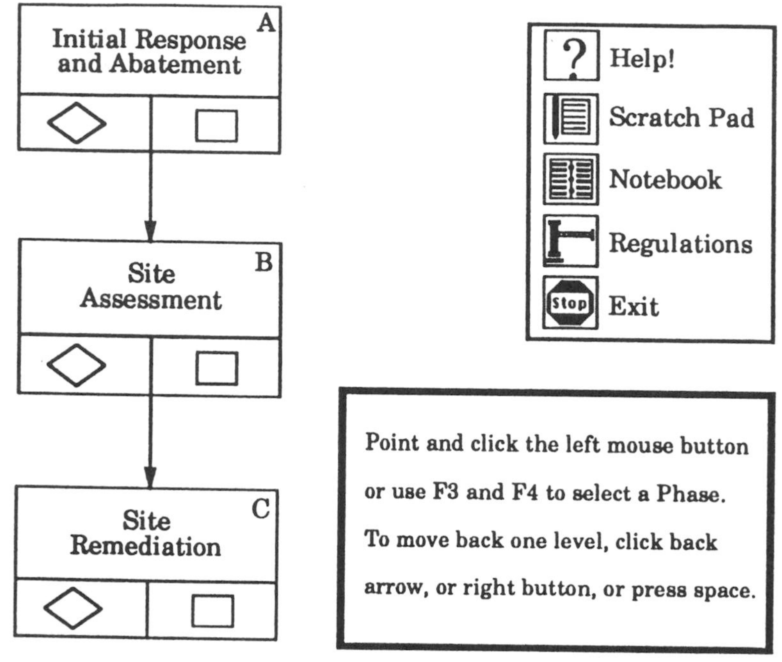

Figure 2. Top tier of decision framework.

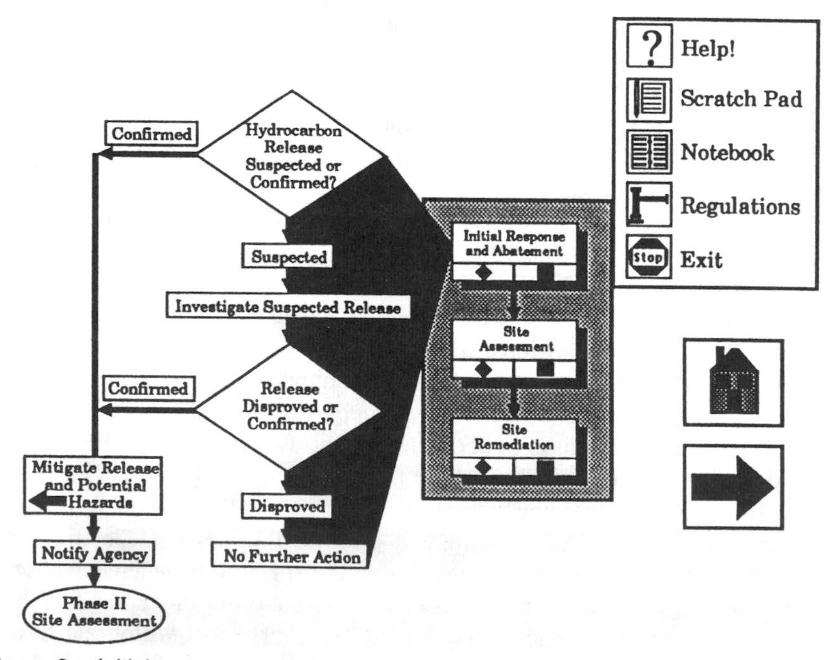

Figure 3. Initial response and abatement decision tree.

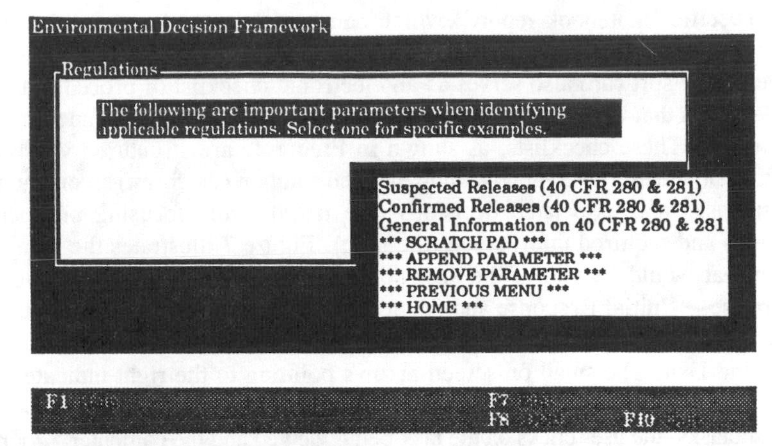

Figure 4. Federal regulatory categories.

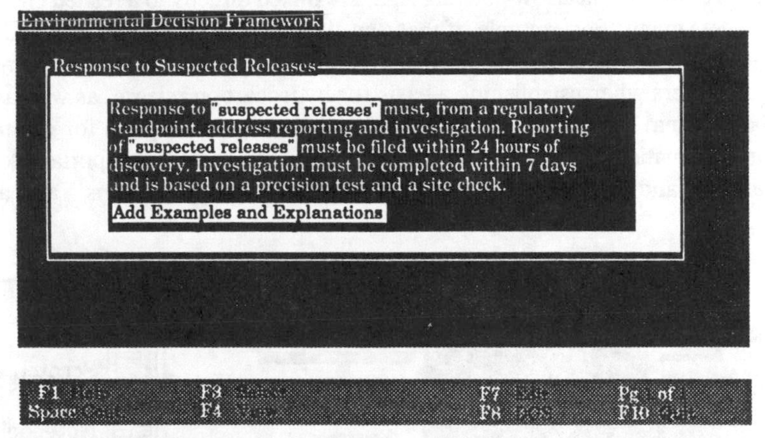

Figure 5. Response to suspected release.

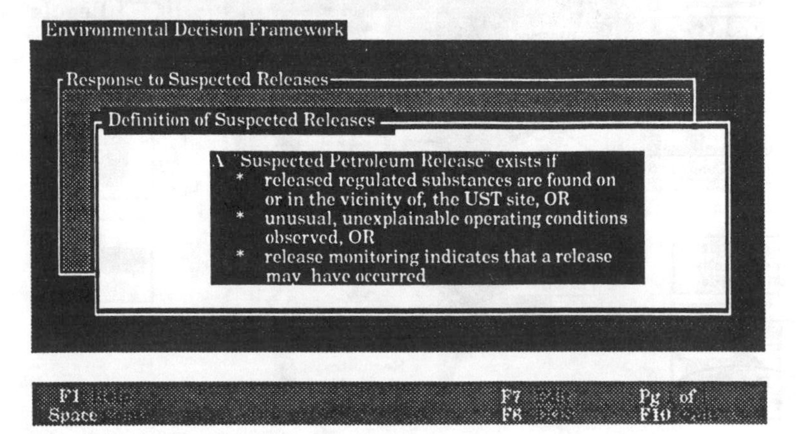

Figure 6. Regulatory definition of suspected release.

a site-specific "notebook report" which can later be edited and archived in a retrievable form for future reference.

The PRDF software also serves as an electronic checklist of procedural steps and decisions that UST project managers can establish for field personnel or hired contractors. These checklists, as shown in Figure 1, are structured via hypermedia to aid the user in comprehending the total solution (the forest) among numerous subtasks and associated decisions (the trees) while focusing on specific decisions and required information (the path). Figure 7 illustrates the computer screen that would be seen if the user clicks on the procedural steps under the first phase—"Initial Response and Abatement."The user is presented with five subtasks as well as the system utilities—Help, Scratch pad, Notebook, Regulations, and Exit. The small on-screen arrows pointing to the right indicate there are additional tiers of subtasks related to the current task being viewed. To access these subtasks, the user clicks on the task being viewed and the computer "zooms" to the next level of detail (Figure 8).

The tasks without an arrow pointing right are linked directly to a related checklist (parameter) menu. For example, if the user clicks on "Review Site for Evidence of Release," a checklist menu appears (Figure 9). This menu lists several important parameters when establishing a basis for hydrocarbon release, as well as the standard system utilities. The user can select a given menu option for examples and/or explanations of how or why the specific parameter is important to the sub-task at hand. For example, the user selects "Piping Problems" and a list

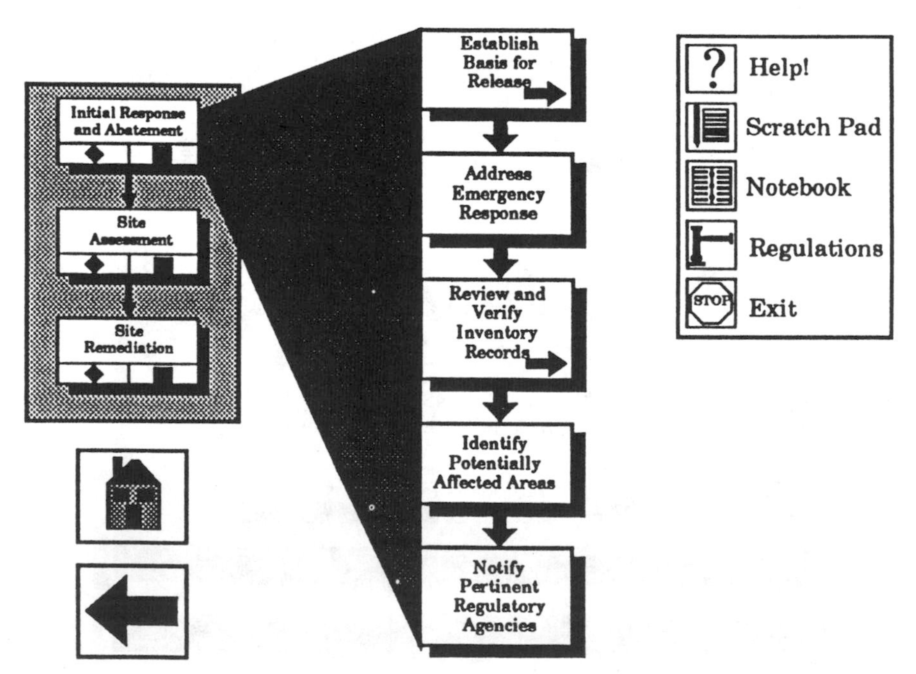

Figure 7. Procedural steps under initial response and abatement.

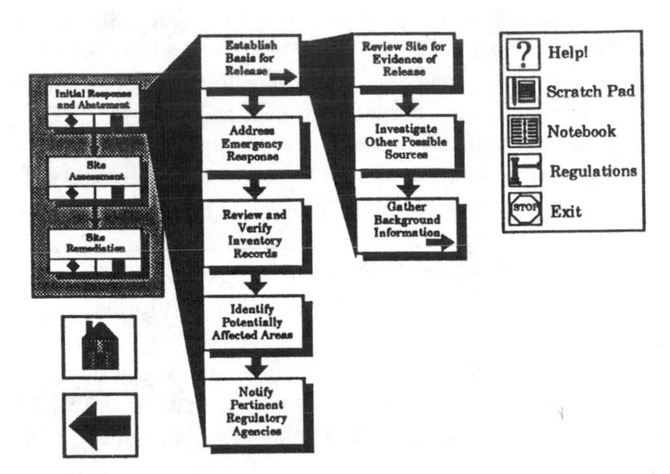

Figure 8. Higher level of subtasks.

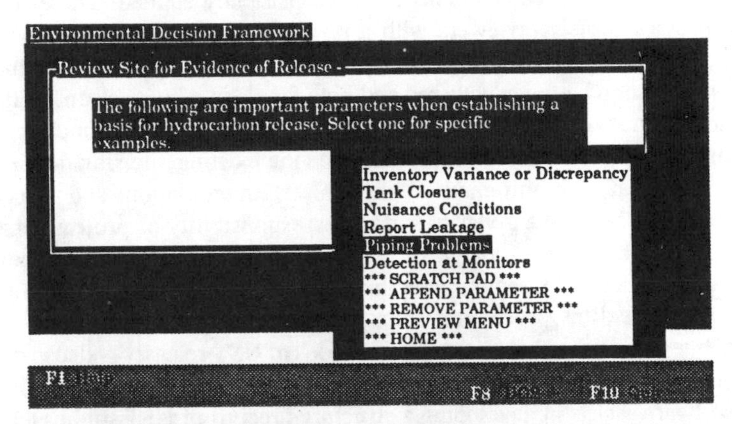

Figure 9. Checklist menu of important parameters.

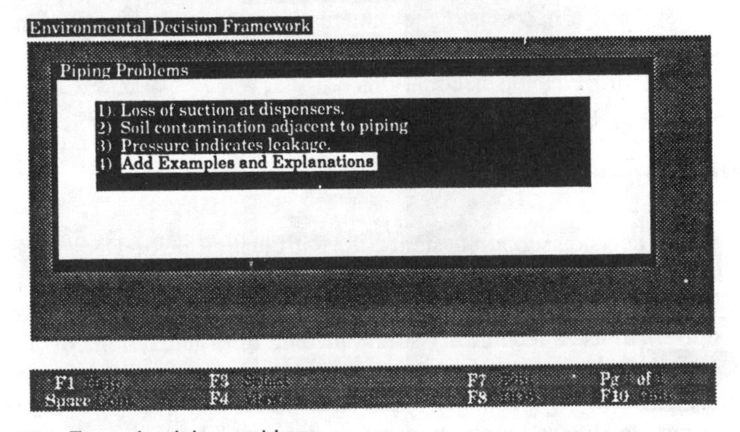

Figure 10. Example piping problems.

of example problems related to piping is displayed (Figure 10). The PRDF software offers experienced users the flexibility to append additional checklist items (parameters) as well as examples and/or explanations of how or why the additional checklist items are important to the task at hand. Thus, environmental managers can easily customize the framework—unaffected by the rigors and costs of reprogramming. For example, the UST project manager (PM) clicks on the regulations icon. A menu of pertinent federal regulation categories appears; however, no state or local regulations have previously been entered into the framework. The PM simply clicks on "APPEND PARAMETER," and types in the name of a site-specific local regulation, for example, California, and this appears on the menu. The PM then clicks on that menu selection; a screen window opens, and the PM can use the full-screen editor to enter the regulation and, if necessary, any interpretation. This information becomes part of the PRDF, and the enhanced software can be disseminated to field personnel equipped with portable field computers.

Framework "scratch pads" and subsequent "notebook reports" can be used to document that correct procedural steps were consistently applied. These software features provide project reviewers with a powerful verification tool. Since each notebook report is similarly structured, reviewers can quickly ascertain what site-specific information has been gathered and how far along a given cleanup effort is.

The novice user who has just perused a regulatory definition can click on the regulatory "scratch pad" and simply describe the existing site situation as it relates to the regulatory requirement. Although certain regulations and procedures may not be applicable to a given site, it is the responsibility of project personnel to document such situations via each "scratch pad" throughout the framework. The compiled "notebook report" can then be used to quickly verify that correct regulations, as well as procedural steps, were consistently applied.

Often, questions arise about past investigations of UST releases and also cleanup efforts undertaken at the time of the remedial investigations. As an archive, the PRDF software assists in developing a structured record of UST studies that contains site-specific information related to initial response and abatement, risk-based site assessment, and if necessary, site remediation. This record is used in building conceptual site models that are extremely helpful in preparing technically defensible documents for cleanup negotiations.

DECISION SUPPORT

Similar to all decision support systems, the Petroleum Release Decision Framework offers a mechanism for the user to learn, analyze, and communicate. The system graphically structures a decisionmaking approach, thus teaching inexperienced users to use a technically sound method of organization. By allowing easy access, via hypertext, to pertinent information (e.g., regulatory standards, appropriate chemical test methods, and other important decision parameters) the framework facilitates comprehensive analysis. And, finally, the graphical interface and automated report generation allows diverse parties (industry and regulators) to communicate better by focusing on pertinent information.

OTHER POTENTIAL APPLICATIONS

As its name implies, the Petroleum Release Decision Framework is designed as a framework so that many other tools could easily be incorporated. The use of hypermedia is an effective means of narrowing the UST problem domain via "zoom" processing, thus making it an ideal front-end to very specific UST expert system modules, models, databases, and geographic information systems. Although the framework software is tailored to meet the needs of UST risk management, the concepts, architectures and object-oriented algorithms could be applied to many other environmental problems; in particular, those relating to waste site cleanups.

CONCLUSIONS

The PRDF decision support software is a unique computer-based approach to facilitate rational decisionmaking for investigation and cleanup of petroleum releases to soils and groundwater. The software offers a consist structure to gathering, compiling, and presenting site-specific data; addressing initial response and abatement, site characterization, and remediation; while enhancing communication among diverse parties (industry and regulators).

CHAPTER 4

The Department of Defense Research Program for Cleanup of Contaminated Soil

Patricia L. D. Janssen, Office of the Deputy Assistant Secretary of Defense (Environment), Alexandria, Virginia

The Defense Appropriations Act of 1984 established the Department of Defense (DoD) Installation Restoration Program (IRP). The program set up a centrally managed account under the Office of the Deputy Assistant Secretary of Defense (Environment) [ODASD(E)] in the Office of the Assistant Secretary of Defense (Production & Logistics) called the Defense Environmental Restoration Account (DERA). In 1989, DERA was appropriated $503 million. The amount is expected to grow over the next few years as we start to enter the remediation phase of many of our sites. The primary purpose of the DoD program is to clean up hazardous waste sites. We utilize existing technology where available and adequate for the task. However, DoD realizes the necessity for new research to adequately assess health risks, minimize environmental effects, and achieve permanent remedial actions for the purpose of protecting the public health and the environment. In addition, the Superfund Amendments and Reauthorization Act of 1986 (SARA) encouraged the use of DERA funds to look at innovative ways to clean up our hazardous waste sites.

The ODASD(E) is a policy and review office within the Secretary of Defense, and it is charged with managing and overseeing the Installation Restoration Program and is responsible for oversight of cleanups and research and development (R&D) projects. The majority of installation restoration research is conducted at six service laboratories. These are the U.S. Army Toxic and Hazardous

Materials Agency, U.S. Army Construction Engineering Research Lab, U.S. Army Biomedical Research and Development Lab, U.S. Air Force Engineering and Services Center, U.S. Navy Civil Engineering Lab, and the U.S. Navy Energy and Environmental Support Activity.

The contaminant most often found at DoD sites is trichloroethylene (TCE). The major DoD soil contaminants are:

Organics—fuels and solvents
Combined Waste—hazardous waste landfills
PCBs/Pesticides—chlorinated hydrocarbons
Heavy Metals—Pb, Cu, Cd, Hg, Ni, Ag
Ordnance—TNT, RDX

ODASD(E) identified a need for increased emphasis in DoD research on innovative hazardous waste cleanup demonstration technologies. A policy memorandum was sent to the armed services in the spring of 1987 instructing them to look for, and propose, demonstration projects. The services have done a good job of identifying demonstration projects which include low temperature thermal soil treatment, soil venting, air stripping, and biodegradation. These projects have real potential for developing waste minimization techniques and for identifying alternative technologies and procedures for the cleanup of hazardous wastes. These specific technologies will be discussed later. However, we are still encouraging the services to continue actively pursuing innovative demonstration technologies. Many of the projects may be suitable for cooperative projects with other federal agencies.

In fiscal year 1989, DoD committed $27.6 million on hazardous waste minimization and cleanup research and development projects. DERA provided $16.0 million (58%) of the money for these projects. The individual service research and development programs provided the remaining $11.6 million. In fiscal year 1989, the services have programmed $28.3 million for research programs. DERA will provide $16.5 million (58%) for these efforts.

The Installation Restoration Technology Coordinating Committee (IRTCC) is a tri-service committee which was established to facilitate the exchange of programmatic and technical information among the DoD components. The IRTCC is working to improve and encourage coordination and communication among the services to ensure that limited R&D dollars are spent wisely. DoD and the services are placing great emphasis on innovative hazardous waste cleanup demonstration projects within the R&D community. We are hoping to identify a few promising programs for basic research in hazardous waste cleanup this year. We hope to establish a peer review process through the IRTCC for DERA funded R&D. The IRTCC published a handbook entitled, "Installation Restoration and Hazardous Waste Control Technologies," in 1988. This handbook provides a reference of current installation restoration and hazardous waste control technologies for use in DoD industrial operations, training and readiness, and for environmental and management staffs. This report is currently being updated to include new information. It should be available by the end of 1990.

The DoD is also working to facilitate technology transfer and cooperative R&D as part of the DoD/Environmental Protection Agency/Department of Energy Hazardous Waste Technology Interagency Working Group. It seeks to increase not only service cooperation but interagency cooperation. These three agencies met in 1988 in Washington, D.C. to identify additional projects of mutual interest. Approximately 138 projects are included in the report. These efforts share innovative technical approaches to common problems, eliminate duplicative efforts, and minimize costs by leveraging individual agency funds. Given the current funding climate, especially for DoD, the efforts of this group will become even more important. This effort will assist federal agencies to get the most research possible with limited R&D dollars. DoD will be sponsoring the next meeting in early fiscal year 1990. We will be striving to increase the cooperative participation by the three agencies.

The various research and development studies being conducted under the DERA program can be divided into several categories. These include the following:

- In situ treatment
- Onsite cleanup
- Containment
- Site assessment
- Alternative technologies to land disposal
- Waste minimization
- Underground storage tanks
- Monitoring
- Risk assessment

This chapter will be concentrating on soil cleanup. A brief description of the various technologies under development follows.

Organic

A study is underway to evaluate underground fuel spill cleanup technologies. Thermal vacuum spray/combustion, chemical oxidation, and air stripper/carbon technologies will be evaluated. The study will compare the effectiveness of each technology. Results will be incorporated into a technology transfer guide.

Pilot-scale field remediation studies are planned to evaluate the use of in situ unsaturated zone bioremediation of jet fuel contaminated subsoils. Demonstration tests are planned for 1991.

Another study will identify the parameters that limit on-site bioremediation of JP-5 contaminated soil. The services are looking at various methodologies for onsite bioremediation of volatile organic compound contaminated soils as well as soil venting.

The Army and Air Force have conducted field demonstrations of low temperature thermal stripping to treat JP4 and TCE contaminated soil. A brochure produced by the U.S. Army Toxic and Hazardous Materials Agency entitled,

"Low Temperature Thermal Stripping of Volatile Organic Compounds from Soil," provides an excellent summary of this technology.

The services are also investigating the feasibility of photochemical oxidation of solvents and oils. The Air Force plans to conduct a demonstration study of radio frequency thermal decontamination of soils in 1990. They are also looking at crossflow air stripping/apron treatment technologies, and subsurface oxygen enrichment for biodegradation for the treatment of organic contaminated soils.

Mixed Waste

Studies are underway to test the feasibility and identify the favorable conditions for biodegradation of mixed hazardous waste. Lessons learned from other studies on bioremediation of organics are being applied here.

Another technology under study is in situ oxygen generation by electrolysis on site-specific soils. The objective is to determine the feasibility of oxygen generation and the effect of electrolysis on the microbial population in a bench-scale study. DoD is also looking at electroacoustical effects on microorganisms.

PCB/Pesticides

The Navy has completed a pilot plan evaluation using potassium-polyethylene glycol to dechlorinate PCBs in the soil. In FY 1990, they plan to scale up the decontamination unit and conduct a field unit demonstration. They are optimistic about developing an affordable treatment system and plan to transfer the technology for use in 1992.

Heavy Metals

The services are evaluating stabilization/solidification processes in hopes of developing affordable technologies and materials for decreasing the mobility and toxicity of non-degradable hazardous wastes. Pilot tests have been performed using sulfide processes and silicate processes. Results show that chemical formulation is waste-specific.

Ordnance

Three technologies are currently under review by the services for treatment of ordnance contaminated soil. Bioremediation of TNT and RDX contaminated soil studies are underway using aerobic bacteria and fungi to determine their potential to degrade TNT and RDX.

Composting studies have shown great potential. However, DoD has had difficulty in identifying degradation products. Finally, preliminary studies are being conducted to determine the feasibility of chemical treatment of explosive soils.

Monitoring and Site Assessment

The cone penetrometer is being studied by all the services for use in site characterization. The cone penetrometer uses fiber optic sensors to characterize soil contamination and plume migration.

The Army is developing a biomonitoring traveler for onsite toxicity assessments. The unit will provide a mobile field test laboratory for onsite chemical analysis. The Army is also studying nonmammalian toxicity models to reduce use of animals for testing.

The Navy is looking at in situ bioindicators for site assessment. They are also researching marine environmental survey monitors and protocols for use at many of their military sites that are located around marine environments.

SUMMARY

The primary purpose of the DoD program is to clean up hazardous waste sites with existing technology where available. Our R&D efforts are focused at this primary objective.

REFERENCES

1. ''Installation Restoration and Hazardous Waste Control Technologies,'' U.S. Army Toxic and Hazardous Materials Agency, 1988.
2. ''Interagency Work Group on Hazardous Waste Technologies,'' HAZWRAP Support Contractor Office, Oak Ridge, TN, December, 1988.
3. ''Low Temperature Thermal Stripping of Volatile Organic Compounds from Soil,'' U.S. Army Toxic and Hazardous Materials Agency.

CHAPTER 5

Environment Canada Research on Land Treatment of Petroleum Wastes

T. L. Bulman, Campbell Environmental Ltd., Suite 93 Havelock Mall, West Perth WA 6005, Australia
Bruce E. Jank, Wastewater Technology Centre, Environment Canada, Burlington, Ontario, Canada, L7R 4A6
R. P. Scroggins, Industrial Programs Branch, Environment Canada, Hull, Quebec

Application of industrial wastes in a repeated fashion to dedicated land areas is practiced by the oil refining industry in Canada as a method of waste disposal. In a 1980 survey of waste disposal in the refinery industry,[1] it was reported that 29.7% of refinery waste was recovered for reuse, while 70.3% was disposed of by other means (Table 1). Application to land was the least expensive of identified disposal methods, and was used for the disposal of the greatest proportion of petroleum waste. Landfilling was used predominantly for general refuse which could not be disposed of by other methods. The industry is also actively pursuing the use of land application for treatment of oil and gas exploration and production wastes.

Land application of waste (also called land treatment of waste or landfarming) has potential benefits such as the reduction of waste mass and toxicity. Potential negative impacts include the contamination of the environment and entry of toxic chemicals into the food chain. Guidelines are required, therefore, for the safe application of suitable wastes to soil in a manner which is cost effective, obviates postapplication cleanup measures, and protects the environment.

Environment Canada is sponsoring several research studies which assess the treatability of oily wastes by land application and the environmental acceptability

Table 1. Methods and Costs of Treatment for Oil Refinery Wastes in Canada

Treatment Method	Waste Treated (%)	Cost per Ton ($/Can)
Land application	17.9	9.10
Landfill	35.7	16.30
Incineration	6.9	71.80
Deep well injection	1.4	35.60
Unspecified	8.5	8.00
Recovery/reuse	29.6	15.20

Source: Reference #1.

of this practice. Oily waste types under evaluation include refinery wastes, conventional oilfield production wastes, heavy oil production waste from enhanced oil recovery operations, and diesel-based drilling muds. Most of these studies are cofunded with industrial groups, notably the Petroleum Association for Conservation of the Canadian Environment (PACE) and the Canadian Petroleum Association (CPA). Other government agencies have also participated in funding, including provincial environment ministries (Alberta Environment, Ontario Ministry of Environment and Ministre de l'Environnement Quebec), the Alberta Environmental Research Trust, the federal Interdepartmental Panel for Energy Research and Development, the Great Lakes Water Quality Board, and the United States Environmental Protection Agency.

The purpose of these studies is to identify wastes which can be applied to land in an environmentally acceptable manner and to provide information on which to base guidelines for the proper application of such wastes to land. The information which has been collected to date has focused on (a) the persistence and fate of oil and toxic constituents of petroleum wastes when applied to soil, (b) potential environmental impacts and risk to human health associated with application to land, and (c) site management techniques which enhance treatment of organic constituents of wastes while protecting environmental quality. The potential for contamination of groundwater, the accumulation of hazardous substances in soil, and effects on plant growth have undergone the most intensive investigation to date. Impingement on air quality has received limited study. A recent study on closure of contaminated industrial sites, however, has included an intensive air monitoring component performed under the leadership of R. Dupont, Utah State University, with funding provided by the United States Environmental Protection Agency. A summary of recent Environment Canada research activities on land treatment of petroleum waste is provided in Table 2 and a brief review follows for each of the project areas.

LAND APPLICATION OF REFINERY WASTES

Research on land application of refinery wastes has been performed at the Wastewater Technology Centre in Burlington, Ontario. This work has focused on the fate of polynuclear aromatic hydrocarbons (PAHs), toxic constituents of refinery wastes, when added to soil. The objectives have been (a) to study the kinetics of loss of PAH constituents from soil and (b) to identify PAH loss mechanisms, including biodegradation, volatilization, leaching, plant uptake, and binding to soil.

Kinetic Studies

Two studies to identify the kinetics for loss of PAHs from soil were performed using open flasks. The first study (Activity 1 in Table 2) involved the incubation of pure chemicals in soil to provide detailed, precise data for kinetic modeling. Unacclimated soil was incubated with a suite of eight PAHs ranging in size from two aromatic rings (naphthalene) to five aromatic rings [benzo(a)pyrene] added as pure compounds at two concentrations. The soil was sampled periodically and analyzed for PAHs. Concentration data were fitted by nonlinear regression to a general power rate model of the form;

$$\frac{dC}{dt} = kC^n \qquad \begin{aligned} C &= \text{concentration} \\ k &= \text{rate constant} \\ n &= \text{reaction order} \end{aligned}$$

The study with pure chemicals[2] provided good estimates of loss kinetics for nonpersistent chemicals. Loss of persistent chemicals tended to be less than sample variability, however, and precision of estimated kinetic parameters was poor. The low molecular weight PAHs, such as anthracene, were quickly lost from soil (90% loss within 200 days), whereas an extended period (greater than 200 days) was required before loss of high molecular weight PAHs, such as benzo(a)-pyrene, was observed. Loss kinetics did not always approximate a single first order reaction over the entire incubation period, but suggested either a two stage reaction or a reaction order greater than one. First or zero order kinetics were sufficient, however, to estimate the time required for loss of up to 90% of each compound.

In a second study (Activity 3 in Table 2), soil was used from a refinery land treatment site which had received oil application over a period of 10 years and had an average residual oil content of 6%. Oil refinery waste was added at a level of 3% oil in soil and the concentrations of 16 constituent PAHs were monitored over time.

Sample variability in the study with refinery waste was much greater than with pure chemical addition, and analytical recovery, as determined by spiked duplicate samples, varied a great deal between samples. The results of the refinery waste kinetic study confirmed that low molecular weight PAHs were less persistent than high molecular weight PAHs, but evaluation of the kinetics of loss was inconclusive[3]. Losses due to individual mechanisms, such as volatilization, biodegradation, and photochemical decomposition, were not assessed separately in these studies.

Table 2. Summary of Environment Canada Funded Research on Land Treatment of Petroleum Waste (1983 to 1989)

Research Activity	Contact	Report Date
1. The Fate of Polynuclear Aromatic Hydrocarbons in Soil	T. Bulman Campbell Environmental Ltd. Suite 93 Havelock Mall Perth, WA 6005 Austrailia (9) 332 2344	1985
2. Land Treatment of Diesel Invert Mud Residues: Phase 1: Laboratory and Greenhouse Feasibility Assessments	R. M. Danielson Kananaskis Centre for Environmental Research University of Calgary Calgary, Alberta, Canada T2N 1N4 (403) 220 3194	1987
3. The Persistence and Fate of Polynuclear Aromatic Hydrocarbons in Refinery Waste Applied to Soil	T. L. Bulman Campbell Environmental Ltd.	1988
4. Land Application of Residual Diesel Invert-based Muds: Phase 2	J. Ashworth Norwest Soil Research Ltd. 9938-67 Ave. Edmonton, Alberta, Canada T6E OP5 (403) 438 5522	Oct 1989[a]
5. Development and Validation of a Method for Establishing Site-Specific Clean-up Criteria	B. E. Jank Wastewater Technology Centre (416) 336-4740	Vol 1, 2, 4: 1988 Vol 5, 6: 1989 Vol 3, 7: Dec 1989[a]
6. Persistence and Fate of Aromatic Constituents of Heavy Oil Production Waste	K. Hosler Wastewater Technology Centre (416) 336 6021	March 1991[a]
7. Disposal of Oilfield Wastes by Land Treatment: Effects on the Environment and Implications for Future Land Use	R. M. Danielson Kananaskis Centre for Environmental Research	March 1990[a]
8. Utilization of Oily Wastes to Improve the Quality of Agriculturally Marginal Sandy Soil	V. Biederbeck Agriculture Canada Research Stn. P.O. Box I030 Swift Current, Saskatchewan, Canada S9H 3X2 (306) 773 4621	June 1991[a]

[a]Estimated completion date.

Fate Studies

The fate of PAHs when applied to soil was studied using ^{14}C labeled chemicals in two types of enclosed incubation systems (Activities 1 and 3 in Table 2). One system consisted of soil incubated in biometer flasks, in which the evolution of PAHs or transformation products into the atmosphere (as CO_2 or volatiles) was monitored continuously. Parent PAHs or products remaining in soil were identified at the end of the incubation as extractable or bound residue fractions. This system provided precise results with both pure chemical additions and with addition of petroleum wastes. This monitoring approach has been used to compare the fate of selected PAHs in the unacclimated soil and the soil from a refinery waste application site used in the kinetic studies.[2,3]

On a slightly larger scale, terrestrial microcosms were used to provide information similar to that from biometer flask studies, but including transport of PAHs through the soil profile and uptake by plants. The microcosms consisted of soil columns, as reconstructed soil profiles or undisturbed soil cores, contained in stainless steel cylinders. The column was designed so that the upper portions of plants did not come into contact with the soil or soil atmosphere. In this way, movement of toxic chemicals into the plant shoot could occur only by translocation. Soil atmosphere was continually monitored for evolved ^{14}C-labeled CO_2 and volatile organic compounds. Leachate and plant shoot tissue were periodically collected and analyzed for ^{14}C. In addition, leachate was analyzed for total PAH content and assessed using microtoxicity tests (for example, Microtox assay, green algae growth inhibition, and SOS Chromotest genotoxic test). At the end of the study, the soil columns were analyzed with depth for ^{14}C.[4]

In both the biometer flask and microcosm studies, biodegradation of PAHs in soil from the refinery waste application site was substantially enhanced over that observed in agricultural soil, and constituted a major loss mechanism. This effect may be due to prior exposure of organisms to aromatic hydrocarbons, the availability of other organic substrates for cometabolism (10% organic carbon, 6% oil), and high overall microbial activity.

Small amounts of PAHs were volatilized from refinery waste treated soil in biometer flask and microcosm studies.[3,4] The proportions of added ^{14}C-labeled organic compounds that were volatilized from soil are summarized in Table 3 (averages for three replicate flasks). Volatilization was greater from soil in microcosm studies than in biometer flasks, possibly because the flow rates used for aeration of the microcosms were greater. The majority of the volatilized compounds were transformation products, however, rather than the parent compound under study. In addition to volatilization of organic compounds, ^{14}C-labeled CO_2 was evolved into the soil atmosphere (7.68%, 23.2%, and 73.4% for benzo(a)pyrene, anthracene and naphthalene, respectively, in biometer flasks, and 68.7% and 64.1% for anthracene and naphthalene, respectively, in microcosms). The remainder of ^{14}C remained in soil as extractable parent compound or nonextractable material which was not identified. In microcosm studies, small amounts (less than 0.01%) of ^{14}C were also observed in leachate and plant material as transformation products.

Table 3. The Proportions of ^{14}C-Labeled Organic Compounds Volatilized from Soil Treated with Refinery Waste and ^{14}C-Labeled PAH's in Laboratory Studies

Study Description				
Type	Duration (Days)	Chemical Added	Volatilized Parent Compound (% of ^{14}C Added)	Volatilized Transformation Product (% of ^{14}C Added)
Biometer Flask[a]	145	benzo(a)pyrene	0.03	0.1
	107	anthracene	0.03	2.4
	68	naphthalene	0.1	1.0
Microcosm[a]	67	anthracene	0.001	7.5
	67	naphthalene	0.02	8.0

[a]Reference #3.

Similar techniques are being used to study the fate of heterocyclic nitrogen substituted aromatics in heavy oil production wastes, and are described in a following section.

LAND APPLICATION OF DIESEL-BASED DRILLING MUD RESIDUE

Studies concerning the land application of diesel-based drilling mud residues were initiated in 1984. Environment Canada is cofunding these studies with the Canadian Petroleum Association in order to assess the environmental impact of applying diesel-based drilling mud residues to soil at drill leases. The use of land treatment for oil field waste disposal would involve a single application of waste with a high priority given to rapid revegetation of the site. Phase 1 was a laboratory and greenhouse study undertaken by the Kananaskis Centre for Environmental Research (Activity 2 in Table 2). The focus of effort was to assess the effects of drilling mud residue application rates on degradation of oil and selected hydrocarbons and to evaluate phytotoxicity. Phase 2 was a field demonstration, principally to evaluate methods for revegetation of land application sites and to field validate the results of the Phase 1 study. The field demonstration was performed by Norwest Soil Research Ltd (Activity 4 in Table 2).

Laboratory/Greenhouse Studies

Incubation studies were performed in which two sources of diesel-based drilling mud residue were applied to two soil types at four levels of application.[5] Soil respiration was monitored to assess total carbon mineralization. In addition, soil was periodically sampled and analyzed in detail for target PAHs. Greenhouse studies were also set up in which crop species were assessed for revegetation of the soils/waste mud mixtures. Tall fescue (*Festuca arundinacea*), Manchar smooth bromegrass (*Bromus inermis*), Manhattan perennial rye grass (*Lolium perenne*), and rambler alfalfa (*Medicago sp.*) were planted and percent germination, shoot, and total plant dry weights were measured.

The composition of the diesel-based drilling mud residue was typical of slightly weathered diesel fuel, with 85% to 90% aliphatic hydrocarbons, 8% to 13% aromatic hydrocarbons and about 2% polar materials. Rates of degradation varied between the two sources of mud and was lower in the finer textured clay soil than in the clay loam. Although the rate of CO_2 evolution increased with increasing levels of mud addition (for 10%, 20%, 30%, and 40% additions on a weight/weight basis), the largest proportion of added carbon was mineralized with the 20% addition. The concentration of normal alkanes decreased rapidly (approximately 50% lost in 30 days). Low molecular weight aromatics, such as naphthalene, also disappeared rapidly (95% lost in 30 days). Concentrations of aromatic hydrocarbons in the 156 to 212 molecular weight range were reduced by 75% after 90 days and by 98% at the end of the 240 day incubation period.

The soil/waste mud mixtures were initially inhibitory to seed germination and plant growth. After 5 weeks of incubation, the mixtures no longer inhibited germination, but plant growth was reduced, probably due to changes in soil physical characteristics and nutrient availability. Laboratory grade charcoal and manure were effective in reducing phytotoxicity, whereas other treatments (garden charcoal, waste charcoal from the natural gas processing industry, peat, sawdust, manure, sewage sludge, cultivation, high rate fertilizer addition, and adapted microorganisms) were ineffective. Alfalfa was the most sensitive species. After 15 weeks of incubation, however, phytotoxicity was no longer evident.

Field Study

A field site was established near a drilling lease in the foothills of Alberta in 1986.[6] Levels of 10%, 20%, and 30% diesel-based mud residue (weight/weight) were applied, and concentrations of oil and aromatic hydrocarbons were monitored. The waste contained approximately 11% total hydrocarbon and 0.8% aromatic hydrocarbons. The site was seeded with a mix of grasses for evaluation of plant growth. Soil amendments of laboratory grade charcoal, spent charcoal granules, and manure were evaluated with intensive cultivation and fertilization practices.

In the first year, plant growth decreased with increasing rate of waste application. Growth was noticeably better where fertilizer had been applied and where fertilizer was applied in combination with a surface mulch of manure (not incorporated into soil). Loss of aromatic compounds was as great as 90% within the four month monitoring period. Total extractable hydrocarbons decreased by approximately 25% (i.e., decreased from 4% to 3% total hydrocarbons) over the same period. Degradation was greatest on the fertilized plots. In contrast with the laboratory study results, the activated charcoal and spent charcoal granules treatments did not appear to have a beneficial effect on plant growth. Volatilization was not measured.

Losses of 1% total hydrocarbon in soil occurred in the second year, after which the rate of loss was very slow. Some plots which received the high level of waste application had low rates of hydrocarbon loss from soil and some downward movement of hydrocarbon within the soil.

LAND TREATMENT OF HEAVY OIL PRODUCTION WASTES

Environment Canada is currently supporting a three part program to evaluate land treatment for heavy oil production wastes through the Interdepartmental Panel for Energy Research and Development (PERD). This program involves two field study projects which focus on monitoring soil, groundwater, and crop quality. The field studies are located in proximity to heavy oil deposits in western Canada. In addition, a laboratory study at the Wastewater Technology Centre is investigating the persistence and fate of heterocyclic aromatic compounds, including degradation, volatilization, leaching, and plant uptake.

Field Studies

Field sites for one of the projects were established in four locations in Alberta in 1982, with support from the Canadian Petroleum Association and the Alberta Environmental Research Trust, to investigate the efficiency of land treatment for the degradation of oil in oilfield sludges. These sites had received oily waste application for a three year period, and oil and metal concentrations had been monitored in soil and groundwater. Environment Canada and the Canadian Petroleum Association are currently cofunding a follow-up study which is assessing the rates of degradation of recalcitrant materials over a longer term (4 to 7 years), and evaluating the effects on groundwater quality, soil physical properties, and plant productivity (Activity 7 in Table 2).[7] The study is primarily directed toward rehabilitating the land treatment sites and returning them to agricultural use.

Results to date indicate that decreased plant productivity was related to residual oil content in the soil. The oil content in soil ranged from 2.6% to 8.6%. Plant growth was good on plots with oil content of 3.5% or less, but poor on plots with oil content of over 5%. Residual oil contained 15% to 18% asphaltenes, 40% to 50% resins, 16% to 23% aliphatics, and 11% to 16% aromatics. Shallow groundwater beneath and downgradient from the land treatment sites contained elevated levels of chloride and nitrate. Two interim reports have been published jointly by the Canadian Petroleum Association and Environment Canada.[7,8] An additional interim report which outlines the 1988 field results is in preparation. A final report will be published following completion of the 1989 field program.

A second field program was established in 1986 in central Saskatchewan (Activity 8 in Table 2). This region supports agricultural activity as well as oil production from heavy oil deposits. Some cultivated soils in the region suffer from low productivity, however, due to poor soil characteristics and susceptibility to erosion. This field study was designed to determine if land application of heavy oil waste could be used in these areas, not only to treat and dispose of waste, but to improve soil aggregation and reduce erosion on marginal cropland. The study involves assessment of the effects of low level applications of heavy oil wastes on soil physical and chemical properties, groundwater quality, and plant growth.

The waste studied was from a nearby Enhanced Oil Recovery (EOR) plant. Preliminary greenhouse trials were used to identify waste application levels which were not detrimental to crop growth. Field plots were then established with application of waste at levels of 1% and 2% oil in the soil. In the first season under field conditions, levels of 1% and 2% oil appeared to reduce crop growth. Fertilization treatments did not enhance crop growth and substantial immobilization of nitrogen and phosphorus was observed. Crop growth was enhanced, however, on portions of the plots which inadvertently received levels of 0.5% as a result of uneven waste application. The size of dry and water stable aggregates was substantially increased with waste application, reducing the susceptibility of the soil to erosion. The study will continue in 1989 and 1990.

Laboratory Studies

Complementary to the Saskatchewan field study, laboratory studies are being performed at the Wastewater Technology Centre[4] (Activity 6 in Table 2) using experimental methods similar to those developed for the study of refinery wastes. Incubations have been completed with a suite of seven heterocyclic aromatic compounds, previously identified as constituents of the heavy oil wastes, which were added as pure compounds to soil from the Saskatchewan site. Loss mechanisms such as volatilization and degradation were not studied separately. Kinetic parameters for losses through all mechanisms (in combination) were estimated. Of the compounds studied, the most rapid loss rates were observed for 2-methylnaphthalene and 9-ethylfluorene (50% loss in 20 days or less). Losses of biphenyl and 4-methylbiphenyl were intermediate (50% in 30 to 40 days) while acridine, dibenzothiophene and carbazole were more persistent (30% loss in 40 days). Incubations with heavy oil waste from the EOR plant are currently underway.

Biometer flask and microcosm studies have also been performed to study the fate of ^{14}C-labeled 2-methylnaphthalene, pyridine, and dibenzothiophene with addition of heavy oil waste to soil. Biometer flask studies were performed to assess the effects of environmental conditions on fate of the waste constituents. Treatments included moisture contents of 40% and 80% (dry weight basis), temperatures of 20° and 35° Celsius, as well as with wet/dry cycles and freeze/thaw cycles. Approximately 20% to 30% of added pyridine was extracted from soil after 56 days of incubation. The majority of ^{14}C remained in soil in a nonextractable form. This soil-bound ^{14}C may have been incorporated into microbial biomass or bound to soil organic matter. The proportion of nonextractable ^{14}C was greater with the 35° treatment than with the 20° treatment, and was substantially increased through wetting and drying. Small amounts (2.1% to 4.4%) of ^{14}C were volatilized as transformation products of pyridine. The proportion of volatilized products was greater with the 35° treatment than with the 20°C treatment. Evolution of ^{14}C as CO_2 ranged from 1.8% to 5.0% and was also highest with the 35° treatment. Loss of ^{14}C as CO_2 with two cycles of freezing and thawing was similar to that with constant conditions of 35°.

The microcosm studies were performed with two soil types, (1) Meota loamy sand (pH 7.5) taken from the Saskatchewan field site as intact cores, and (2) an acidic McLaurin sandy loam (pH 4.5) from Mississippi. Carbon-14 labeled 2-methylnaphthalene and dibenzothiophene were added, in combination with heavy oil waste, to the top 15 cm of soil in a 70 cm soil column. The distribution of ^{14}C as parent compound and metabolites in the microcosm systems is summarized in Table 4. The 2-methylnaphthalene was predominantly degraded to CO_2, whereas the dibenzothiophene appeared to be degraded to stable intermediates or irreversibly bound to soil organic matter. Irreversible binding of metabolites was also greater for 2-methylnaphthalene in the acidic McLaurin soil. Breakthrough of ^{14}C-labeled metabolites occurred following leaching of the Meota soil columns with four pore volumes of water; however, negligible amounts of ^{14}C were detected in McLaurin soil leachate. The results of these laboratory studies are being used to assess methods of enhancing degradation and to evaluate mathematical models for prediction of contaminant transport through soil.

Table 4. The Distribution of ^{14}C in Terrestrial Microcosm Systems After 161 Days Incubation with Heavy Oil Waste in Soil

	dibenzothiophene	2-methylnaphthalene	
	Meota	Meota	McLaurin
	% added ^{14}C		
Volatile			
(parent)	0.01	0.3	0.6
(metabolite)	0.2	1.8	1.4
CO_2	7.7	58	56
Soil			
(parent)	7.1	0.06	1.9
(metabolite)[a]	60	8.2	25
Leachate			
(metabolite)	14	8.3	0.1
Total recovery	89	77	85

[a]Not extractable from soil residue.

DECOMMISSIONING OF CONTAMINATED PETROLEUM SITES

A National Steering Committee for Industrial Site Decommissioning was established in 1986 under the direction of the Canadian Council of Environment Ministers to establish national guidelines for the decommissioning of industrial sites. Decommissioning includes the cleanup of soils which have become contaminated through spills, leaks, and improper waste disposal as well as through controlled landfarming activities. Decommissioning is a complex problem because cleanup activities are specific to industry type; products and by-products; the age of the plant; its location (geography, geology, hydrogeology, and climate of the site); waste management practices; and the proposed future use of the site. A project was initiated which involved two major components: (a) the development of a multimedia (air, water, soil) model for use as an aid in establishing site-specific cleanup criteria, and (b) a monitoring program at a field test site for evaluation of transport and fate algorithms in the model (Activity 5 in Table 2).

Model Development

The multimedia model is known as AERIS (Aid for Evaluating the Redevelopment of Industrial Sites). This model links exposure assessment (multimedia pathways models) with toxicity assessment as part of an overall risk evaluation procedure. Information about the site being studied, the environmental behavior of a substance in site soil, and the characteristics of a future site user are used in model calculations. Development of the AERIS model has been based on previous work that developed site-specific cleanup requirements for two decommissioned oil refinery sites in Ontario, Canada.[9] The original work has been enhanced by incorporating information for various Canadian environments, organic and inorganic substances, and including algorithms for detailed estimation of transport in the soil system. An "expert system" shell was used to facilitate the transfer of information between the model user, the model's database, and computational procedures.

A demonstration version of AERIS was initially developed[10,11] for further evaluation. Individual algorithms that estimate environmental fate and concentrations have been verified and compared to measured data. The model has been calibrated against the risk assessment procedure upon which it was based using the data and recommendations obtained for the decommissioned Canadian oil refinery sites. Additional comparisons will be made with independently derived risk-based criteria for other decommissioned sites as such information becomes available. Sensitivity of model predictions to variation in input parameters is assessed with each use of the model. AERIS model validation and improvement will be undertaken as an ongoing process involving the model developers and users.

The advantage of the AERIS model is that many model runs can be performed in a relatively short period of time and at reduced cost as compared to current risk assessment methods. This allows sensitivity analysis to be performed on cleanup criteria, i.e., assessment of a number of "what if" scenarios. In consideration of the uncertainty in the risk assessment process in general, the comparison of several risk assessment scenarios is highly desirable.

Field Evaluation

Movement of contaminants in the soil's unsaturated zone is a critical factor in the potential release of soil-based contaminants into the environment and is therefore a critical component of transport and fate pathways analysis in the AERIS model. To date, limited work has been completed in the modeling of the unsaturated zone of soil, particularly when compared to the efforts expended in groundwater and air quality modeling. The U.S. Environmental Protection Agency (EPA) has recently developed the Regulatory and Investigative Treatment Zone (RITZ) model[12] to predict the proportions of a contaminant which will be degraded, volatilized, and leached in soil, based on soil and waste characteristics, kinetic parameters of degradation, and volatilization and phase partitioning. A cooperative study between EPA and Environment Canada was implemented to field evaluate the transport algorithms which are components of AERIS, RITZ, and other transport models.

The objectives of the field program were (a) to acquire site information to use as input parameters for environmental pathways analysis, and (b) to collect sufficient information on the concentrations of chemicals in soil, water, and air phases over time at the field site to allow comparison with pathways predictions. The field program was conducted on a landfarm site at the Texaco Canada refinery, Nanticoke, Ontario. Refinery waste was applied to plots at two application levels. Soil sampling was performed periodically at three depths using a coring device. Water sampling was performed at two depths using both porous cup and glass brick sampling devices. A significant component of the field program was extensive monitoring of volatile emissions before and after waste application and tillage operations. The air monitoring program was performed by Dr. R. Dupont, Utah State University.

Data collection from the field site has been performed over a two year period to evaluate transport and fate pathways predictions for a variety of models. Preliminary results indicate that the transport models evaluated provided a qualitative estimate of contaminant movement and degradation in soil. The contaminant most susceptible to leaching was predicted and observed to be phenol. The most persistent and immobile contaminant was predicted and observed to be 2-methylnaphthalene. In general, contaminants at the field site have not moved appreciably within the soil profile, preventing a quantitative evaluation of the models at this time. A discussion of preliminary project results[13] can be obtained from H. Campbell at the Wastewater Technology Centre.

SUMMARY

Land application of waste in a controlled manner is a cost-effective method of treating waste. Studies performed by Environment Canada have indicated that there is potential for degrading most constituents of petroleum waste through appropriately controlled land application. Waste application to soil may be limited, however, by persistence of some waste constituents, transport of unidentified transformation products, salinity, and the phytotoxicity of residual oil. Some waste types may be unsuitable for land treatment.

Environment Canada is currently in the process of collecting information on which to base policy decisions regarding the land treatment of waste. Critical to the development of policy and guidelines for land treatment of waste is the determination of acceptable concentrations of compounds in environmental media. One approach being investigated by Environment Canada to determine environmentally acceptable concentrations is the use of computer modeling. A significant research need in this area is the toxicological assessment of a large number of compounds. Research is also continuing to identify and model contaminant pathways and to investigate land treatment techniques which are protective of human health and the environment.

REFERENCES

1. Petroleum Association for Conservation of the Canadian Environment, "Canadian Petroleum Refining Industry Waste Survey," PACE Report 80-4 (1980).
2. Bulman, T. L., K. Hosler, P. J. A. Fowlie, S. Lesage, and S. Camilleri, "The Fate of Polynuclear Aromatic Hydrocarbons in Refinery Waste Applied to Soil," PACE Report 88-1 (1988).
3. Bulman, T. L., S. Lesage, P. J. A. Fowlie, and M. D. Webber, "The Persistence of Polynuclear Aromatic Hydrocarbons in Soil," PACE Report 85-2 (1985).
4. Hosler, K. R., T. L. Bulman, and P. J. Fowlie, "The Persistence and Fate of Aromatic Constituents of Heavy Oil Production Waste Applied to Soil-Interim Report," April 1990.
5. Visser, S., R. M. Danielson, and E. Peake, "Land Treatment of Diesel Invert Mud Residues: Laboratory and Greenhouse Feasibility Assessments," Report to Canadian Petroleum Association and Environment Canada (1987).
6. Ashworth, J., R. P. Scroggins, and D. McCoy, "Feasibility of Land Application as a Waste Management Practice for Disposal of Residual Diesel Invert-based Muds and Cuttings in the Foothills of Alberta," presented at the International Conference on Drilling Wastes, April, 1988.
7. Danielson, R. M., N. Okazawa, W. J. Ceroici, E. Peake, and D. Parkinson, "Disposal of Oilfield Wastes by Land Treatment: 1986 Studies on the Effects on the Environment and Implications for Future Land Use," Report to Canadian Petroleum Association and Environment Canada (1987).
8. Danielson, R. M., N. Okazawa, and W. J. Ceroici, "Disposal of Oilfield Wastes by Land Treatment: 1987 Studies on the Effects on the Environment and Implications for Future Land Use," Report to Canadian Petroleum Association and Environment Canada (1988).
9. Ibbotson, B. G., D. M. Gorber, D. W. Reades, D. Smyth, I. Munro, R. F. Willes, M. G. Jones, G. C. Granville, H. J. Carter, and C. E. Hailes, "A Site-Specific Approach for the Development of Soil Clean-up Guidelines for Trace Organic Compounds," *Proc. 2nd Con. on Environmental and Public Health Effects of Soils Contaminated with Petroleum Products*, University of Massachusetts, Amherst, Massachusetts (1987).

10. SENES Consultants Ltd., "Contaminated Soil Cleanup in Canada—Volume 2 Interim Report on the Demonstration Version of the AERIS Model" prepared for the National Steering Committee, Industrial Site Decommissioning (1988).
11. SENES Consultants Ltd., "Contaminated Soil Cleanup in Canada—Volume 5 Development of the AERIS Model Final Report" prepared for the National Steering Committee, Industrial Site Decommissioning (1989).
12. Nofziger, D. L., J. R. Williams, and T. E. Short, "Interactive Simulation of the Fate of Hazardous Chemicals During Land Treatment of Oily Wastes: RITZ User's Guide," Robert S. Kerr Environmental Research Laboratory, US EPA, Ada, Oklahoma (1987).
13. Bulman, T. L., K. R. Hosler, B. Ibbotson, D. Hockley, and M. J. Riddle, "Development of a Model to Set Clean-up Criteria for Contaminated Soil at Decommissioned Industrial Sites," *Contaminated Soil '88; Proceedings of the Second International TNO/BMFT Conference on Contaminated Soil, Volume* 1, Kluwer Academic Publishers, Norwell, MA (1988), pp. 299–308.

CHAPTER 6

An Update on a National Survey of State Regulatory Policy: Cleanup Standards

Charles E. Bell, Paul T. Kostecki, and **Edward J. Calabrese,** School of Public Health, University of Massachusetts, Amherst, MA

When the U.S. Environmental Protection Agency's (EPA) Office of Underground Storage Tanks (UST) released its Final Rules[1] for UST technical requirements and state program approvals, it was clear that each state was to be responsible for developing their own programs and policies for cleanup of petroleum hydrocarbon contaminated soils (PCS). Such an approach would allow, among other things, for each state to develop and implement a program that would emphasize and take into consideration local climatic, geologic, and demographic conditions. All that was required was that individual state programs be as, or more, stringent than current EPA requirements. In an effort to keep abreast of regulatory developments in this area, an ongoing national state survey has been conducted for the purpose of cataloging states' approaches and management strategies for investigation and cleanup of PCS sites.

Survey efforts focused on a number of different aspects of state regulatory programs including: research activities relating to PCS; cleanup levels for soils; relationship to existing air and water quality standards and programs; classification of PCS as a hazardous, nonhazardous, industrial and/or solid waste; rules and regulations governing treatment and disposal options; analytical and field screening protocols; funding mechanisms for cleanup; permitting programs; treatment technologies in use; and standard operating procedures for site investigation and corrective action.

Both the regulated and regulatory communities have been particularly interested in the development and application of cleanup levels or numbers for PCS. This chapter focuses on this topic by providing a state-by-state overview of regulatory cleanup objectives and an examination of the use of contamination levels as a screening tool, action level, cleanup standard, and remediation goal for PCS.

OVERVIEW

Potential respondents were identified from a list of participants of past surveys conducted on this topic.[2] Very often, two or more agencies within a state have overlapping jurisdiction or are responsible for different aspects of PCS site cleanups. As a result, as many as five regulatory representatives from different departments within a state may have been contacted and interviewed.

Once identified, participants were provided with an overview of the material to be covered in the survey, and arrangements were made to conduct the interview over the phone at a later date. This provided an opportunity for regulators to collect reference materials and prepare their responses. At the time of the interview, questionnaires were filled out on behalf of the participants and subsequently mailed to participants for verification and validation. Respondents were asked to sign and return the questionnaire along with any relevant literature or documentation.

In order to compare and contrast information from different states the terminology used to describe the use of contamination levels must be defined. For the purposes of this chapter, a soil cleanup standard is one comparable to an air or water quality standard, established by law as a rule. Action levels are most often used to advise responsible parties (RPs) if further corrective action is required. Guidance levels function to direct and dictate the regulatory response by the agency, and remediation goals refer to acceptable endpoint contamination concentrations to be determined on a site-by-site basis.

Much of the controversy surrounding the use of cleanup numbers centers on their use or interpretation as fixed standards, and concerns that the parameters measured do not adequately characterize the contaminant present and its potential risk to public health. Survey data from 40 states has revealed that virtually all agencies contacted use total petroleum hydrocarbon (TPH) as one means to measure the extent of soil contamination resulting from a spill or tank leak of petroleum product. Additional analytical measurements of benzene, toluene, ethylbenzene, total xylenes (BTEX), polynuclear aromatic hydrocarbons (PAHs), methyl tert-butyl ether (MTBE) and 1,2 dichloroethane may or may not also be required. Thirty-four states also use TPH either as a guidance level for cleanup or as a site-specific remediation goal.

CLEANUP OBJECTIVES BY STATE

The following section provides a brief description of some aspects of each state program in relation to their cleanup objectives. The name of the agency or agencies refers to those from whom the information was collected and/or those having jurisdiction in this area. The contact date reflects the time the information was collected, updated, or verified. The guidance and limits of soil cleanup and related information sections indicate whether numbers are being used, for what parameters, under what conditions, and any factors that may influence their current or future application. The references listed are regulations or in-house documents from which the information was derived and/or which may provide more detail on the information presented.

Alabama

Name of agency: AL Department of Environmental Management (ADEM), Water Division

Contact date: 01/18/90

Guidance and limits of soil cleanup: Analysis of soil samples for TPH and the elevation of the groundwater table are determined as part of a preliminary site investigation or closure assessment. Sites where depth to groundwater is 5 feet or more below the base of the tank excavation, and TPH concentrations are 100 ppm or less for each sample, or TPH concentrations are 10 ppm or less for every sample irrespective of groundwater conditions, are considered satisfactory and no further action will be required.

Related information: If the conditions identified above cannot be met, the Department may require a secondary investigation which may include additional sampling, analyses, and determination of the full lateral and vertical extent of soil and groundwater contamination. Based on this information, the RP may be required to develop and submit a corrective action plan in response to the contamination. Corrective action limits (CALs) for petroleum contaminated soils are 100 ppm TPH. Not surprisingly, state landfills will normally not accept contaminated soil for disposal in excess of 100 ppm TPH. Alternative levels may be established by the Department based on (a) site-specific factors or (b) if, after implementation of the CAL, the concentration of contaminant in groundwater no longer decreases with continued treatment. A risk assessment must be performed for proposed alternate CALs which may include both exposure and toxicity assessments to characterize the cumulative risks to both the public and environment.

Documents: ADEM Administrative Code R., Water Quality Program. Sections 335-6-15.26 to 335-6-15.34

Alaska

Name of agency: AK Department of Environmental Conservation (DEC), LUST (Leaking Underground Storage Tank) Program

Contact date: 02/05/90

Guidance and limits of soil cleanup: Alaska continues to use the CA Leaking Underground Fuel Tank (LUFT) manual as an assessment tool for guidance in selecting remediation goals. However, a working group of DEC staff members are in the process of drafting regulations tailored more specifically to Alaska's climatic, demographic, and hydrogeologic conditions. Proposed rules include recommended guidelines of 100 ppm TPH for diesel and other middle distillates, and 0.5 ppm benzene, 2 ppm toluene, 0.7 ppm ethylbenzene and 10 ppm xylenes as target values for all sites.

Related information: From the agency perspective, establishment of soil guidance values provides a framework for policy that is enforceable, as compared to current policy in which site-specific levels are left up to the discretion of a regional supervisor. Soil standards are under development in response to a perceived need by the agency from both public and private sectors. They provide a target for industry and offer the potential for minimal involvement by the agency for tank closures, particularly in the case of home heating oil tanks. Included in the proposed rules are soil sampling guidelines adopted from the state of Oregon.

Documents: AK Department of Environmental Conservation, Interim SCRO Soil and Groundwater Cleanup Standards, August 12, 1988.

Arizona

Name of agency: AZ Department of Environmental Quality (ADEQ)

Contact date: 05/01/89

Guidance and limits of soil cleanup: Health-based guidance levels of 130 ppb benzene, 200 ppm toluene, 68 ppm ethylbenzene, 44 ppm xylenes, and 100 ppm TPH are currently used as suggested cleanup levels for contaminated soils in Arizona. A risk assessment must be performed by the RP to justify alternative, site-specific cleanup levels.

Related information: The health-based guidance levels proposed by the Arizona Department of Health are not enforceable soil cleanup levels. Consequently,

ADEQ relies on voluntary compliance by RPs for cleanup. Experience has shown the vast majority of RPs do comply rather than electing to perform a detailed risk-based analysis to arrive at less stringent remediation goals. The health-based guidance level for benzene was based on a literature review of soil ingestion studies in children. Toluene, ethylbenzene, and xylene levels were derived from recommended groundwater cleanup and action levels of 2 ppm, 680 ppb, and 440 ppb, respectively (an attenuation factor of 100).

Documents: None available.

Arkansas

Name of agency: AR Department of Pollution Control and Ecology, UST Program

Contact date: 07/26/89

Guidance and limits of soil cleanup: There are no formal cleanup standards or guidelines that are followed, and remediation goals are determined on a case-by-case basis.

Related information: The legislature appropriated monies in July 1989 for a state UST program. The initial approach has been to apply federal requirements where appropriate, in lieu of development of their own reporting and operating procedures for site investigation and corrective action. The agency does not normally perform site inspections as part of investigation procedures at this time.

Documents: None available.

California

Name of agency: CA State Water Resources Control Board, Department of Health Services

Contact date: 01/13/90

Guidance and limits of soil cleanup: The California LUFT Field Manual provides a general approach for site cleanup based on the leaching potential of compounds or groups of compounds in soil. It is intended to be used as a tool for the determination of site-specific cleanup levels, as opposed to the establishment of state-wide remediation goals. Estimated contaminant concentrations that can be left in place without threatening groundwater are: 10 to 1000 ppm TPH for gasoline; 100 to 10,000 ppm TPH for diesel; 0.3 to 1 ppm benzene; 0.3 to 50 ppm toluene; 1 to 50 ppm xylene; 1 to 50 ppm ethylbenzene.

Related information: California classifies PCS as hazardous if it exceeds 1000 ppm TPH concentration, in which case the LUFT manual procedures may not apply. Regulatory agencies, including county and municipal offices, may consider additional or alternative factors such as potential groundwater and land use when characterizing a site. The LUFT manual also provides responsible regulatory agencies with guidance for site investigation, risk assessment, and remediation while functioning as a tool for screening out sites which may or may not require further study. Other agencies, including local air quality management districts, Office of the State Fire Marshall, and Office of Emergency Services, have jurisdiction for toxic air pollution control, fire, and explosion hazards, and have developed procedures for tank closures and fuel leak sites.

Documents: Leaking Underground Fuel Tank (LUFT) Field Manual, State Water Resources Control Board, State of California, Sacramento, CA, October 1989.

Colorado

Name of agency: CO Department of Health (DOH), Water Quality Control Division

Contact date: 02/07/90

Guidance and limits of soil cleanup: Cleanup levels are determined on a site-by-site basis. Specific remediation goals are often to background or a level that would protect groundwater resources and minimize potential offsite impacts.

Related information: Colorado recently passed UST program legislation, intended to be no more stringent than current EPA regulations, which state that the Department of Health can be involved in corrective action for site cleanup. The Water Quality Control Division of the DOH has established limits of 5 ppb benzene, 680 ppb ethylbenzene, and 2420 ppb toluene for drinking water purposes. Corrective action plans submitted by RPs must address both environmental and public health concerns, particularly with respect to potential offsite impacts. An owner or operator may elect to perform a health-based risk analysis in order to make that determination. The DOH reviews and approves all corrective action plans. In those cases for which no agreement can be reached between the department and the owner or operator, a UST Advisory Committee made up of public officials, industry representatives, and public interest groups makes recommendations for acceptable corrective action.

Documents: State of Colorado, House Bill No. 1299. Amendment to Section 1. Article 20 of Title 8, Colorado Revised Statutes, Underground Storage Tanks.

Connecticut

Name of Agency: CT Department of Environmental Protection (DEP), Water Compliance Unit

Contact date: 01/09/90

Guidance and limits of soil cleanup: There are no soil cleanup guidelines or action levels at present. Site-specific cleanup levels are those determined necessary to protect groundwater resources.

Related information: The UST Group within the DEP Hazardous Management Unit regulates underground storage tanks and the handling of petroleum products. Soil cleanup levels are determined based on the potential for groundwater contamination and its use. The Department of Health Services uses public health code regulations and action levels for volatile organics to determine the potablility of drinking water supplies. Connecticut is currently using 1 ppb for benzene and 1 ppm for toluene as action levels for groundwater contamination.

Documents: CT Department of Health Services, Public Health Code Regulation 19-13-B102. CT Department of Health Services, Volatile Organics and Inorganics Action Levels, January 1988.

Delaware

Name of agency: DE Department of Natural Resources and Environmental Control, Division of Air and Waste Management

Contact date: 07/10/89

Guidance and limits of soil cleanup: No guidance or action levels are reported at this time.

Related information: The Division of Air and Waste Management is reportedly in the process of developing a technical document which may include soil cleanup guidelines or remediation goals. Cleanup levels may be incorporated into the existing corrective action plan process for UST sites.

Documents: None available.

Florida

Name of agency: FL Department of Environmental Regulation, Bureau of Waste Cleanup

Contact date: 09/18/89

Guidance and limits of soil cleanup: Soils that result in a TPH reading greater than 500 ppm using an organic vapor analysis (OVA) instrument with a flame ionization detector are considered excessively contaminated and must be remediated. Soils with vapor readings between 10 and 500 ppm TPH are deemed contaminated and may require remediation, depending upon the concentration, the potential effect on groundwater, and soil type. Soils with vapor readings between zero and 10 ppm TPH are considered clean.

Related information: Current guidelines allow for "initial remedial actions" (IRA) to be taken by RPs at sites to expedite the removal of excessively contaminated (>500 ppm) soils and free product. At sites where only kerosene or diesel fuel contamination is known to exist, an OVA reading higher than 50 ppm is considered justification for timely removal of the material. The responsible party must notify the Bureau of Waste Cleanup verbally within 24 hours and in writing within 3 days in order to be eligible for reimbursement for cleanup costs associated with IRAs. This approach is intended to eliminate the source of contamination, thereby minimizing further potential impact while a more detailed site assessment and remedial action plan can be developed to address the entire problem. Treatment options for soils generated by IRAs are, in order of preference, incineration, landfilling, and land framing. Treated soils may be disposed or used as clean fill, road bed material or incorporated into asphalt mix, depending on the residual concentration of BTEX and TPH left in the soil. There is pending legislation and rule making that may transform current cleanup guidelines into enforceable rules.

Documents: "Guidelines for Assessment and Remediation of Petroleum Contaminated Soils," FL Department of Environmental Regulation, January 1989.

Hawaii

Name of agency: HI Department of Health

Contact date: 01/09/89

Guidance and limits of soil cleanup: No guidance or action levels reported.

Related information: The agency currently relies on federal guidelines (RCRA) and state water quality standards to evaluate site-specific environmental and public health risks and determine cleanup requirements.

Documents: None available.

Illinois

Name of agency: IL Environmental Protection Agency (IEPA), Division of Land Pollution Control

Contact date: 01/11/90

Guidance and limits of soil cleanup: At sites where groundwater contamination is not a factor, soil cleanup objectives are 0.025 ppm for benzene, and the sum of individual BTEX values should not exceed 16.025 ppm. All visibly contaminated soils, as well as those exhibiting petroleum odors must also be removed. These objectives do not apply in cases of offsite releases, when the capacity of excavation equipment is a limiting factor, or when excavation limits (i.e., a public road) have been encountered. Groundwater impacted sites are handled on a case-by-case basis.

Related information: A Cleanup Objectives Team (COT) and Coordinated Permit Review Committee (CPRC) have been developed by IEPA to address site-specific cleanup objectives for cases such as offsite releases and those involving groundwater contamination. The COT provides technical support (risk assessment) and recommends numerical cleanup objectives to the CPRC based on protection of the environment and the public health. The CPRC reviews the COT's recommendations (risk management) while taking into account other factors including financial, policy, and legal aspects of the case, and either accepts or modifies those recommendations accordingly. Final cleanup objectives are then incorporated into the CAP for the site.

Documents: Guidance Manual for LUST Cleanups in Illinois. IL Environmental Protection Agency, September 1989.

Indiana

Name of agency: IN Department of Environmental Management

Contact date: 11/15/89

Guidance and limits of soil cleanup: No guidance or action levels at present. Cleanup determined on a site-by-site basis.

Related information: The Department is considering the use of a 100 ppm TPH action level as a screening tool for site investigations.

Documents: None available.

Iowa

Name of agency: IA Department of Natural Resources, Department of Water, Air and Waste Management

Contact date: 02/05/90

Guidance and limits of soil cleanup: Iowa uses a soil vapor analysis level of 10 ppm as a screening tool and action level for further investigation. Soils greater than 100 ppm TPH as determined by laboratory soil tests are considered contaminated and must be remediated. Final remediation goals are based on site-specific conditions and determined on a case-by-case basis.

Related information: The Department is currently developing a more detailed, comprehensive soil cleanup policy which will take into account recently established groundwater and surface water cleanup guidelines.

Documents: None available.

Kansas

Name of agency: KS Department of Health and Environment, Bureau of Environmental Remediation

Contact date: 11/06/89

Guidance and limits of soil cleanup: Contaminant levels for soils in excess of 1.4 ppm benzene, 8 ppm 1,2-dichloroethane (for leaded gasoline releases) and 100 ppm TPH are contaminated and require treatment. TPH is defined as the sum of individual concentrations of toluene, xylene, ethylbenzene, and methyl tert-butyl ether (MTBE).

Related information: The Department has established a field testing protocol for screening contaminated soils to allow for rapid remediation and closure of sites. Field screening of gasoline contaminated soils is performed using head space testing with drager sample tubes. Sensory methods (sight and smell) are used for diesel contaminated soils, with laboratory analysis being used where significant contamination is evident.

Documents: Technical Guidance Procedures for Petroleum Storage Tank Site Evaluations. KS Department of Health, January 1989.

Kentucky

Name of agency: KY Dept. of Environmental Protection, Hazardous Waste Branch, UST Program

Contact date: 06/20/89

Guidance and limits of soil cleanup: The agency requires all contaminated sites to be returned to background levels or detection limit.

Related information: Collection of both background and suspected or confirmed contaminated soil samples is required at all tank removals. Gasoline spills are analyzed for BTEX; diesel fuel contamination requires analysis for PAHs; and waste oil contamination levels are measured for TPH.

Documents: Guidelines for Site Investigations of Leaking Underground Storage Tank Sites in Kentucky. KY Dept. of Environmental Protection, January 1990.

Maine

Name of agency: ME Department of Environmental Protection, Bureau of Oil and Hazardous Materials

Contact date: 02/15/90

Guidance and limits of soil cleanup: No soil cleanup guidelines at present. Remediation goals established on a case-by-case basis as a function of the potential for surface and groundwater contamination, and related public health concerns.

Related information: Cleanup objectives of 20 to 50 ppm TPH as measured by field instrumentation (H-Nu) are often considered acceptable for the purpose of protecting the environment and public health. Pending legislation (June 1990) may provide a rule-making process that may serve to further standardize site investigation procedures.

Documents: None available.

Maryland

Name of agency: MD Department of the Environment, UST Program

Contact date: 02/15/90

Guidance and limits of soil cleanup: All free-phased product must be removed from hydrocarbon contaminated soil sites. The department may also determine that dissolved contamination must be reduced to levels both feasible and cost beneficial.

Related information: Remediation of dissolved (groundwater) contamination must be performed until an asymptotic level of contamination is reached or the department indicates remediation may be discontinued. Sites where subsurface recovery systems are used may discharge wastewater effluent at concentrations not to exceed 5 parts per billion (ppb) benzene, or the sum of individual BTEX concentrations not to exceed 100 ppb.

Documents: Workshop Materials—Consultants Day. Maryland Department of the Environment, February 15, 1990.

Massachusetts

Name of Agency: MA Department of Environmental Protection

Contact date: 02/06/90

Guidance and limits of soil cleanup: Soils contaminated with gasoline in excess of 1800 ppm volatiles, waste oils, or oil residuals with a weight/weight concentration greater than 300 ppm TPH, are regulated as hazardous waste. Volatiles are measured as total organic headspace vapors expressed as benzene. Cleanup guidelines of 100 ppm TPH or 10 ppm total volatiles have been established as remediation goals, but are not fixed standards. The agency may set site-specific alternative levels where it deems appropriate.

Related information: Treated soils may be returned to the original excavation site, depending upon the type of area and concentration of contaminant remaining in the soil. High environmental impact areas include highly populated residential areas, and those where known or potential drinking water supplies can occur. "On-site reuse" of these materials may be allowed if contamination levels do not exceed 10 ppm total headspace volatiles (as benzene) or 100 ppm TPH. Low environmental impact areas include commercial, industrial, and sparsely populated areas which may have high background contamination and/or little potential for future use of groundwater resources. Contaminated soils less than 100 ppm headspace volatiles or 300 ppm TPH may be permitted for storage in these areas.

Documents: Management Procedures for Excavated Soils Contaminated with Virgin Petroleum Oils. MA Dept of Environmental Protection, Policy #WSC-89-001, June, 1989.

Michigan

Name of agency: MI Department of Natural Resources

Contact date: 01/29/90

Guidance and limits of soil cleanup: At present responsible parties are required to cleanup all contaminated soils to background levels.

Related information: Recently proposed rules provide for one of three options as remediation goals an RP may select for cleanup: type (A)—reduce concentrations to levels not to exceed background or detectable limit; type (B)—cleanup to concentrations that do not pose an unacceptable risk on the basis of standardized exposure assumptions and acceptable risk levels; type (C)—cleanup to concentrations that do not pose an unacceptable risk based on a site-specific assessment of risk. Acceptable cleanup criteria to be included in any corrective action plan are: individual concentrations of 10 ppb for BTEX in gasoline contaminated soils; individual concentrations of 10 ppb BTEX, 300 ppb for PNAs, and 100 ppm TPH for diesel and kerosene contaminated soils.

Documents: Proposed Administrative Rules—Environmental Response Act (1982 PA 307 as Amended). MI Dept. of Natural Resources, February, 1990.

Minnesota

Name of agency: MN Pollution Control Agency (MPCA)

Contact date: 12/12/89

Guidance and limits of soil cleanup: PCS are to be cleaned to background or less than 1 ppm TPH.

Related information: Approximately 85% of PRs accept responsibility for reported releases and agree to perform a remedial investigation and implement a corrective action plan. In cases where the RP cannot be identified or refuses to cooperate, the MPCA may elect to perform the investigation, develop, and implement a corrective action plan itself. Although the MCPA could enforce an order for cleanup via the Attorney General's office, the current policy is intended to ensure timely remediations as opposed to potentially protracted litigation and delays in cleanup.

Documents: None available.

Mississippi

Name of agency: MS Department of Natural Resources, Bureau of Pollution Control

Contact date: 02/26/90

Guidance and limits of soil cleanup: Cleanup levels of 100 ppm BTEX for soils contaminated with gasoline, and 100 ppm TPH for diesel and/or waste oil spills are currently applied. In cases where a site is contaminated with both gasoline and diesel fuels, the more stringent of 100 ppm BTEX or 100 ppm TPH would apply.

Related information: More stringent cleanup goals may be required in areas considered sensitive in terms of environmental receptors, such as geologic recharge areas and public or private wells.

Documents: None available.

Nevada

Name of agency: NV Department of Conservation and Natural Resources, Division of Environmental Protection

Contact date: 08/25/89

Guidance and limits of soil cleanup: Soils with TPH readings in excess of 100 ppm may or may not have to be removed, depending on site-specific conditions. Removed soils may require treatment such as land farming prior to their disposal.

Related information: Decisions regarding soil removal are based on site-specific information including: depth, quality, and use of groundwater; distance to nearest well; soil type and permeability; annual precipitation; type and age of contaminant; migration potential; present and future land use.

Documents: Hydrocarbon Spill and Remedial Action Policy. NV Division of Environmental Protection, October, 1987.

New Hampshire

Name of agency: NH Department of Environmental Services

Contact date: 09/12/89

Guidance and limits of soil cleanup: Remediation goals of 10 ppm TPH and 1 ppm total BTEX for gasoline contaminated soils and 100 ppm TPH and 1 ppm total BTEX for all other petroleum products are used.

Related information: The agency may set more stringent cleanup goals based on site-specific conditions (i.e., well head protection areas, sensitive wildlife habitats, heavily populated residential and commercial areas). Remediation goals were established using the California LUFT Field Manual leaching potential analysis for gasoline and diesel. Data from New Hampshire case histories were used in the model resulting in these goals.

Documents: Policy for Management of Soils Contaminated from Spills/Releases of Virgin Petroleum Products. NH Dept. of Environmental Services, Multi-Media Oil Contamination Task Force, June, 1989.

New Jersey

Name of Agency: NJ Department of Environmental Protection, Division of Hazardous Site Mitigation

Contact date: 09/11/89

Guidance and limits of soil cleanup: Soils contaminated in excess of 3% (30,000 mg/kg) are classified as hazardous and (if not treated) require disposal in a hazardous waste facility. Contaminated soils greater than 100 ppm TPH usually require cleanup based on the extent of volatile (VO) and semivolatile (SVO) or base-neutral constituents present.

Related information: If the levels of VOs and SVOs present are near or below 1 ppm and 10 ppm, respectively, and soil contamination is at least 6 inches below the surface, no remediation may be required. The rationale for this is that in the absence of any significant amount of the mobile constituents, the remaining petroleum hydrocarbons are considered long chained, relatively immobile compounds that do not pose a significant environmental or human health threat. Current policy is undergoing critical review to investigate use of alternative parameters of measurement (PNAs, VOCs) for site investigation.

Documents: None available.

New Mexico

Name of agency: NM Health and Environment Department, Environmental Improvement Division, UST Bureau

Contact date: 02/09/90

Guidance and limits of soil cleanup: Recently proposed action levels for soils contaminated with gasoline or lighter hydrocarbons call for remediation to proceed if one of the following conditions are met: concentrations exceed 100 ppm using field headspace methods (PID/FID); laboratory analysis indicates the sum of all detected aromatics is greater than 50 ppm, or benzene is greater than 10 ppm. For diesel sites, soils in excess for 100 ppm TPH must be remediated.

Related information: At sites contaminated with more than one petroleum product, the strictest action level would apply. Soils are considered highly contaminated if, when placed on filter paper, hydrocarbon saturates the paper. Soils exhibiting a very strong odor, gross staining, or high moisture content may also be termed highly contaminated. It is normally recommended these soils be excavated and thin spread to promote aeration. Alternatives to thin spreading such as bioremediation or incineration must be submitted for agency approval.

Documents: UST Bureau Soils Policy (draft). NM Health and Environment Department, February, 1990.

New York

Name of agency: NY State Department of Environmental (NYSDEC) Conservation, Bureau of Spill Response

Contact date: 02/01/90

Guidance and limits of soil cleanup: Guidance levels for soils are currently under development.

Related information: Proposed cleanup criteria for soils will be determined for individual chemicals and compounds based on the potential human health risk, water quality considerations, and the protection of fisheries and wildlife. Actual soil guidance values (Cs) are to be calculated using the water partition theory equation (Cs=f*Koc*Cw), where Cw is the most stringent New York State Department of Health drinking water or groundwater standard in parts per billion, Koc the contaminant-specific partition coefficient, and an assumed organic content (f) of 2.5%.

Documents: Methodology for Soil Guidance Module (draft). NYSDEC, Bureau of Spill Response, September, 1989.

North Carolina

Name of agency: NC Department of Natural Resources and Community Development, Division of Environmental Management

Contact date: 09/18/89

Guidance and limits of soil cleanup: No guidelines or formal standards at present. Sites normally required to be cleaned to background or detection limit.

Related information: Laboratory analysis of soil samples required for TPH, BTEX, and naphthalene, depending on the nature of the contaminant. Landfills will not accept excavated soils in excess of 100 ppm TPH.

Documents: None available.

North Dakota

Name of agency: ND State Department of Health, Division of Waste Management

Contact date: 08/01/89

Guidance and limits of soil cleanup: No cleanup guidelines at present. Each site handled subjectively on a case-by-case basis.

Related information: Soil action levels for TPH and BTEX are under consideration. The Department does not promote land treatment or bioremediation as the most suitable treatment technology available, but does have guidelines for its use at agency approved landfills. However, landfill owners or operators can refuse to accept contaminated soils.

Documents: ''Guidelines for Proper Land Treatment of Petroleum Product Contaminated Soils.'' North Dakota Department of Health and Consolidated Laboratories, January 24, 1989.

Ohio

Name of Agency: OH Department of Commerce, Division of State Fire Marshall, Bureau of UST Regulations

Contact date: 08/18/89

Guidance and limits of soil cleanup: All contaminated soils must be removed as part of corrective actions and cleaned to background or analytical detection limit.

Related information: Alternatively, health-based cleanup standards may be proposed by the RP on a site-specific basis. The proposal must be able to demonstrate that the contaminants remaining in the soil will not adversely impact the environment and that exposure to this material will not threaten human health or local fauna. The exposure assessment must address dermal, inhalation, and ingestion exposure pathways.

Documents: "Corrective Action Policy and Procedures." Appendix C1, OH Department of Commerce, State Fire Marshal's Office, September, 1988.

Oklahoma

Name of agency: OK Corporation Commission, UST program

Contact date: 08/29/89

Guidance and limits of soil cleanup: Temporary action levels of 50 ppm TPH and 10 ppm total BTEX have been established for tank closures and sites where the contamination is limited to soils.

Related information: State water quality standards will be applied at sites where groundwater or surface waters have been impacted. Soil action levels are being used to advise RPs if further corrective action may be required. RPs may justify alternative levels based on site-specific factors including water quality and soil conditions.

Documents: None available.

Oregon

Name of Agency: OR Department of Environmental Quality, Environmental Cleanup Division

Contact date: 08/18/89

Guidance and limits of soil cleanup: A matrix of numeric soil cleanup standards based on TPH levels are used at sites where contamination is restricted to soils. Levels range from 40 to 130 ppm TPH for gasoline and 100 to 1000 ppm TPH for diesel fuel.

Related information: A matrix score is calculated as the sum of 5 parameter scores based on site-specific evaluation criteria: depth to groundwater; mean annual precipitation; native soil type (in terms of permeability); sensitivity of the uppermost aquifer; potential receptors as determined by the distance to the nearest well, and the number of people at risk. Results are compared to a table of required numeric soil cleanup standards based on the level of TPH as measured by EPA Method 418.1. The soil cleanup standards or matrix are part of a more comprehensive set of cleanup rules in which initial emergency response, abatement measures, and site characterization procedures must first be performed in order to determine their potential application. Sites involving groundwater contamination normally require a more detailed investigation and development of a corrective action plan in which alternative cleanup levels may be set. In the case of very large or complex releases, the agency may require the remedial action plan be developed under the Hazardous Substance Rules (OAR 340-122-010 to 110).

Documents: "Cleanup Rules for Leaking Petroleum Systems and Numeric Soil Cleanup Levels for Motor Fuel and Heating Oil," OAR 340-122-101 to 340-122-335. Environmental Quality Commission, State of Oregon, August 1989.

Pennsylvania

Name of agency: PA Department of Environmental Resources (PADER), Bureau of Waste Management

Contact date: 07/27/89

Guidance and limits of soil cleanup: No established requirements for cleanup. Each site handled on a case-by-case basis.

Related information: Currently a computer modeling system is under development named the Risk Assessment/Fate and Transport (RAFT) Modeling System which may be applied to (but not limited to) hydrocarbon contaminated soil sites for the purpose of addressing the question of "How Clean is Clean?" It is intended for use as a tool in the determination of site-specific cleanup levels based on the principles of environmental fate and exposure, exposure assessment, and risk assessment. It may also serve to provide consistency in the development of regulatory cleanup strategies and a means to better understand the relationship between contaminant concentrations in the environment and the potential risk associated with exposure to that contamination.

Documents: User's Manual for Risk Assessment/Fate and Transport (RAFT) Modeling System. PADER, Bureau of Waste Management, October 24, 1989.

Rhode Island

Name of agency: RI Department of Environmental Management, Division of Groundwater

Contact date: 04/01/89

Guidance and limits of soil cleanup: Determination of site-specific cleanup levels are primarily based on the nature of the contaminant and whether the site is located on residential or commercial property. A corrective action level of 50 ppm TPH is most often used on residential properties.

Related information: In addition to TPH measurements, analysis of soil samples may also be required for VOCs in the case of gasoline spills, naphthalene for diesel, and PAHs for heavier fuels (No. 4) as part of the site investigation.

Documents: None available.

South Carolina

Name of agency: SC Department of Health and Environmental Control, Groundwater Protection Division

Contact date: 09/21/89

Guidance and limits of soil cleanup: No soil cleanup guidelines at present. Remediation requirements handled on a site-by-site basis.

Related information: Initial site characterization requires laboratory analysis of soil samples for both TPH and BTEX. The Solid and Hazardous Waste Bureau oversees offsite disposal of petroleum contaminated soils. Most landfill operators will not accept soils in excess of 10 ppm total BTEX. As a result, much of the material requires pretreatment (incineration) prior to its disposal in a landfill.

Documents: None available.

Tennessee

Name of agency: TN Department of Health and Environment, Division of Underground Storage Tanks

Contact date: 12/08/89

Guidance and limits of soil cleanup: Soil cleanup levels range from 10 to 500 ppm total BTX and 100 to 1000 ppm TPH, depending on the permeability of the soils and whether the groundwater present below a site is classified as a drinking water or nondrinking water supply.

Related information: Drinking water supplies are defined as any aquifer or water source that meets the requirements of primary and secondary state drinking water standards, and yields at least one-half gallon per minute. For sites where natural background levels of petroleum exceed the required levels for cleanup, the RP may only be required to clean to the natural background level. Once having treated contaminated soils at a site for an extended period of time so that asymptotic levels have been reached for removal of the contaminant, the RP may request establishment of an alternative site-specific cleanup standard from the agency.

Documents: Petroleum Underground Storage Tanks Program. Rules 1200-1-15-.01 through 1200-1-15-.07 (draft). Technical Standards and Corrective Action Requirements for Owners and Operators of Petroleum Underground Storage Tanks. TN Department of Health and Environment, October, 1989.

Texas

Name of agency: TX Department of Health (TDH), Water Commission

Contact date: 08/29/89

Guidance and limits of soil cleanup: Soils do not require remediation if they are less than 100 ppm TPH or total BTEX is less than 30 ppm and are located in areas where groundwater is not considered threatened. All other sites are handled on a case-by-case basis.

Related information: The Division of Solid Waste Management has released requirements and conditions for the disposal of petroleum contaminated soils in municipal landfills based on the type of contaminant: gasoline contaminated soils with a total BTEX concentration less than 500 ppm and TPH concentration less than 1000 ppm may be accepted without specific TDH authorization; diesel contaminated soils less than 1000 ppm TPH may be accepted; fuel oil, aviation gasoline, and unknowns are reviewed on a case-by-case basis.

Documents: None available.

Utah

Name of agency: UT Bureau of Solid & Hazardous Waste Management

Contact date: 03/18/89

Guidance and limits of soil cleanup: No soil cleanup guidelines or policy at present. Remediation goals established on a site-by-site basis.

Related information: Site investigations require laboratory analysis of soil samples for BTEX and TPH. Corrective action plans may take into account economic factors as they relate to site-specific conditions when determining remediation goals.

Documents: None available.

Vermont

Name of agency: VT Department of Environmental Conservation, Hazardous Material Division

Contact date: 04/15/90

Guidance and limits of soil cleanup: In cases where contamination is limited to soils, an action or ''trigger'' level of 20 ppm volatile constituents, as measured using a photo ionization device (PID), is used to determine what soils must be excavated and treated onsite or removed for offsite disposal. Remediation goals are handled on a case-by-case basis.

Related information: Initial corrective actions taken or required by the Department are based on the presence or absence of potentially sensitive receptors such as drinking water wells or surface waters. Excavated soils destined for landfill disposal must first be land spread at the landfill site for a minimum of six weeks before they may be used as cover material. Treated soils may then be placed in the landfill if concentrations are less than 100 ppm volatile constituents using a PID.

Documents: None available.

Virginia

Name of agency: VA State Water Control Board, VA Department of Waste Management

Contact date: 12/12/89

Guidance and limits of soil cleanup: Contaminated soils greater than 100 ppm TPH must be removed from a site.

Related information: One hundred ppm TPH is also used as a disposal guideline for landfilling excavated material. Land farming (bioremediation) is commonly used as an offsite treatment option. The agency is interested in developing a policy for the offsite treatment of contaminated soils so as to minimize the necessity for ultimate disposal of these soils to a landfill.

Documents: None available.

Washington

Name of agency: WA State Department of Ecology, Hazardous Waste Investigations and Cleanup Program

Contact date: 01/20/90

Guidance and limits of soil cleanup: Current regulations require cleanup to 100 ppm TPH for gasoline and 200 ppm TPH for diesel fuel releases, and individual compounds (BTEX) must not exceed 100 times the state drinking water standard in soils.

Related information: Recently proposed rules would allow regional administrators to apply a set of compliance cleanup levels for soils: 0.1 ppm for benzene; 4.0 ppm for toluene; 3.0 ppm for ethylbenzene; 2.0 ppm for xylenes; 100 ppm TPH for gasoline releases; and 200 ppm TPH for diesel. Alternatively, they may choose to establish site-specific levels using a Leaching Potential Analysis and Risk Appraisal approach to estimate the levels of BTEX and TPH that can safely be left in place without threatening groundwater resources. The Risk Appraisal approach was developed by the California LUFT Task Force and adapted for use in Washington. The leaching potential analysis focuses on site characteristics that may most significantly influence the downward migration of contaminants. They include depth to groundwater, subsurface fractures, precipitation, and any man-made conduits. If BTEX or TPH values from samples exceed those derived from the leaching potential analysis, additional site investigation may be required in the form of a General Risk Appraisal. This involves the use of two computer models (SESOIL, and AT123D from the CA LUFT manual) to estimate acceptable BTEX concentrations that may be left in place, while incorporating both general (environmental fate and chemistry) and site-specific information.

Documents: Draft Interim Underground Petroleum Storage Tank Removal and Remediation Guidelines. WA Department of Ecology, Hazardous Waste Investigations and Cleanup Program, April, 1989.

Wisconsin

Name of agency: WI Department of Natural Resources, Bureau of Solid and Hazardous Waste Management

Contact date: 02/05/89

Guidance and limits of soil cleanup: An informal action level of 10 ppm TPH has been used as a basis for further corrective action.

Related information: Laboratory analysis of soil samples required as part of site investigation based on a characterization of the contaminant as gasoline, diesel, etc. The Department of Industry, Labor, and Human Relations has site assessment criteria requiring TPH analysis at tank closures. Written guidelines and procedures for RPs for cleanup are under development.

Documents: None available.

Wyoming

Name of agency: WY Department of Environmental Quality, Water Quality Division

Contact date: 06/19/89

Guidance and limits of soil cleanup: Satisfactory cleanup is achieved when contaminant concentrations are less than 10 ppm TPH for soils located in areas where the depth to groundwater is less than 50 feet, and less than 100 ppm TPH where depth to groundwater is greater then 50 feet.

Related information: Remediation goals are to be consistent with current state water quality regulations. Oil and grease content of groundwater must not exceed 10 ppm.

Documents: Wyoming Oil and Hazardous Substances Pollution Contingency Plan. WY Department of Environmental Quality, November, 1989.

REFERENCES

1. "Underground Storage Tank Technical Requirements and State Program Approval; Final Rules," 40 CFR Parts 280 and 281, Environmental Protection Agency, September, 1988. 40 CFR Parts 280 and 281, *Federal Register,* Vol. 53. No. 185, 37082–37247.
2. Bell, C. E., P. T. Kostecki, and E. J. Calabrese. *Petroleum Contaminated Soils,* Volume 2 (Chelsea, MI: Lewis Publishers, 1989), p. 73–94.

PART III

Analytical and Environmental Fate

CHAPTER 7

Misapplications of the EP-Tox, TCLP, and CAM-WET Tests to Derive Data on Migration Potential of Metals in Soil Systems

James Dragun, John Barkach, and **Sharon A. Mason,** The Dragun Corporation, Berkley, MI, 48072-1634

During the late 1970s, the U.S. Environmental Protection Agency (EPA) developed the Extraction Procedure Toxicity Test Method (EP-Tox). This test method is the predecessor to the Toxicity Characteristic Leaching Procedure (TCLP) and serves as the technical basis for the TCLP. In the early 1980s, the state of California adopted the California Assessment Manual Waste Extraction Test (CAM-WET).

These tests are utilized to classify a waste as being either hazardous or nonhazardous. The classification depends upon the amount of inorganic chemical that can be extracted from the waste by the acidic extractant solution containing an organic ligand.

During the past few years, individuals, consulting firms, and governmental agencies have attempted to utilize these test methods to estimate the leaching potential of metals from soils. These groups have theorized that the greater the concentration of metal that can be extracted from the soil, the more mobile the metal must be; the more mobile the metal, the more likely that the metal will migrate to groundwater. More recently, some individuals, consulting firms, and agencies have theorized that the concentration of metals in soil potentially able to migrate to groundwater, such as lead (Pb), zinc (Zn), chromium (Cr), and copper (Cu), is directly proportional to the amount of these metals extracted from soil by EP-Tox, TCLP, and CAM-WET.

EP-Tox, TCLP, and CAM-WET test methods can be used only for the classification of hazardous waste produced by waste generators; they cannot be utilized to determine the migration and degradation potential of any chemical in a soil system. This chapter discusses the reasons why this is the case.

GENERAL FORMS OF INORGANIC CHEMICALS IN SOIL

The total concentration of any chemical, C_{Total} in a soil is equal to:

$$C_{Total} = C_{Fixed} + C_{Ads} + C_{Water} \quad (1)$$

where:

C_{Fixed} = Concentration of fixed chemical comprising part of the structure of clay and soil minerals, in milligrams (mg) chemical/kg soil
C_{Ads} = concentration of chemical adsorbed onto the surface of soil minerals and onto organic matter exchange sites, in mg chemical/kg soil
C_{Water} = concentration of chemical in soil water or groundwater in equilibrium with C_{Ads}, in mg soluble chemical/kg soil.

C_{Fixed} represents the "immobile" fraction of C_{Total}. C_{Ads} and C_{Water} represent the potentially mobile portion of C_{Total}. The ratio C_{Ads}/C_{Water} is defined as the distribution or adsorption coefficient, K_d.[1] This is the coefficient most commonly utilized by predictive equations and groundwater transport models that estimate chemical migration in soil and groundwater systems.

There are three important facts that should be understood concerning the parameters listed in Equation 1.

First, analytical data derived from the chemical analysis of the total metal content of soil (i.e. C_{Total}) relays no information regarding C_{Fixed}, C_{Ads}, and C_{Water} other than the magnitude of their combined concentrations. In other words, if a laboratory report states that a soil contains 125 mg/kg total Pb, this datum cannot reveal if 0.1% of this concentration is potentially mobile (i.e., $C_{Ads} + C_{Water}$) or if 99% is potentially mobile.

At background concentrations, the relative magnitudes of the parameters listed in Equation 1 generally are:

$$C_{Fixed} \gg C_{Ads} > C_{Water} \quad (2)$$

The greater part of C_{Total} exists as C_{Fixed} and is immobile. However, this relative ranking may or may not change as C_{Total} increases above the background concentration.

Second, background concentrations represent the total concentration of a chemical present after the soil was formed and weathered. This concentration gives no information on the loading capacity of a soil. The loading capacity can be defined as the maximum amount of chemical that can be added to soil which does not cause water migrating through this soil to contain a harmful concentration of that chemical. In other words, knowing that a soil contains 125 mg/kg total background Pb will not reveal if soil will or will not completely convert an additional loading of 500 mg/kg Pb into an immobile form (C_{Fixed}).

Soil cleanup standards that specify the excavation or treatment of soil containing concentrations of a chemical over a background concentration are usually based on an incorrect premise that the background concentration of a chemical in soil represents a maximum concentration of a chemical which the soil can immobilize. The background concentration only represents the total concentration present after the soil was formed and had undergone some degree of weathering; it gives no indication of the maximum concentration of a chemical which a soil can immobilize, i.e., the loading capacity of the soil.

Third, there is no "universal" analytical method or extractant which is applicable for all forms of chemicals in all soils. The test method employed is dependent upon the individual chemical to be tested, the parameter to be tested (e.g., C_{Fixed} versus C_{Ads}) and the soil type. A number of established, accepted laboratory methods exist for determining the magnitude of C_{Total}, C_{Fixed}, C_{Ads}, and C_{Water} in soil.[2-6] Typically for metals, C_{Total} can be measured by wet ashing with a mixture of perchloric, nitric, or sulfuric acids. For metals, C_{Ads} and C_{Water} can be determined by using mineral acids (e.g., 0.1 N HCl), organic acids, and chelating agents (e.g., EDTA, DTPA); hot water extractions can be utilized for elements that exist as anions (e.g., B, Mo, Se).

C_{Fixed} can be further fractionated.[7] The analytical method which should be utilized varies with the metal of concern and the form of the metal to be analyzed. For example, the forms of fixed copper (Cu) are expressed in Equation 3.

$$C_{Fixed} = C_{sac} + C_{saom} + C_{sao} + C_{oo} + C_{bo} + C_{ml} \tag{3}$$

where:

C_{sac} = concentration of Cu specifically adsorbed onto clay
C_{saom} = concentration of Cu specifically adsorbed onto soil organic matter
C_{sao} = concentration of Cu specifically adsorbed onto oxides
C_{oo} = concentration of oxide occluded Cu
C_{bo} = concentration of biologically occluded Cu
C_{ml} = concentration of mineral lattice Cu.

The various extractants and the forms of Cu extracted by each are listed in Table 1.

Table 1. The Forms of Cu Analytically Determined by Sequential Extraction Utilizing Various Extractants[a]

Form of Copper	Extractant				
	$CaCl_2$	Acetic Acid	Potassium Pyrophosphate	Oxalate + UV	Hydrofluoric Acid
C_{Water}	+	+	+	+	+
C_{Ads}	+	+	+	+	+
C_{sac}	–	+	+	+	+
C_{saom}	–	P	P	+	+
C_{sao}	–	P	P	+	+
C_{oo}	–	P	P	+	+
C_{bo}	–	–	–	P	+
C_{ml}	–	–	–	–	+

[a]Reference #7.
+ = Form is extracted by the extractant.
P = Form is partially extracted by the extractant.
– = Form is not extracted by the extractant.

What Happens When EP-TOX and Other Extractants Are Utilized on Soils?

Test methods such as the TCLP, EP-Tox, the CAM-WET, as well as others that are utilized to determine the amount of extractable chemical from wastes, when applied to soils, provide a value, $C_{Extract}$, in which:

$$C_{Extract} = aC_{Fixed} + bC_{Adsorbed} + C_{Water} \tag{4}$$

where:

$C_{Extract}$ = concentration of a chemical extracted from a soil; $C_{Total} > C_{Extract}$
a,b = fractions

Since a and b are not determined, $C_{Extract}$ provides no information regarding the magnitude of C_{Fixed}, $C_{Adsorbed}$, and C_{Water}, information which is needed to determine the potential migration and transformation of a chemical in soil.

The technical support documents prepared by the U.S. EPA for these tests do not state that these tests should be utilized to determine the migration and degradation potential of any chemical in a soil system.

The scientific literature reveals that soils react with inorganic and organic chemicals. For example, metals can be immobilized in soils via several soil-chemical reaction pathways:[1,8,9]

- chemisorption
- irreversible penetration of soil-mineral lattice structures
- metal precipitation with subsequent formation of new soil-minerals such as insoluble soil oxides, oxyhydroxides, phosphates, etc.
- metal occlusion by formation of soil oxides, oxyhydroxides, phosphates, etc.

In addition, organic chemicals can be transformed in soils via several reaction pathways,[1,10,11] which include:

- Microbial degradation
- Aqueous oxidation
- Aqueous reduction
- Aqueous hydrolysis
- Surface catalyzed oxidation
- Surface catalyzed reduction
- Surface catalyzed hydrolysis
- Bound residue formation

The TCLP and the EP-Tox do not take these reactions into account. In fact, when soils are exposed to the extractants utilized by these methods, gross alterations will occur in soil mineralogy, in naturally occurring soil-chemical reactions, and in soil physical/chemical properties. Gross soil alterations result from the interaction of the extractant with soil and will (a) dissolve some soil minerals which serve as the basic "building blocks" of soil, (b) impede crystallization and formation of aluminum hydroxides and other soil minerals while causing structural distortions in newly formed minerals, (c) perturb hydrolytic reactions of aluminum, and (d) desorb, via mass action, metals and organic chemicals from soil adsorption sites which may not normally be desorbed.[12-15]

Because the TCLP and Ep-Tox extractants cause gross alterations in the chemical and mineralogical properties of soil systems, the data derived from these test methods, when soil is utilized as the solid phase, cannot be extrapolated to actual field conditions.

Scientific Literature Has No Information Supporting the Use of These Test Methods to Determine the Leaching Potential of Metals in Soils

The authors of this chapter have conducted a detailed review of the scientific literature. The purpose of this review was to determine if any scientist or scientific organization has published any data or information showing that EP-TOX, TCLP, and CAM-WET can generate meaningful data on the leaching potential of any metal in soil.

The authors searched 55 technical journals published between 1976 and 1989, over 50 conference proceedings, 30 books, and over 40 U.S. EPA reports. No data, no information, and no papers were found showing that these test methods can generate useful data on the leaching potential of metals in soils. In summary, the scientific literature does not support the use of these test methods to generate meaningful data on the leaching potential of metals in soils.

Scientific Literature Contains Other Test Methods That Can Generate Meaningful Data on the Leaching of Chemicals in Soils

The scientific literature contains at least seven standardized laboratory test methods addressing the leaching potential of chemicals in soils.[16-24] In addition, the scientific literature contains at least 60 methods addressing the chemical analysis of soil, and over 40 methods addressing the biological properties of soil.[4-6,25] Some of these test methods were developed over 100 years ago and are still scientifically valid and acceptable.

Because acceptable laboratory test methods now exist, these methods should be utilized to determine the potential migration, degradation, and transformations of metals in soils.

SUMMARY AND CONCLUSIONS

During the late 1970s, the U.S. Environmental Protection Agency (EPA) developed the Extraction Procedure Toxicity Test Method (EP-Tox). The EP-Tox test method is the predecessor to the Toxicity Characteristic Leaching Procedure (TCLP) and serves as the technical basis for the TCLP. In the early 1980s, the state of California adopted the California Assessment Manual Waste Extraction Test (CAM-WET).

These tests are utilized to classify a waste as being either hazardous or nonhazardous. However, during the past few years, individuals, consulting firms, and other governmental agencies have attempted to utilize these test methods to estimate the leaching potential of metals from soils. These groups have theorized that the greater the concentration of metal that can be extracted from the soil, the more mobile the metal must be; the more mobile the metal, the more likely that the metal will migrate to groundwater.

The EP-Tox, TCLP, and CAM-WET must not be utilized to determine the migration and degradation potential of any chemical in a soil system. These test methods can be used only for the classification of hazardous waste produced by waste generators for handling and/or disposal.

These test methods are not suitable for the estimation of the leaching potential of metals from soil for five reasons.

First, these tests provide no information regarding the magnitude of C_{Fixed}, $C_{Adsorbed}$, and C_{Water}, the parameters which determine the potential migration and transformation of a chemical in soil. Second, the technical support documents provided by the U.S. EPA for these tests show that the agency does not recommend that these tests be utilized to determine the migration and degradation potential of any chemical in a soil system. Third, the scientific literature contains substantial data showing that these test methods alter soil properties and soil reactions. Fourth, the scientific literature has no information supporting the use of these test methods to determine the leaching potential of metals in soils. Fifth, the scientific literature contains other test methods that can generate meaningful data on the leaching of chemicals in soils.

REFERENCES

1. Dragun, J. *The Soil Chemistry of Hazardous Materials.* (Silver Spring, MD: Hazardous Materials Control Research Institute, 1988).
2. Baker, D. E., and L. Chesnin. "Chemical Monitoring of Soils for Environmental Quality and Animal and Human Health," *Advances in Agronomy* 27:305–374 (1975).
3. Baker, D. E., and M. C. Amacher. *The Development and Interpretation of a Diagnostic Soil-Testing Program.* Bulletin 826. June, 1981. University Park, PA: College of Agriculture, The Pennsylvania State University.
4. Hesse, P. R. *A Textbook of Soil Chemical Analysis.* (New York: Chemistry Publications, 1972).
5. Jackson, M. L. *Soil Chemical Analysis-Advanced Course.* 2nd Edition, 8th Printing. (Madison, WI: University of Wisconsin, 1973).
6. Page, A. L., R. H. Miller, and D. R. Keeney, Eds. *Methods of Soil Analysis. Part 2. Chemical and Microbiological Properties.* Second Edition. (Madison, WI: American Society of Agronomy, 1982).
7. McLaren, R. G., and D. V. Crawford. "Studies on Soil Copper. I: The Fractionation of Copper in Soils," *Journal of Soil Science* 24:172–181 (1973).
8. Marshall C. E. *The Physical Chemistry and Mineralogy of Soils.* Volume 1—Soil Materials. (New York: John Wiley & Sons, 1964).
9. Lindsay, W. L. *Chemical Equilibria in Soils.* (New York: John Wiley & Sons, 1979).
10. Dragun, J., and C .S. Helling. "Soil- and Clay-Catalyzed Reactions: I. Physicochemical and Structural Relationships of Organic Chemicals Undergoing Free-Radical Oxidation," in *Land Disposal of Hazardous Waste.* Proceedings of the Eighth Annual Research Symposium. EPA-600/9-82-002, 1982.
11. Alexander, M. "Biodegradation of Chemicals of Environmental Concern," *Science* 211:132–138 (1981).
12. Clark, C. J., and M. B. McBride. "Chemisorption of Cu(II) and Co(II) on Allophane and Imogolite," *Clays and Clay Minerals* 32: 300–310 (1984).
13. Miller, W. P., D. C. Martens, and L. W. Zelazny. "Effect of Sequence in Extraction of Trace Metals from Soils," *Soil Sci. Soc. Am. J.* 50:598–601 (1986).
14. Pohlman, A. A., and J. G. McColl. "Kinetics of Metal Dissolution from Forest Soils by Soluble Organic Acids," *J. Environ. Qual.* 15:86–92 (1986).
15. Wang, M. K., J. L. White, and S. L. Hem. "Influence of Acetate, Oxalate, and Citrate Anions on Precipitation of Aluminum Hydroxide," *Clays and Clay Minerals* 31:65–68 (1983).

16. Helling, C. S., and J. Dragun. ''Soil Leaching Tests for Toxic Organic Chemicals,'' in *Test Protocols for Environmental Fate and Movement of Toxicants*. (Arlington, VA: Association of Official Analytical Chemists, 1981).
17. U.S. EPA. Office of Pesticides and Toxic Substances. ''Soil Thin Layer Chromatography,'' Test Guideline #CG-1700, in *Chemical Fate Test Guidelines*. EPA 560/6-82-003. Washington, DC: U.S. EPA, 1982a.
18. U.S. EPA. Office of Pesticides and Toxic Substances. ''Sediment and Soil Adsorption Isotherm,'' Test Guideline #CG-1710, in *Chemical Fate Test Guidelines*. EPA 560/6-82-003. Washington, DC: U.S. EPA, 1982b.
19. U.S. EPA. Office of Pesticides and Toxic Substances. ''Thin Soil Layer Chromatography,'' Support Document #CS-1700, in *Chemical Fate Test Guidelines*. EPA 560/6-82-003. Washington, DC: U.S. EPA, 1982c.
20. U.S. EPA. Office of Pesticides and Toxic Substances. ''Sediment and Soil Adsorption Isotherm,'' Support Document #CS,1710, in *Chemical Fate Test Guidelines*. EPA 560/6-82-003. Washington, DC: U.S. EPA, 1982d.
21. U.S. EPA. Office of Toxic Substances. ''Soil Thin Layer Chromatography,'' in Proposed Environmental Standards; and Proposed Good Laboratory Practice Standards for Physical, Chemical, Persistence, and Ecological Effects Testing. *Federal Register* 45(227):77332–77365, 1980a.
22. U.S. EPA. Office of Toxic Substances. ''Adsorption,'' in Toxic Substances Control Act Premanufacture Testing of New Chemical Substances. *Federal Register* 44(53):16257–16264, 1980b.
23. U.S. EPA. Office of Toxic Substances. ''Soil Thin Layer Chromatography,'' in *Support Document for Test Data Development Standards*. EPA-560/11-80-027. Washington, DC: U.S. EPA, 1980c.
24. U.S. EPA. Office of Toxic Substances. ''Adsorption,'' in *Technical Support Document for Guidance for Premanufacturing Testing: Discussion of Policy Issues. Alternative Approaches and Test Methods*. Washington, DC: U.S. EPA, 1979.
25. Black, C.A., Ed. *Methods of Soil Analysis*. Part 2. Chemical and Microbiological Properties. (Madison, WI: American Society of Agronomy, 1965).

CHAPTER 8

Fingerprinting Petroleum Products: Unleaded Gasolines

Thomas L. Potter, Mass Spectrometry Facility, Massachusetts Agricultural Experiment Station, University of Massachusetts, Amherst, Massachusetts

At most petroleum storage tank facilities there are at least several tanks onsite containing the same or different types of products. It is also common for tanks or pipelines to be located on nearby properties. Thus, when a product release is detected there may be many potential sources, including both on and offsite structures. Under these circumstances source identification is essential for the selection and implementation of appropriate corrective actions and ultimately for assigning responsibility. To this end, chemical analysis of products may play a critical role. Ideally, through chemical fingerprinting the release can be traced directly to its source.

A variety of analytical schemes have been published for petroleum product fingerprinting,[1-3] although few have focused on gasolines. Those that have, typically begin with a simulated distillation analysis and proceed as necessary to more complex analyses involving identification of additives such as octane boosters and hydrocarbon profiling by gas chromatography and/or mass spectrometry.[4-6] In general, methods and procedures based on high resolution gas chromatography-mass spectrometry (HRGC/MS) have shown the greatest success.[6,7]

In this chapter, HRGC/MS analysis of a set of unleaded gasolines is discussed with respect to product fingerprinting. It is shown that computer-based pattern recognition techniques like cluster analysis can facilitate data analysis. Nevertheless, uncertainty in results may be relatively large and the burden of proof beyond the limits of data that can reasonably be obtained.

ANALYSIS

Instrumentation and analytical conditions were based on U.S. Environmental Protection Agency (EPA) Method 8270.[8] This HRGC/MS technique represents the "state of the art" for trace organic analysis. Modifications were that injections were made using the "split" mode, toluene-d8 was used as an internal standard, and alternate chromatographic conditions were utilized.

Use of split injection permitted direct injection of the products into the gas chromatograph without the use of dilution solvents. By avoiding dilution solvents, the entire chromatographic profile of each product was obtained. Figure 1 provides a typical total ion current chromatogram.

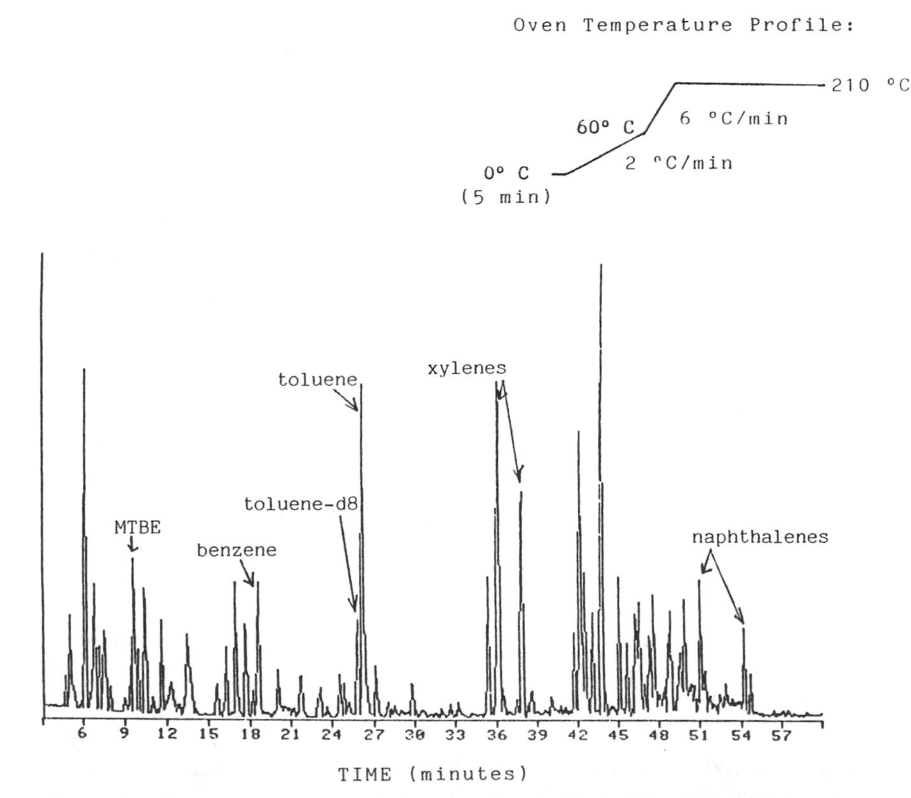

Figure 1. Total ion current chromatogram of an unleaded regular gasoline.

The alternate internal standard, toluene-d8, was selected since it has physical-chemical properties which are similar to many gasoline constituents and can be quantified in gasolines under the analytical conditions described. The toluene-d8 was spiked directly into each gasoline sample at the rate of 0.5% by volume. Note that all quantitation was based on relative response to the internal standard and a series of relative response factors determined from the replicate analyses of a standard hydrocarbon mixture. This approach substantially improved the precision of concentration measurements (Table 1).

Table 1. Coefficient of Variation for the Concentration of Selected Compounds in a Gasoline Using Internal and External Standardization[a, b]

	Coefficient of Variation (%)[c]	
Compound	Internal	External
Methyl-tert-butyl ether	4.7	13.1
Benzene	3.3	14.4
Pentane, 2,2,4-trimethyl	1.9	15.7
Toluene	1.1	18.2
Ethyl-benzene	4.2	19.6
Xylenes, para and meta	3.2	18.8
Xylene, ortho	4.6	15.7
Naphthalene	3.1	16.6

[a]Data from three replicate analyses of an unleaded regular gasoline (product M).
[b]Analysis by HRGC/MS with split injection and quantitation based on the internal standard, toluene-d8.
[c]Coefficient of variation is an expression of the standard deviation divided by the mean and reported as relative percent.

Differences in chromatographic conditions were in the GC oven temperature profile. A lower initial temperature and slower oven temperature program rates were used than are specified in method 8270. These conditions were selected to allow better separation of the constituents in the products. For example, the alternate conditions permitted complete resolution of the octane booster, methyl tert-butyl ether (MTBE), from various five carbon alkanes. This improved MTBE quantitation by eliminating interferences from the heavy isotope ions (C13) of these compounds.

The alternate chromatographic conditions did not, however, permit complete separation of all constituents. This is reflected in the fact that isomer specific identifications of the C8-alkanes (Table 4) were not attempted. Such identifications were beyond the limits of the analytical conditions.

RESULTS AND INTERPRETATIONS

The sample set was a series of six regular unleaded gasolines purchased at service stations operated in the Amherst, Massachusetts area. In addition, a "weathered" sample was prepared by 50 % evaporation of one of the products. Concentration

Table 2. Concentration of Selected Compounds in Six Unleaded Regular Gasolines[a]

	Concentration (Grams per Liter)					
	Product					
Compound	M	Ge	A	T	S	G
MTBE[b]	23.0	50.9	36.0	7.2	13.6	<.1
Cyclohexane	1.1	1.3	2.5	1.3	1.7	1.9
Isooctane[c]	35.4	44.1	56.1	64.2	64.5	26.6
Benzene	12.7	9.6	14.2	9.4	20.2	28.7
Toluene	47.8	36.8	41.6	37.5	68.0	68.2
Ethyl-benzene	15.8	13.3	10.4	14.2	17.5	16.2
Xylenes	67.0	60.7	58.9	63.2	78.4	67.3
Naphthalene	6.7	8.0	7.1	8.7	6.2	2.8

[a]Analysis by HRGC/MS with split injection and quantitation based on the internal standard, toluene-d8.
[b]MTBE is the octane booster, methyl tert-butyl ether.
[c]Isooctane is the common name for 2,2,4-trimethyl pentane.

Table 3. Concentration of Selected Compounds in an Artificially Weathered Unleaded Regular Gasoline[a]

	Concentration (grams per liter)	
Compound	Product (T)	50% Evaporated (T50)
MTBE[b]	7.2	0.8
Benzene	9.4	4.9
Cyclohexane	1.3	2.9
Isooctane[c]	64.2	80.7
Toluene	37.5	59.6
Ethyl benzene	14.2	24.6
Xylenes	63.3	128
Naphthalene	8.7	24.5

[a]Analysis by HRGC/MS with split injection and quantitation based on the internal standard, toluene-d8.
[b]MTBE is the octane booster, methyl tert-butyl ether.
[c]Isooctane is the common name for 2,2,4-trimethyl pentane.

Table 4. Concentration of C8-Alkanes in Five Unleaded Regular Gasolines and an Artificially Weathered Sample[a]

	Concentration (grams per liter)					
	Products					
Peak No.[b]	M	Ge	A	T	S	T50[c]
1	60.9	101	128	147	148	178
2	9.4	10.5	11.2	15.3	15.8	23
3	9	10.8	11.3	14.7	13.7	20
4	13.9	23.5	31	34.2	43.4	54
5	25.7	27.5	24.1	26.9	26.8	39
6	21.1	22.5	19.3	22.6	22.7	33

[a]Analysis by HRGC/MS with split injection and quantitation based on the internal standard, toluene-d8.
[b]Peak 1 was identified as 2,2,4-trimethyl pentane.
[c]T50 was prepared by 50% evaporation of product T.

data for selected constituents in each sample has been summarized in Tables 2, 3, and 4.

Additives Analysis

The utility of additives analysis is that presence or absence of a unique compound may allow a definitive product identification. To this end, the data summarized in Table 2 show that five out of the six products contained the octane booster, MTBE. Within the sample set the product that did not contain this compound was easily distinguished from the others. The converse is that among the products which contained MTBE, detection of this additive provided little information regarding product source. Note that no other unique compounds were detected at a detection limit of .01% by weight.

This is not to say that distinctive additives were absent. The problem is that most additives are used at low product concentrations and that "a priori" knowledge of additive concentration and structure is required before methods can be developed for their analysis. In general, this information is proprietary and not available to the analyst.

Hydrocarbon Profiling

In the absence of unique additives, fingerprinting usually turns to hydrocarbon profiling. This approach involves identifying a unique pattern for each product based on the relative concentrations of various constituents. Techniques for data interpretation range from simple visual comparison of chromatograms to computer-based pattern recognition techniques. An example of the latter is cluster analysis.

The basic concept of cluster analysis and related techniques is that a product can be represented as a point in a space whose dimensionality is equivalent to the number of number parameters for which concentration data is available.[9] Within this space, the distance between two points is a relative measure of similarity, and points that are close to each other are said to group or "cluster." Operationally, what is accomplished through computation is the generation of alternate variables which allow the data to be represented in fewer dimensions while retaining most of the inherent variability within the data set.

Results of the cluster analysis of the Table 2 data are presented graphically in Figure 2 (only data for the products which contained MTBE were used). Computations were performed on a Zenith 386 workstation using the software package, "Spring-stat."[10]

That the points on this two-dimensional plot are well separated is indicative of the unique chemical character of each product. It also indicates that the subset of gasoline hydrocarbons on which the cluster analysis was based effectively describes the chemical variability of these products. These compounds are representative of at least four major refinery streams which are combined in varying proportions to yield products meeting performance specifications.[11]

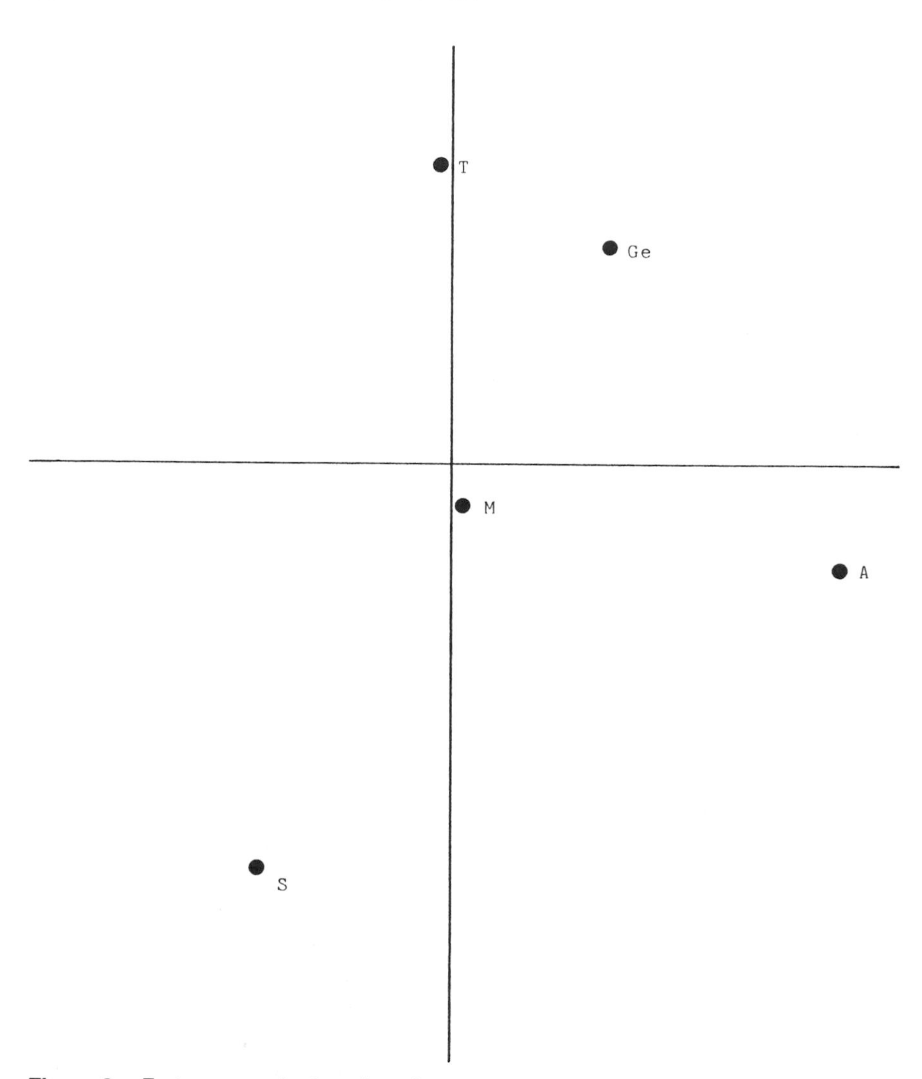

Figure 2. Factor score plot based on the concentrations of nine principal constituents in five gasolines.

The expectation is that by collecting similar concentration data for an unknown, and performing a cluster analysis with the reference sample data (Table 2), a product similarity index can be derived. The more similar the composition of an unknown and product, the closer they will cluster, and the more likely their source is related.

Unfortunately, even with the high precision measurements obtained in this study, this approach was not found to be an effective predictor of sample source for the artificially weathered sample (Table 3). In fact, Figure 3 shows that T50 failed

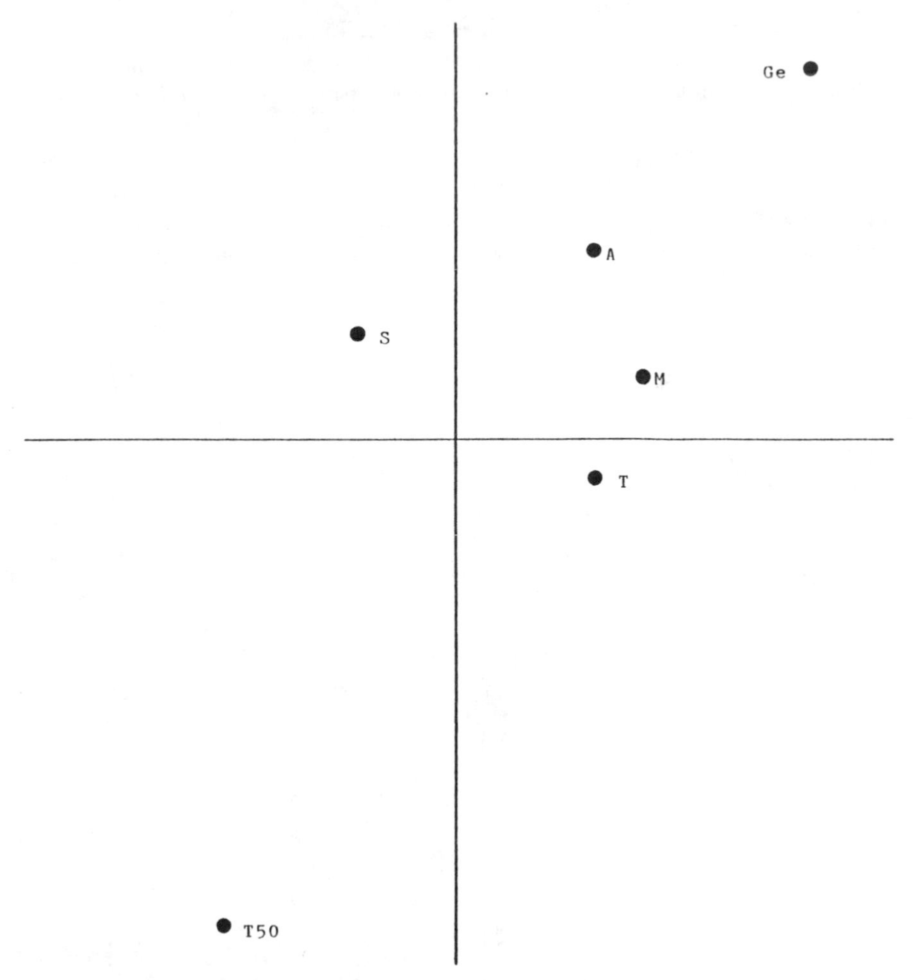

Figure 3. Factor score plot based on the concentrations of nine principal constituents in five gasolines and an artificially weathered product.

to cluster with any of the products, and that product T (its source) was more like the other products than it is T50.

The breakdown in this product matching system occurred because the Table 2 compounds evaporate from gasolines at different rates. For example, the Table 3 data show that after 50% evaporation, the product concentration of MTBE and benzene had decreased, whereas the concentration of isooctane and other compounds had increased.

Given this behavior, alternative cluster analyses based on the relative concentrations of various isomers were attempted. Isomeric groups evaluated included C2-benzenes, C1-naphthalenes, C3-benzenes, and C8-alkanes. Each of these groups are well represented in gasolines, and their concentration in the product phase tends to increase as weathering advances. The rational for examining isomers is that aqueous solubilities and vapor pressures of isomers exhibit relatively small differences.[12]

In short, it was found that the only isomeric group which had potential for fingerprinting were the C8-alkanes. Indeed, with the C8-alkane data, the weathered and unweathered samples, T50 and T, clustered closely (Figure 4). The limitation is that only two clusters were observed for the five products; thus, an exact match was beyond the limits of the data.

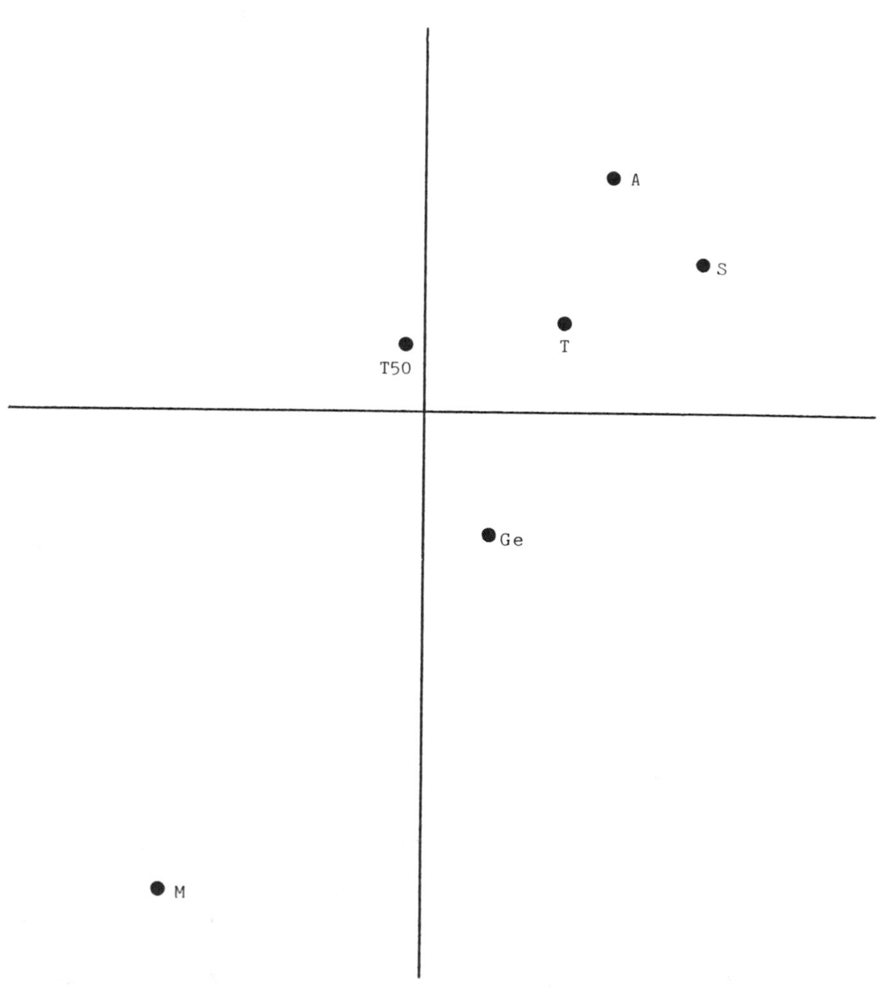

Figure 4. Factor score plot based on the C8-alkane concentrations in five gasolines and an artificially weathered product.

What was found for the other isomers was that the relative distributions of isomers within a group were nearly identical. This is reflected in the fact that gasolines are highly engineered products. Their composition is a function of refinery and blending processes which promote a "leveling" effect among products.[11] With less refined products the relative distributions of various isomers is usually more distinctive. Calculation of isomeric concentration ratios is one of the principal tools of crude oil fingerprinting.[2,3]

CONCLUSIONS

The examples cited illustrate that unleaded gasolines can be fingerprinted through additives analysis and hydrocarbon profiling, but there are significant limitations. The occurrence of (or the ability to detect) unique additives in unleaded products is apparently rare and, weathering can obscure much of a product's unique chemical character.

Other obstacles to effective gasoline fingerprinting is that, unlike the situation described in this work, product reference samples which can be directly related to a release are rarely if ever available, and product composition is not constant. Product mixing (co-mingling) in pipeline and other distribution systems, product switching (i.e. trading among refiners and blenders) are equally if not more important problems.[11,13]

This serves to emphasize that uncertainty in fingerprinting results may be large. The work described in this chapter was both expensive and time-consuming, and "state of the art" analytical technology was utilized. Yet a definitive match could not be obtained for the weathered sample. Simply, the burden of proof may be beyond the limits of data which can reasonably be obtained.

ACKNOWLEDGMENTS

Financial support for this work was provided by the Massachusetts Agricultural Experiment Station, University of Massachusetts, Amherst, MA.

REFERENCES

1. ASTM Standard Method, D3327-78, Comparison of Waterborne Petroleum Oils by Gas Chromatography, 1978 *Annual Book of ASTM Standards,* Part 31, American Society for Testing Materials, Philadelphia, PA., 1978.
2. Butt, J. A., D. F. Duckworth, and S. G. Perry, eds. *Characterization of Spilled Oil Samples* (New York: John Wiley & Sons, 1986).
3. Bentz, A., "Who Spilled the Oil," *Anal. Chem.* 50: 655A (1978).
4. Rygle, K., "Methods for 'Free' Product Analysis," in M. Kane, ed., *Manual of*

Sampling and Analytical Methods for Petroleum Hydrocarbons in Ground Water and Soil. Publication 841-44490, American Petroleum Institute, Washington, DC (1987).

5. Youngless, T. L., J. T. Swansiger, D.A. Danner, and M. Greco. "Mass Spectral Characterization of Petroleum Dyes, Tracers, and Additives," *Anal. Chem.* 57:1894–1902 (1985).
6. Sleck, L. W., "Fingerprinting and Partial Quantification of Complex Hydrocarbon Mixtures by Chemical Ionization Mass Spectrometry," *Anal. Chem.* 51:128–132 (1979).
7. Sutton, D. L., "Component Analysis by High Resolution Gas Chromatography," in M. Kane, ed., *Manual of Sampling and Analytical Methods for Petroleum Hydrocarbons in Groundwater and Soil,* Publication 841-44490, American Petroleum Institute, Washington, DC (1987).
8. "Test Methods for Evaluating Solid Waste", 3rd ed., U.S. Environmental Protection Agency, Doc.: SW-846, 1987.
9. Hopke, P., "An Introduction to Multivariate Analysis of Environmental Data," in B. Natusch and P. Hopke, eds., *Analytical Aspects of Environmental Chemistry* (New York: John Wiley & Sons, 1983).
10. *SpringStat,* version 2.2, Spring Systems Inc., Chicago, Illinois.
11. Whitmore, I., "Identification of Spilled Hydrocarbons," 79-in M. Kane, ed., *Manual of Sampling and Analytical Methods for Petroleum Hydrocarbons in Groundwater and Soil,* Publication 841-44490, American Petroleum Institute, Washington, DC (1987).
12. Mackay, D., and W. Y. Shiu. "A Critical Review of Henry's Law Constants for Chemicals of Environmental Interest," *J. Phys. Chem. Ref. Data* 10:1175–1199 (1981).
13. Anderson, E.V. "Fuel Ethanol Wins Two Regulatory Decisions," *Chem. Eng. News* Feb. 2, 1987, pp. 14–15.

CHAPTER 9

Relationships Between Chemical Screening Methodologies for Petroleum Contaminated Soils: Theory and Practice

S. A. Denahan, B. J. Denahan, W. G. Elliott, W. A. Tucker, M. G. Winslow, Environmental Science & Engineering, Inc., Gainesville, Florida
S. R. Boyes, GeoSolutions Inc., Gainesville, Florida

The contamination of the environment by petroleum products from surface spills and leaking underground storage tanks (USTs) is a leading environmental problem today. Identification, assessment, and remediation of petroleum contaminated sites is a major task facing civilized society. Thousands of these sites exist in the country. Traditional assessment technology for the definition of the spatial extent of contamination involves the collection and laboratory analysis of samples of the various environmental media. The economic pressures engendered by the extremely large number of petroleum sites to be dealt with has spawned the development of several screening methodologies, which can provide a cost-effective alternative to the traditional sampling and laboratory analysis.

This chapter discusses one of the most useful of these methodologies, the measurement of vapor phase petroleum compound concentrations in the vadose zone. Because most petroleum products have a significant component of volatile compounds in their makeup, they are well suited to vapor phase analysis. Whenever petroleum products are released into the environment a vapor phase component will be present. Most commonly, a three phase equilibrium is established, with petroleum compounds present as vapor phase, liquid phase, and also adsorbed to the solid phase, mainly organic soil components. Several different methodologies have been employed to sample and analyze these vapor phase components. These methodologies can be divided into two broad categories: in situ methods and head space methods.

The in situ methods provide a mechanism for collecting a sample of the vapor phase components in situ without the need for the collection of a soil sample. There are two primary variations of this technique (active and passive), with the principal differences being how the sample of the vapor phase material is obtained. In the active technique, a sample of vapor is extracted from the vadose zone by means of a hollow probe and a vacuum pump. In the passive technique, a collection device containing an adsorbent medium is buried in the soil for a specified period of time and then exhumed with whatever vapor phase components have been adsorbed. The passive technique has several apparent disadvantages in that it requires two and possibly more field efforts, and the chemical data are not acquired in real time. Additionally the passive technique relies on the natural flux of vapor phase components to move to the collector, and the rate of this natural movement is often very slow and also variable, depending on soil permeability, moisture content, temperature, and atmospheric pressure. Only active methods will be discussed in this chapter.

The head space methodologies require the collection of a soil sample from the zone of suspected contamination. This sample is then placed into a closed container (various types are used) and the vapor phase components concentration in the head space above the soil is measured directly using a portable instrument. Certain states have mandated the use of this methodology for the field determination of the extent of petroleum contaminated soils.

INSTRUMENTATION

Two principal types of portable analytical instruments are commercially available for the analysis of the vapor phase components of petroleum contaminated soils, flame ionization detectors (FIDs), and photo ionization detectors (PIDs). There are significant differences between the responses of these two types of instruments. These differences and the practical implications of them are one of the principal topics of discussion in this chapter.

Flame Ionization Detectors (FID)

The Flame Ionization Detector (FID) in today's commercially available portable screening instruments is a nonspecific detector designed to measure total organic compounds present in the vapor phase. Vapor phase components are drawn through a probe into an ionization chamber containing an oxygen (air)-hydrogen flame where chem-ionization of organic molecules occurs:

$$CH + O -> CHO^{+} + e^{-}$$

The ions and electrons pass between electrodes to which a voltage is applied, decreasing the resistance and causing a current to flow in the external circuit. The FID should respond to all organic molecules containing an ionizable carbon atom. Clean air components, H_2O and CS_2 should yield no effective response. The FID's response to different organic molecules differs only in the total number of ionizable carbon atoms in the molecules. A substituent (e.g., Cl) on an individual carbon atom in a molecule generally reduces the ionization compared to hydrogen. When measuring petroleum hydrocarbon contamination in the vapor phase, the detector response will, in general, increase with increasing carbon number. The detector response should also be approximately the same for alkanes, alkenes, and aromatics with the same number of effective (ionizable) carbon atoms. The screening FID responds quantitatively only to the individual compound to which it is calibrated. Response to an unknown compound or mixture of compounds, as in the case of petroleum hydrocarbons, can only be nonqualitative and semiquantitative in nature.

Photoionization Detectors (PID)

The Photoionization Detectors (PID) in today's commercially available portable instruments that are designed to screen vapor phase contaminants is a nonspecific detector that measures total ionizable compounds, both organic and inorganic. Unlike the FID, the PID is a nondestructive detector (i.e., molecules exit the detector chemically unaltered from their original state). Vapor phase components are drawn through a probe into an ionization chamber which is separated by a window from an ultraviolet (UV) lamp. The UV lamp emits photons with a specific energy (several interchangeable lamps of varying intensities are available for most commercial instruments). The photons are transmitted through the window into the ionization chamber. Molecules that have lower ionization potentials than the energy of the radiated photons will absorb a photon and become ionized:

$$AB + \text{photon} -> AB^{+} + e^{-}$$

The ionization chamber also contains two electrodes to which a voltage is applied. The ionized molecules and electrons flow into the electric field and the resulting increase in current is proportional to the concentration of ionized molecules in the vapor phase. The PID does not respond to the components of clean air or H_2O because they have ionization potentials higher than the energy of any commercially available UV lamp (8.4, 9.5, 10.2, 10.6, and 11.7 eV). Standard lamps that are most commonly used in field screening (10.2 and 10.6 eV) of petroleum hydrocarbon contaminants will respond well to numerous inorganic gases such as CS_2, H_2S, pH_3, and NH_3. All compounds manifest a unique ionization potential, but all ionizable compounds are not detected by the PID equivalently. In general, the lower the ionization potential of the compound, the greater the PID's response to it. Within a homologous series, the detector response will generally increase with increasing carbon number and is different between homologous series. Although most petroleum hydrocarbon compounds will be detected by the PID equipped with a standard lamp (10.2 or 10.6 eV), the response increases with increasing unsaturation of the molecule; benzene > cyclohexene > cyclohexane and for different types of petroleum hydrocarbons sensitivity increases as follows: aromatics > alkenes > alkanes, and cyclic compounds > noncyclic compounds. Like the FID, the screening PID responds quantitatively only to the individual compound to which it is calibrated. Response to an unknown or unknowns can only be nonqualitative and semiquantitative.

FIELD SAMPLE COLLECTION PROCEDURES

This section describes the methods used to collect field data using various detection techniques. The spacing between sample stations should be chosen with consideration of the potential size of the contaminated zone to be detected. The interval between soil gas stations can be greater for potentially larger contaminated areas and conversely smaller for potentially small areas.

In Situ Soil Gas Measurements

To obtain in situ samples from the vadose zone, a soil gas probe, consisting of a hollow steel tube, is driven into the soils to a desired depth. The soil gas probe is evacuated, and a representative sample is collected. The desired depth of the soil gas probe should be determined based on the physical soil characteristics of the study area. The tip of the soil gas probe should be driven into the soil far enough to isolate the sampling port from ambient air. In sandy soils the sand tends to close around the soil gas probe. In silty or clayey soils the hole will tend to remain open; therefore, care must be used to assure a tight seal. All soil gas probes for a given study should be driven to approximately the same depth below ground surface. A depth of 3 to 5 ft. below ground surface is generally sufficient. The probe can be driven with a slide hammer or a pneumatic hammer.

At each sample location, a soil gas probe constructed of 0.5 in. diameter galvanized steel tubing should be driven to its sampling depth and capped. For sampling, the cap is removed, and the soil gas probe is attached by Teflon® tubing to a 1 liter Tedlar® purge bag. The purge bag is placed inside a laboratory desiccator; a vacuum pump is then connected to the desiccator and used to evacuate the space inside the desiccator surrounding the bag. This negative pressure causes the soil gas to be drawn into the purge bag as it fills the evacuated volume. Prior to sampling, approximately two liters of gas should be evacuated to purge the ambient air from the sampling train and to draw in gases from the soil. One bag should be dedicated to purging operations.

After purging the sampling train, a new labelled bag is placed in the desiccator and used to collect the sample for analysis. Once the sample is collected, the sample bag is removed from the desiccator, and the contents are analyzed using one of several techniques previously described. If more than one method of analysis is employed, a single sample bag of gas should be used to supply soil gas to all instruments for a given sampling location. In this way, all data from a specific soil gas probe will be generated from a single sample of a specific concentration.

Gas to be analyzed by field gas chromatograph (GC) is removed from the sample bag using a gas-tight syringe and injected into the GC for quantification. Tygon® tubing is used to convey the sample from the capture bag to the other field instruments such as the FID or PID. This is done by connecting the Tygon tubing directly from the sample bag to the field instrument and allowing the instrument to draw in the soil gas. The instrument reading is then entered in the field notebook.

Dynamic trapping, another method of sampling soil gases for analysis, relies on adsorption of the contaminant on a medium of activated charcoal. Prior to sampling, each charcoal adsorption tube is thermally desorbed and stored in double-walled containers to prevent contamination. The soil gas probes are installed and evacuated. Sampling is accomplished by connecting a thermal desorption tube in-line to the soil gas probe and vacuum pump. Soil gas is evacuated through the tube for 15 minutes at a rate of 100 cubic centimeters per minute, for a total of 1500 cubic centimeters of gas sampled. Quality assurance samples may be collected by pumping ambient air through the thermal desorption tubes. After sampling is completed, each tube is sealed in a double-walled container, transported to the laboratory and thermally desorbed, then analyzed using EPA Method 602 for volatile aromatic compounds.

Head Space Measurements

Head space measurements are obtained by collecting a soil sample and placing it in a sealable container without aerating the sample. The container is then sealed and the temperature of the contents allowed to equilibrate for a minimum of five minutes. After the sample equilibrates, the gases in the headspace of the container are analyzed. If a GC is used, the sample can be removed from the container using a syringe and injected into the GC for analysis. If a FID or PID is used, consideration should be given to the technique used to remove the sample from the container, the type of container used (rigid or flexible) and the instrument being used (destructive or nondestructive). If a rigid container is used and the sampling port of the instrument is inserted into the sample container, only the initial moments of the instrument reading will reflect the true concentration of the contaminants in the soil gas. This occurs because as the instrument pumps the gases out of the container the sample is diluted.

There are several ways to avoid compromising the analytical results. If a nondestructive instrument such as a photoionization detector (PID) is used, the soil gas can be recirculated back into the sample container by use of the vent port. A Tygon® tube can be attached to the vent port of the instrument and inserted back into the container. The sample will enter the sampling port, circulate past the ionizing lamp and out the vent port back into the jar. This method does not introduce air into the system, nor does it destroy the sample. If the field instrument uses a flame ionization detector (FID), the sample is destroyed, and recirculating the gases through the instrument will not provide satisfactory results. In this case, a flexible container should be used. The container will collapse as the gas is removed, maintaining the integrity of the sample.

Ambient Screening

Ambient screening is the simplest method of collecting soil gas data on soils as they are obtained from a bore hole. Utilizing this method provides real time information concerning vertical distribution of the contaminant in the vadose zone. Although this method is very quick and inexpensive it often can prove difficult to exactly reproduce the results. The screening is usually conducted as follows, as soils are removed from the bore hole, a field screening instrument such as a PID is moved along the soil sample. Any positive response above the background is recorded, along with the sampling depth interval. To increase the likelihood of detecting any contaminants, the soil sample is sometimes scored or split open to increase the rate of release of volatile soil gases. This screening must be completed immediately after the soils are obtained to assure the best possible results.

Several environmental factors, given below, influence the results of this screening technique and should be considered while collecting and interpreting data:

1. If soil temperatures have not equilibrated, variations in ambient temperatures will influence the results.
2. Variations in wind speed can skew the data; obviously high ambient wind speeds will tend to disperse the volatiles coming from the soil samples.
3. Sample moisture content will influence the response of some instruments.

Collection of Samples for Laboratory Analysis

The collection of representative samples for laboratory analysis of volatile compounds requires preparation and planning prior to sampling. Care should be taken throughout the sampling effort to minimize aeration of the soils. Sampling for volatile compounds consists of taking an undisturbed sample, and quickly isolating the sample from the ambient conditions. The sample should be sealed into a container with minimum or no headspace, chilled, and taken to a laboratory for analysis. This often presents a problem in the field, as packing a soil sample into a rigid container with no headspace requires considerable manipulation of the soil, with the resulting loss of volatiles.

THEORETICAL CONSIDERATIONS

Uncontaminated soil is a three phase system of solids, liquid, and air. Soil solids are predominantly minerals such as silica or calcite that do not interact chemically with petroleum hydrocarbons but also include organic matter, a phase that preferentially absorbs petroleum hydrocarbons. As a result, it is sometimes useful to think of soils as comprised of an inert mineral phase, organic matter, water, and air.

Units of Measurements

In the assessment of multimedia contaminant fate and transport, confusion regarding the units of measurement and appropriate methods of conversion to permit intercomparison of contaminant concentrations occasionally obscures more substantial phenomena. The following discussion is provided to clarify such issues and to facilitate communication and understanding.

Contaminant concentrations in soils are most commonly expressed on a mass/mass basis. In laboratory analysis for volatiles, a soil sample at "field moisture" is weighed and then extracted to determine the total mass of contaminant that had been present in soil, liquid, and vapor phases at the time of collection. A corollary portion from the same sample is weighed before and after drying to determine its moisture content. Then the total mass of contaminant that had been present in the bulk sample (all three phases) is referenced to the mass of a dry soil (bulk soil minus water).

On a mass/mass basis this concentration may be expressed as μg/g or mg/kg, or ppm by weight.

Contaminant concentrations in water are commonly reported on a mass/volume basis, as in mg/L. The metric mass and volume scales are linked via the specific gravity of water: 1 L of water has a mass of approximately 1 kg. Consequently, the concentration expressed in mg/L, is numerically equivalent to a concentration in mg/kg, so either of these units may be referred to as ppm.

Furthermore, the bulk density of most soils is between 1.5 and 2.0 kg/L; not so different from the specific gravity of water. As a result it is not too misleading to compare soil concentrations and water concentrations without rigorously considering the difference between mass/mass versus mass/volume as routinely reported.

This comfortable situation does not prevail when considering soil gas concentration data. The standard reporting convention for vapor phase concentrations is in volume/volume, which is precisely equivalent to moles/mole based on the Ideal Gas Law. One ppm of a contaminant in air would represent, for example, one μL/L or one micromole of contaminant per mole of air.

Air is predominantly a mixture of nitrogen (79%) and oxygen (21%), and behaves like a gas with a molecular weight of 29. (This is similar to the determination of the atomic weight of an element based on the fractional abundance of its isotopes.) A mole of a gas at standard temperature and pressure occupies a volume of 22.4 liters. As a result the density of air is 29 grams per 22400 milliliters or 1.29×10^{-3} gm/mL: air is approximately 800 times lighter than water and more than 1000 times lighter than bulk soil.

To convert from contaminant concentrations in air, routinely expressed as ppm by volume, to a mass per volume basis, one must consider the molecular weight of the contaminant. Benzene has a molecular weight of 78. At STP, then pure benzene vapor has a density of 3.5×10^{-3} gm/mL (78/22400). If the concentration of benzene in air is 1 ppm, then 1 L of air contains 3.5 μg of benzene. By contrast, if 1 L of water contained 3.5 μg of benzene, the concentration might be expressed as 3.5 ppb.

Equilibrium Partitioning in the Three Phase System

Organic contaminants are readily exchanged between soil phases, and tend to partition between phases according to the chemical potential gradients across phase boundaries. Water and air move relatively slowly through soil pores, allowing time for contaminants to partition between phases, establishing geochemical equilibrium. There are complications in predicting or estimating what this equilibrium distribution will be in real soils: organic compounds dissolved in the soil solution such as humic and fulvic acids alter the nature of the aqueous solution that is often simplistically assumed to be pure water. The solid phase conceptualized as soil organic carbon is very complex and may differ from one soil to the next. Nonetheless, a simplified model has proven to be useful in soil contaminant data interpretation and transport modeling. This model, presented concisely by Jury et al.[1] and applied to petroleum hydrocarbons by Tucker et al.[2], considers soil to consist, geochemically, of water, air, and organic carbon. The contaminant is assumed to distribute itself among these phases so as to establish chemical equilibrium. Furthermore, this equilibrium condition is assumed to be the same in all soils. The relative amounts of contaminants sorbed to solids, dissolved in water, or in vapor form depends solely on the contaminant physical chemical properties and the relative amounts of each of these phases in the particular soils.

This equilibrium condition is well defined based on soil organic carbon content, moisture content, bulk density or porosity, and the organic carbon adsorption coefficient and Henry's law constant of the contaminant. With knowledge of the contaminant concentration in bulk soil, or any single phase such as soil gas, the contaminant concentration in other phases can be estimated. The simplified model is expressed as follows:

$$C_s = KC_w \quad (1)$$

$$C_a = HC_w \quad (2)$$

where

C_s is the concentration of contaminant on solid phases (mass/mass, dry weight)
K is the soil adsorption coefficient (vol/mass)
Cw is the concentration in the soil moisture (mass/vol)
Ca is the concentration in the soil air (mass/vol) and H is the Henry's law constant expressed in dimensionless concentration units.

The soil adsorption coefficient is given by K=Koc foc where Koc is the partition coefficient between soil organic carbon and water (vol/mass) and foc is the organic carbon content of the soil (mass/mass).

Based on this equilibrium partitioning relationship, the concentration in the bulk soil is comprised of contribution from different phases. Concentration in bulk soils are referenced to the dry mass of the soil. As shown by Jury et al.[1], the concentration in bulk soil can be related directly to the concentration in any single phase. For this application the relationship to soil air is most relevant.

$$C_{st} = (Kocfoc/H + nw/\ bH + na/\ b)\ Ca \quad (3)$$

where

$$nw + na = n \quad (4)$$

nw is the water filled porosity (vol/vol)
na is the air filled porosity (vol/vol)
n is the total porosity (vol/vol)
b is the soil bulk density (dry)
C_{st} is the contaminant concentration in bulk soil according to the standard reporting convention, i.e. referenced to dry weight.

A few other standard soils relationships assist in application of this equation;

$$b\ (1-n)\ 2.65 \quad (5)$$

assuming the solid phase is predominantly silica or calcium carbonate. If soil moisture 0 is expressed as mass of moisture divided by dry mass of soil, then

$$nw = 0\ b \quad (6)$$

An example of how this equation may be applied is as follows. A soil gas sample has been taken from a sandy soil with a moisture content, 0=8%, foc=0.6%, and b=1.59.

These soil properties imply

$$n = 1 - \frac{1.59}{2.65} = 0.40 \quad \text{from Equation 5}$$

$$nw = 0.13 \quad \text{from Equation 6}$$

$$na = 0.27 \quad \text{from Equation 4}$$

Assume the soil gas contained 500 ppm, benzene. Referring to the section on Units of Measurement, the benzene vapor concentration, is 1.75 mg/L. Assume Koc is 100 and H is 0.235. We would, consequently, expect bulk soil sample from the same location to contain

$C_{st} = [100(0.006)/0.235 + 0.13/1.59\ (0.235) + 0.27/1.59]\ 1.75\ mg/L = 5.4$ mg/kg = 5.4 ppm = benzene by substituting the corresponding value in Equation 3.

Although a direct relationship between soil gas and bulk soil concentration is expected, the numerical relationship is complex and not a direct comparison. Soil properties and contaminant properties will affect the comparison. Generally soil concentrations of petroleum hydrocarbons, expressed in ppm (mass/mass) will be substantially less than soil gas concentrations in ppm (volume/volume).

Aging of Petroleum Products in the Environment

Petroleum products are complex mixtures of many petroleum hydrocarbons. Specific products tend to predominantly contain compounds with similar boiling points, as dictated by the distillation processes used in refining. Upon release of a petroleum product to the environment, certain fractions or groupings of similar compounds may quickly be lost via volatilization, dissolution/leaching, or biodegradation. This leaves behind a residue of more persistent fractions of the original mixture, or weathered product. A very important fate process affecting many refined products is volatilization. In fact, the prominence of this process accounts for the success of the soil gas monitoring techniques.

For example, a representative gasoline mixture may contain 70% paraffinic compounds and 30% aromatics. Over time the proportion of aromatics may increase since the lower molecular weight paraffins are extremely volatile and may not persist. The weathering proceeds, the aromatics and higher molecular weight paraffinic constituents are expected to persist and increase in fractional abundance relative to their contribution to fresh product. The weathering process is very complex and not readily predictable. Nonetheless, the interpretation of vapor monitoring must recognize that as product weathers in the ground, the vapor "signal" may fade, and the relationship between soil gas and bulk soil contaminant concentration may steadily change. Furthermore, monitoring devices that have differential sensitivities to compounds of different molecular weight or structure (aromatic versus paraffinic) may exhibit variable relative sensitivities to fresh product versus weathered product.

THE REAL WORLD: WHAT WORKS AND WHAT DOESN'T

FID vs PID Detectors

In this section we will present some real data collected in the course of site assessment and remediation activities at several sites in Florida. We will examine the relationships among the various types of data and explain those relationships in terms of the theory presented previously.

Table 1. Comparison of Field VOC Detection Methods for Soils Containing Adsorbed Gasoline Surrounding Subsurface Storage Tanks

		Central Florida Site 1		
ID	**PID (ppm)**	**OVA (ppm)**	**GC (ppm)**	**Method SW-846 8020**
26	471	>1000	3477	
27	187	300	245	
28	65	86	101	
29	501	>1000	2869	
30	490	>1000	10400	
31	547	>1000	10227	
32	469	>1000	6101	
33	458	>1000	13370	BTEX = <20 ug/kg
		North Florida Site		
ID	**PID (ppm)**	**OVA (ppm)**	**GC (ppm)**	**Method SW-846 8020**
2	122	160	320	
3	443	>1000	11630	
4	389	>1000	18800	
6	411	>1000	9297	
7	380	680	800	
8	460	>1000	6497	BTEX = 336 ug/kg
9	4	5	4.1	
10	280	400	322	
11	9	6		
		Central Florida Site 2		
ID	**PID (ppm)**		**GC (ppm)**	
1	1.6		0.6	
2	47		39.8	
3	1.3		0.3	
4	528		10480	
7	41		38	
8	71		110	
9	6.7		3.5	
10	1.8		0.4	
11	68		54	
12	131		151	
13	432		13539	
14	140		513	

PID = Photovac TIP®—calibrated to 100 ppm isobutylene.
OVA = Foxburough OVA®—calibrated to 100 ppm methane.
GC = FID & PID serial detectors, GC run isothermally at 75°C and calibrated with 100 ppm methane.
Method SW-846 8020 = Laboratory Analysis.

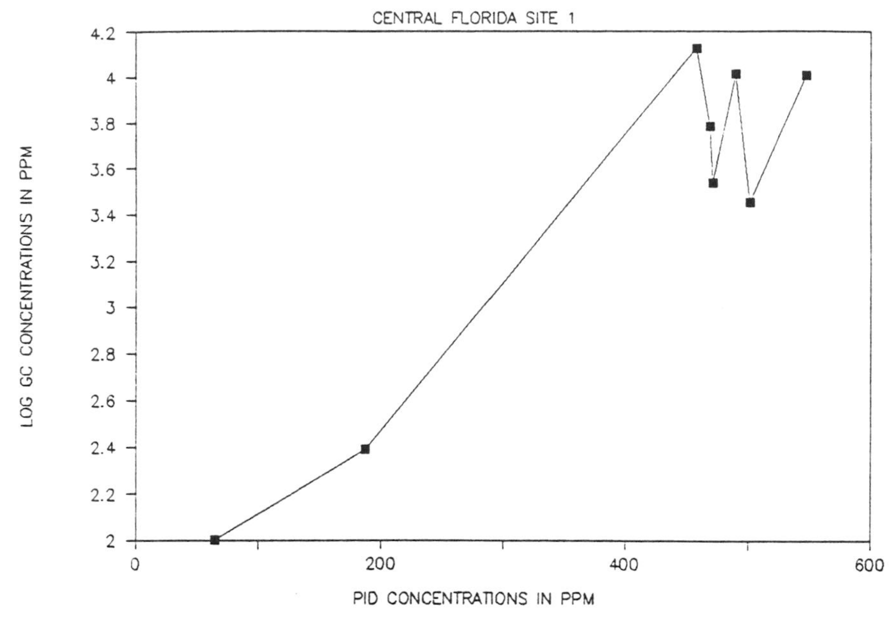

Figure 1. PID vs LOG GC concentrations (central Florida site 1).

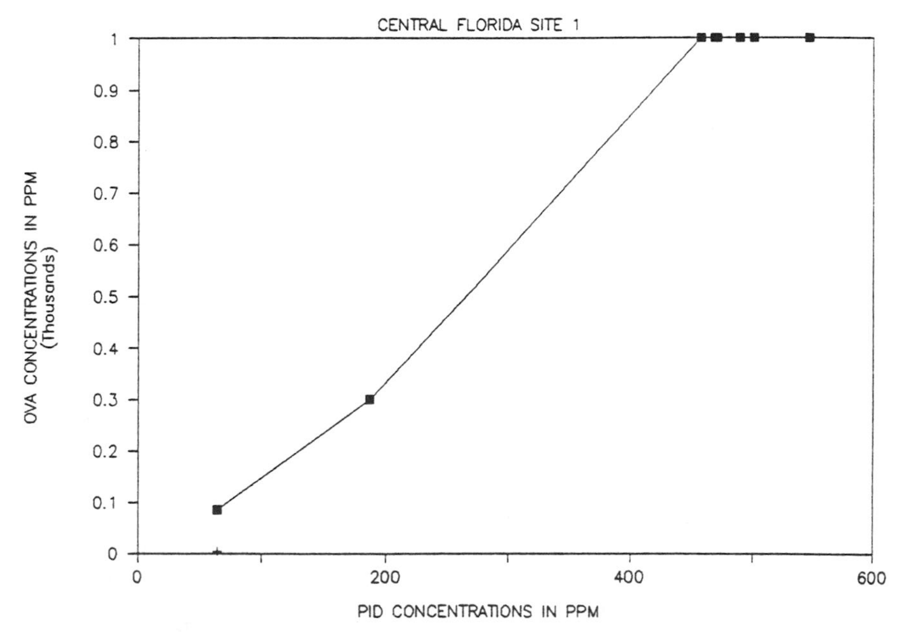

Figure 2. PID vs OVA concentrations (central Florida site 1).

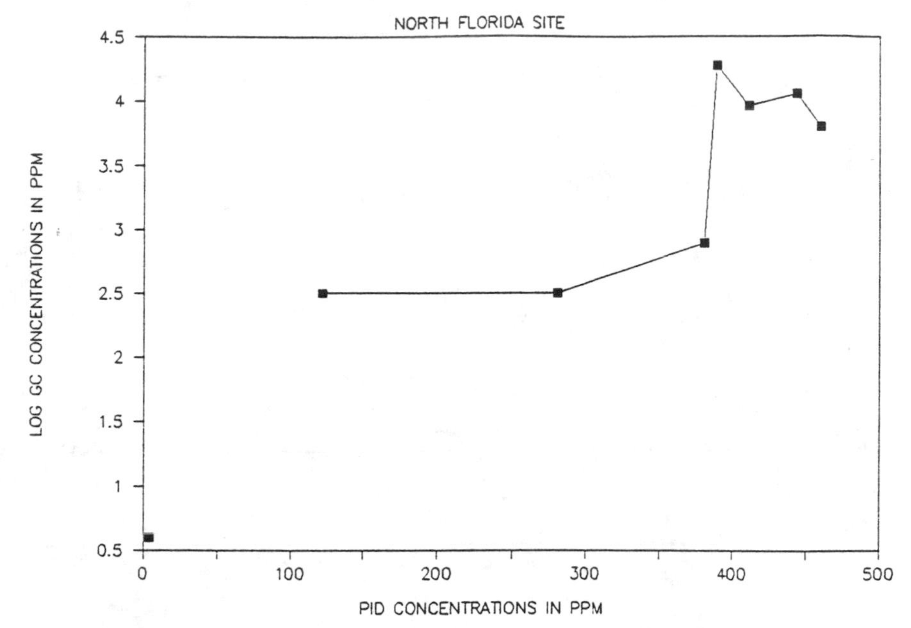

Figure 3. PID vs LOG GC concentrations (north Florida site).

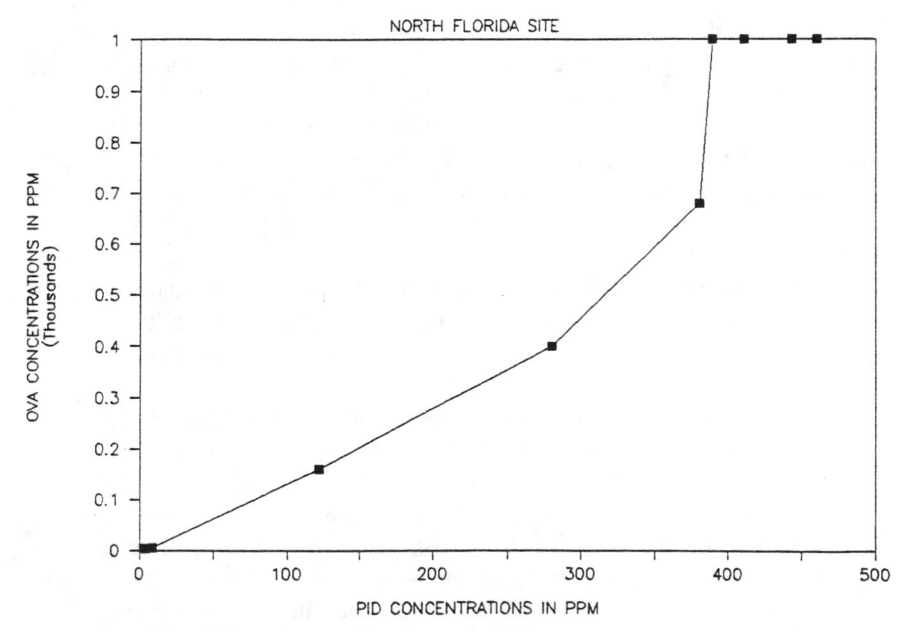

Figure 4. PID vs OVA concentrations (north Florida site).

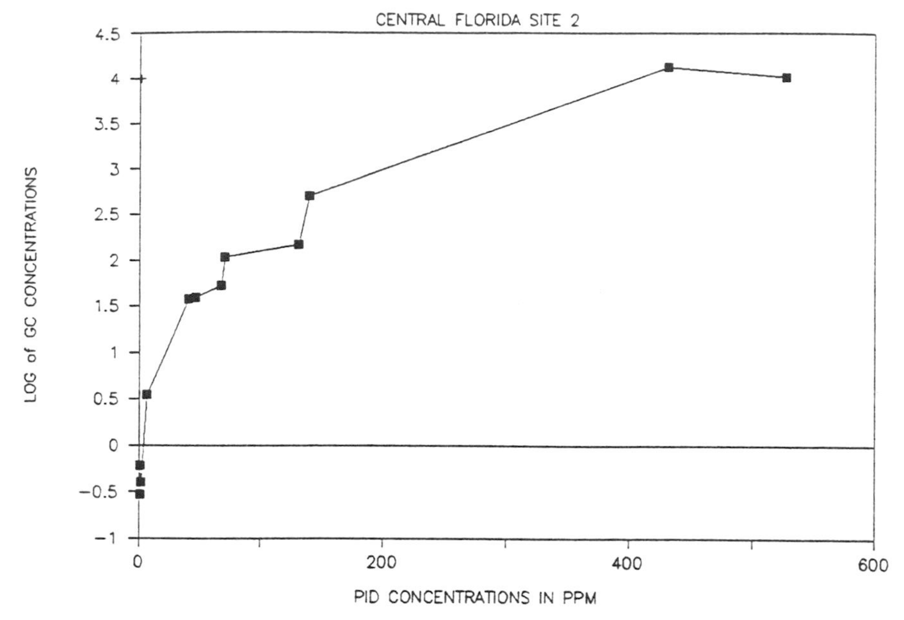

Figure 5. PID vs LOG GC concentrations (central Florida site 2).

Table 1 and Figures 1 through 5 provide a comparison of soil quality data developed during the excavation of contaminated soils from three former gas stations in Florida. The samples were collected following the removal of underground gasoline storage tanks at each of the facilities. The table compares the results of three different field methods (Photovac TIP I, a PID, Foxborough OVA, an FID and gas chromatography, using a PID and FID in series) used to quantify relative total contaminant levels in each sample. Two of the samples subsequently were analyzed for benzene, toluene, ethylbenzene, and xylenes (BTEX) by EPA method SW-846-8020 in the laboratory.

The Foxborough OVA (FID) was calibrated to 100 ppm methane, and the Photovac TIP I (PID) was calibrated to 100 ppm isobutylene. Both were calibrated in accordance with the methods outlined in their respective operating manuals.

The gas chromatograph was run isothermally at 75°C and calibrated to the peak height of a 100-ppm methane standard. Duplicates and controls were run for at least 10% of the analyzed samples. Calibration was verified after analysis was completed. Serial detection by both flame- and photo-ionization detectors on the GC allowed for the exclusion of methane from the calculated results.

Sandy soils containing adsorbed gasoline were excavated from around the subsurface tanks. The source of hydrocarbons detected in the material originated from tank overfill that had occurred over a period of more than 20 years. The adsorbed hydrocarbons in the material had been subjected to weathering by volatilization, biodegradation, and solution into infiltrating water. The samples were collected and placed in clean, 16-ounce, wide-mouth glass jars that were capped with an aluminum foil septum. The samples' temperature was reduced to approximately 20×C. After approximately five minutes, the septum was perforated, allowing sampling of the head space by the three methods (GC, OVA, and PID) to determine relative concentrations of vapor phase hydrocarbons.

The results presented in Table 1 indicate variability between methods of field analysis. The variation is attributable to the detection method and the relative suite of contaminants (weathered, adsorbed gasoline) present in the samples. Each method appears to indicate a relative difference in concentration between samples; therefore, each can be used in the field to discriminate successfully between soils that contain adsorbed hydrocarbons and those that do not. Inspection of Figures 1 through 5 shows that for each site the variation between the responses of the FID and PID instrument is systematic. Considering the calibration procedures employed, one would have to expect nonequivalence of response. The FID was calibrated to methane, a single carbon compound to which the detector is not very sensitive. The PID, on the other hand, was calibrated to isobutylene, a compound which is about in the middle of the sensitivity range for the instrument. This calibration procedure has the effect of making the FID significantly more sensitive than the PID, even though the PID is generally more sensitive to petroleum compounds.

Additional factors to consider when choosing the PID or FID for screening include the portability of the instrument. The PID and FID are both portable, with rechargeable nickel-cadmium batteries powered with A/C power options. However, the FID utilizes a hydrogen flame detector running from compressed hydrogen gas; if the FID is required for extended field efforts, potentially hazardous compressed hydrogen gas must be kept onsite.

The ability of an instrument to detect methane is another consideration. The FID has the ability to detect methane in soil gas vapor, while the PID does not. Since methane occurs in both gasoline and natural soils, the possibility exists that misleading soil gas "hits" actually attributable to natural conditions may be measured using the FID.

The FID's destructive analysis of samples is a factor which is undesirable where repeat analysis of a sample or simultaneous analysis by multiple techniques is required. In these instances a PID would be the preferred instrumentation.

Soil Gas Methods vs Direct Chemical Analysis

Both the in situ and headspace soil gas screening methods are important screening tools for assessing soil contamination quickly and inexpensively. Both methods also permit onsite analytical results, allowing continuous adjustment of the scope and focus of the investigation while it is ongoing. However, a measured soil gas vapor concentration typically is not interpreted to represent the true analytical concentration of petroleum in soils.

The question of whether the soil gas techniques are as effective as direct chemical analysis of soil samples at detecting low levels of soil petroleum contamination was investigated by comparing headspace and/or in situ measurements for specific samples/locations with results of chemical analyses on selected soil samples.

The data presented in Table 1 and Figure 6 indicate that the correlation between soil gas screening and direct chemical analysis does exist, but is not a direct relationship; this was expected due to the differences in instrument response, sensitivity, calibration procedures, environmental conditions, and other factors discussed earlier which influence the measured soil gas vapor concentration. The relationship between soil gas concentration and laboratory analytical concentration appears to have a lower threshold, as shown in Figure 6. Soil gas headspace concentrations measured at around 300 ppm or less were below the EPA Method 8020 total VOA (TVOA) detection limit for soil samples analyzed. When soil gas vapor concentrations are above around 300 ppm, the direct chemical analysis also detected TVOAs.

This situation indicates that soil gas techniques are very good at detecting petroleum components in the vapor phase at low concentrations, below the saturation

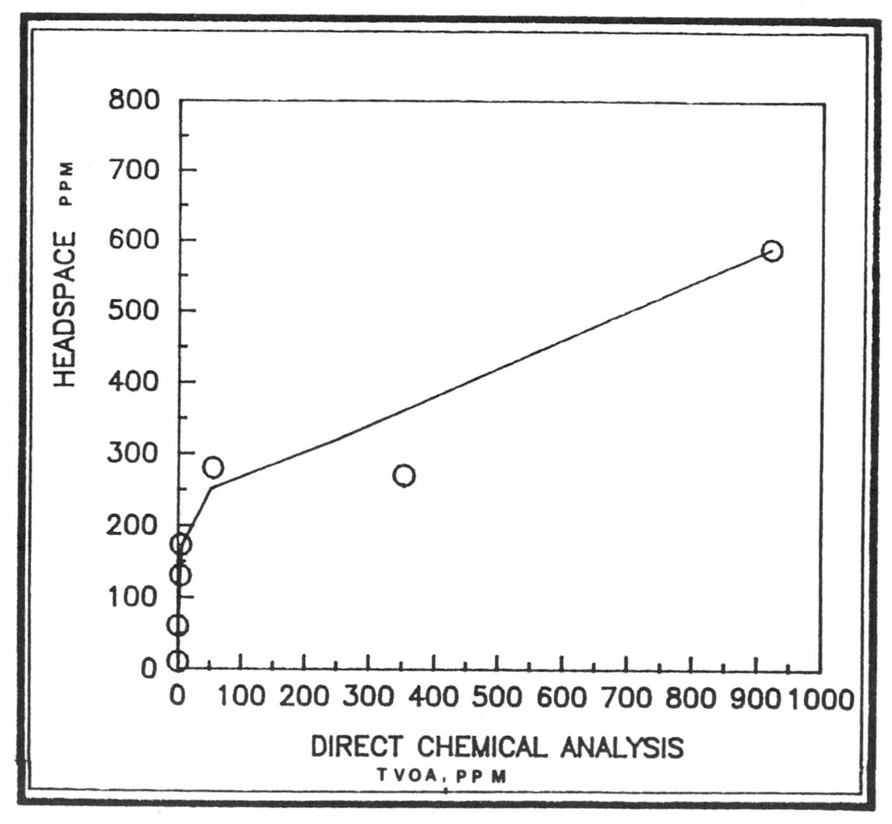

Figure 6. Correlation between headspace and direct chemical analysis.

point required for residual liquid to be detected by direct chemical extraction and analysis. When volatile petroleum components are present as vapor in the soil pores, soil gas techniques can be used effectively to track this vapor plume back to the spill or leak site (source). If the spill has reached phase equilibrium, the soil gas plume may extend well beyond the zone of soil saturation.

For this application, soil gas techniques prove to be more sensitive and much faster than direct laboratory soil analysis for determining whether petroleum products are present.

The Florida Case

The Florida Department of Environmental Regulation (FDER) has incorporated into its regulations a headspace screening procedure. The intent of this is to provide a quick procedure which may be used in the field to determine the extent of severe petroleum contamination in soils. The definition of "excessively contaminated soil" in the regulation is specified as 500 ppm. The procedure is described only briefly in the regulation and has required clarification through the issuance of a guidance document. Calibration procedures are not specified in either the regulation or the subsequent guidance, other than to follow manufacturers' recommendations. The regulation recommends the use of an FID instrument, but allows the use of a PID if the response can be demonstrated to be equivalent. As we have seen from the preceding discussion, instrument response is dependent on many factors, and the variety of equivalences should surprise no one.

REFERENCES

1. Jury, W. A., W. F. Spencer, and W. J. Farmer. "Behavior Assessment Model for Trace Organics in Soil: I. Model Description," *Journal of Environmental Quality,* 12(4):558–564 (1983).
2. Tucker, W. A., C. T. Huang, J. M. Bral, and R. E. Dickinson. "Development and Validation of the Underground Leak Transport Assessment Model," *Proceedings of Petroleum Hydrocarbons and Organic Contaminants in Ground Water: Prevention, Detection, and Restoration,* National Water Well Association, 1986.

CHAPTER 10

Electrochemiluminescent Optrode Development for the Rapid and Continuous Monitoring of Petroleum Contamination

Brian G. Dixon, John R. Deans, and **John Sanford,** Cape Cod Research, Inc., Buzzards Bay, MA
Bruce J. Nielsen, Engineering & Services Laboratory, Tyndall Air Force Base, FL

The problem of petroleum contamination of soils and water is one that has been well-publicized and one that is of pressing environmental concern throughout the world. This problem has many facets to it, including the type of pollutant encountered, its detection, removal, and destruction. Before any remedial action can be taken at a site it would be useful to know the identities of the offending chemicals and be able to measure their concentrations as well.

A wide variety of analytical techniques have been investigated over the years to accomplish these tasks. The development of these techniques has been accompanied by the demand for detecting ever lower concentration levels of pollutants. Detection of part per million concentrations is now routine and part per billion measurements are now required. This means that the next generation of instrumentation will be expected to accurately measure concentrations in the low, parts per billion range.

The state of the art analytical techniques for detecting trace levels of organic compounds in water are primarily gas chromatography (GC) with a flame ionization or photoionization detector and tandem gas chromatography-mass spectrometry (GCMS). These standard methods, although accurate and dependable, suffer from a number of shortcomings. In general, laboratory based instruments are not easily transportable and are expensive, both to purchase and to maintain. They also require fairly involved training of technical personnel, which is expensive. Laboratory instruments are somewhat delicate and not well adapted to use in the field. Those devices which are portable do not possess the ability to measure concentrations lower than the parts per million range.

The last several years have witnessed the development of prototype analytical techniques that have shown some promise. Perhaps the closest related work to the research carried out in this project is the remote laser-induced fluorescence assay for groundwater contaminants.[1-5] Various laser-fiber optic systems have been designed and constructed that are capable of detecting trace levels of contaminants in groundwater. In terms of use in the field, lasers are delicate since they require precisely aligned optics and shock sensitive equipment. Lasers and their associated hardware are also, in general, quite expensive. Most of these laser systems are dependent upon the use of excitation light in the ultraviolet region of the spectrum which can rapidly degrade the glass of fiber optic cables. This then requires the use of quartz fibers which are much more expensive and brittle. Currently available fiber optic-laser systems also suffer from high attenuation problems in the ultraviolet region of the excitation beam.

Another approach which is fieldable, but not remotely applicable, are microbial assays which rely on a bioluminescent technique that has been shown to be effective in detecting toxic organics at the ten to hundred ppb concentration.[6,7] While such microbial tests represent an improvement in the simplicity of monitoring dissolved organics, they too have flaws. Microbiologically based systems possess the attendant problem of keeping microorganisms alive and functioning, and they are more qualitative than quantitative assays. These drawbacks limit the remote sensing applicability of microbial techniques and the ability of such assays to distinguish between toxic substances. What is needed, therefore, is a luminescent-based optrode, with the sensitivity of bioluminescence, that can function well remotely.

The impetus for the research effort about to be described came from the Environics Division of the Air Force Engineering and Services Center (AFESC), located at Tyndall AFB, Florida. This division is responsible for investigating improved methods for subsurface contaminant monitoring. The Air Force Installation Restoration program has now identified over 2800 chemically contaminated sites on Air Force installations due to spills or past disposal practices. The majority of these sites have JP-4 jet fuel as the primary contaminant. Fuel leaks and spills are by far the most frequent sources of soil and groundwater contamination on these installations. The number of fuel spill sites is expected to increase as the number of underground tanks and pipelines are inspected under new tank regulations. Cape Cod Research was funded, by the Air Force, under the auspices of the Department of Defense-Small Business Innovation Research (SBIR) program to carry out a feasibility study involving the use of fiber optic optrode technology to detect aromatic hydrocarbons in water. Of particular concern was benzene detection at concentrations in the 1–5 ppb (parts per billion) range. Benzene is a known carcinogen with a groundwater standard in this range.

RESEARCH OBJECTIVES AND FUNDAMENTALS

The primary objective of this research was to develop an electrochemiluminescence (ECL) optrode (by definition an optrode is an optical electrode) system that would be capable of detecting aromatic hydrocarbons in water at concentrations in the low, parts per billion range. A second objective of this research was to use the feasibility studies to design a device which would be durable, dependable, easily transportable, and relatively low in cost. Other objectives were to establish the feasibility for using the novel ECL technique to qualitatively and quantitatively measure the concentrations of aromatic hydrocarbons in water.

Electrochemiluminescence (ECL) is a blend of the two distinct disciplines of photochemistry and electrochemistry. In effect it combines all of the rich and useful features of photochemistry, such as the phenomena of fluorescence and phosphorescence, with the radical cation/anion and electron transfer capabilities inherent in electrochemistry. In essence, ECL involves the use of electrodes and the application of small voltages to create photochemical excited states.

Aqueous Electrochemiluminescence System

Both organic solvent and aqueous based luminescent systems are well known. Although much less studied, aqueous luminescent systems are, in fact, far more common. Examples of such systems include the firefly, luminescent bacteria, and a multitude of other bioluminescent organisms.[8-10] These naturally occurring luminescent systems are also almost invariably quite efficient as well. For example, the quantum yield of the firefly luminescence approaches 100%, a yield that is unmatched in synthetic, chemiluminescent systems. In general, water and oxygen quench excited states very efficiently, which has meant that the vast majority of laboratory studies have been carried out in rigorously dried organic solvents and in the absence of oxygen.[11-13] Over the last ten years or so, research into aqueous based transition metal complexes in luminescence applications has been intense, primarily because of their promise for solar energy harvesting applications. In an ongoing series of excellent studies, Bard has demonstrated the capabilities of ruthenium bipyridyl complexes to be sensitive ECL agents in aqueous systems.[14-17] The most studied of these ruthenium complexes is tris (2, 2'-bipyridyl) ruthenium (II) chloride (hereafter referred to as $Ru(bipyr)_3$), which is a crystalline, orange, water-soluble complex that possesses a rich and varied chemistry and the octahedral structure shown below:

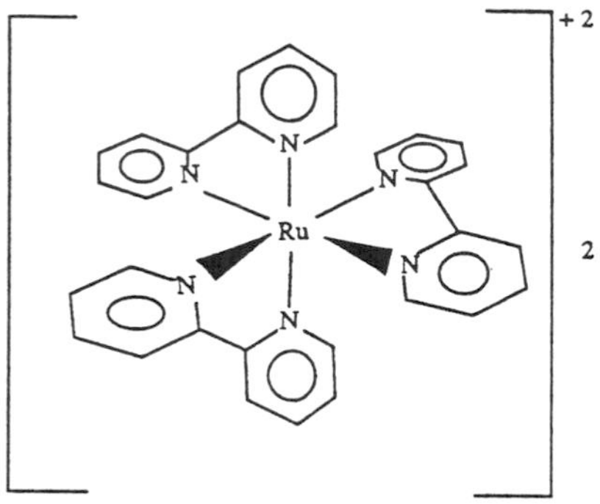

Among the interesting properties of this complex is the fact that it gives efficient ECL in aqueous solution. This is clearly an advantage for the application of detecting petroleum in soils or water. The ECL photochemistry of this compound is complex, but basically involves emission from a $d-\pi^*$ excited state. Of immediate interest here is the fact that its ECL emission occurs in the visible range of the electromagnetic spectrum, specifically ~500 to 700nm with a maximum at ~600nm. This means that silicon-based photodetectors and nonquartz fiber optic cables can be used which are much less costly and more readily available. Of equal importance is the fact that the subject complex can be immobilized within a polymer or onto its surface, and thus the required ECL carried out in the solid state. The mechanism of the luminescence, depicted below, involves a reversible charge-transfer excited state complex between the Ru^{+2} and the bipyridyl ligands.

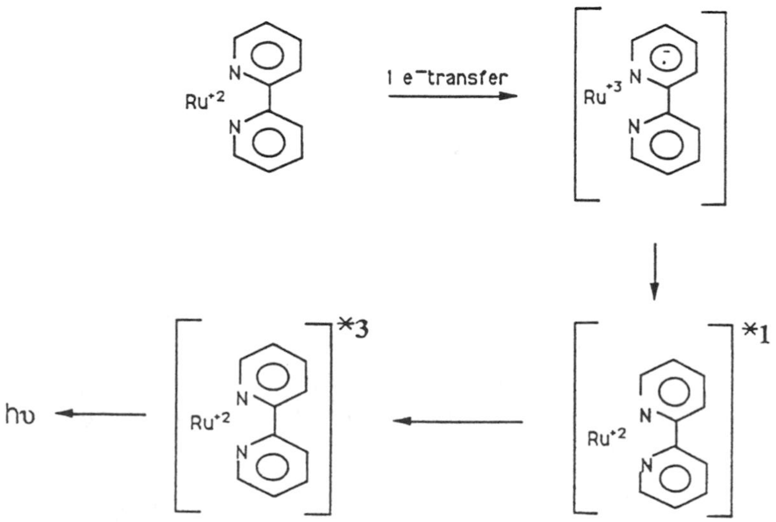

The quantum efficiency of this process is dependent upon, among other factors, the nature of the solvation of the complex. This phenomenon is used to great advantage in the described work.

Photomultiplier Detection System and ECL Studies

Both homogeneous and solid state ECL laboratory experiments were carried out with the Ru(bipyr)$_3$ complex under model conditions to establish the scope of the technology. Benzene was used in all of the experimental studies as a model compound for aromatic hydrocarbon pollutants that are present at fuel spill sites. In all of the experiments to be discussed, the benzene solutions were carefully prepared daily for a given set of experiments. All of the experiments were reproduced at least in duplicate. Control experiments were also performed with the various components individually to assure that the observed phenomena were real.

General Laboratory Techniques

Sensitive photodetection experiments were carried out with a photomultiplier-monochromator apparatus model (Model 77250 and Model 77762, respectively, the Oriel Corp.). This instrument consisted of a sensitive photomultiplier tube whose operational detection range is from 200nm to 800nm, a monochromator, and attached fiber optic cables. A polypropylene mold was machined to hold the quartz cell in which the electrochemical experiments were actually carried out. This mold also was modified so that the end of the fiber optic cable could be placed flush to the side of this quartz cell.

Experiments were carried out using the tris (2, 2'-bipyridyl) ruthenium (II) chloride in the presence and absence of trace levels of benzene. These experiments were carefully performed to assure that extraneous impurities would not be introduced into the test solutions. All glassware was first acid washed, rinsed with deionized water, and thoroughly dried prior to use. New and carefully cleaned syringes were used as well. In the experiments about to be described, the stock solutions were prepared immediately prior to the performance of the experiments.

Effect of Voltage

There is little doubt that the experimentally observed ECL was due to the Ru(bipyr)$_3$ complex. The mechanism of this ECL involves a charge transfer $(d-\pi^*)$ transition with the emitting excited state being a triplet. The crucial electron transfer occurs at a voltage of $\sim$0.95V vs SCE. Virtually no luminescence was detectable below this voltage while, at or above it, an increase of up to 40% in total intensity was observed. In addition, this emission corresponded to the well-known characteristic spectrum expected for this complex. The appropriate control experiments were carried out to ensure that Ru(bipyr)$_3$ ECL was actually being observed. These experiments included running all of the components of the stock solution independently and together, but in the absence of the Ru(bipyr)$_3$, at 0 and 1.000V and confirming that no luminescence changes were observed.

Homogeneous (Aqueous) Phase ECL Studies

Figure 1 is composed of two plots, 1a and 1b. Figure 1a is a plot of the luminescence of a 1×10^{-4} M Ru(bipyr)$_3$ solution as a function of voltage applied to the solution and in the absence of benzene. The baseline corresponds to zero applied voltage. As can be seen, at an applied voltage of 0.950V there is an immediate response due to Ru(bipyr)$_3$ electrochemiluminescence. This response was found to be very reproducible and quite fast. Removal of the applied voltage resulted in an immediate drop in intensity to the background light level. Figure 1b is a plot of the luminescence after the injection of enough stock benzene solution to yield a final benzene concentration of 25 ppb. It should be noted that the volume of such an injection is insignificant in terms of the total volume of solution. As can be seen there was a significant increase in the total detected luminescence. This result was found to be very reproducible and the electrode response was very quick as well.

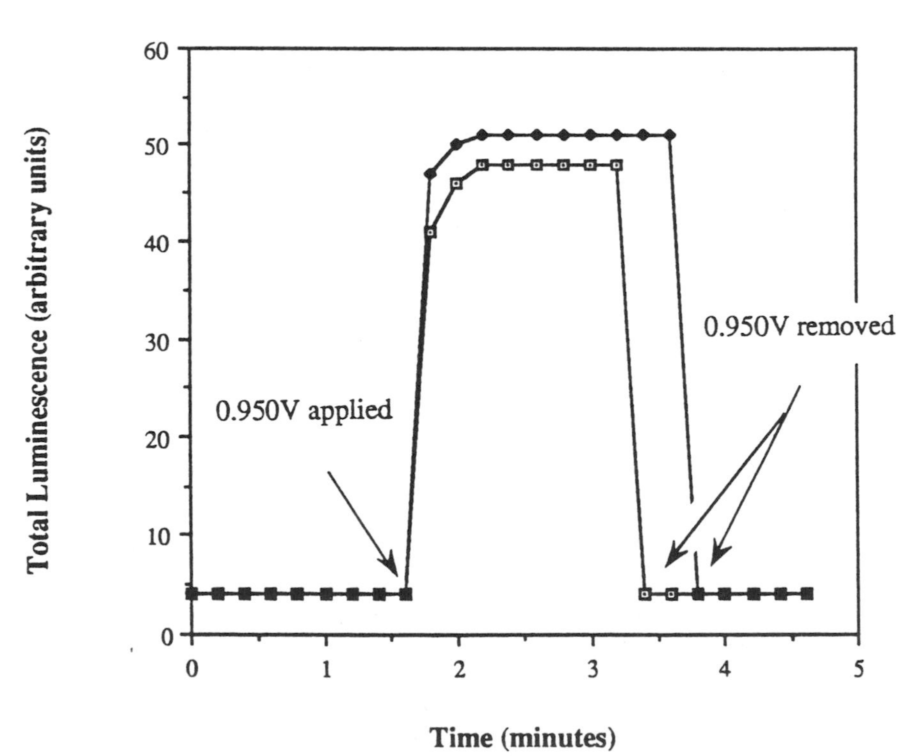

Figure 1. Electrogenerated chemiluminescence of the Ru(bipyr)$_3$Cl$_2$ with and without benzene in water.

Ruthenium Complex Emission Spectrum

An ECL spectrum of the $Ru(bipyr)_3$ complex was obtained by using the monochromator to scan from ~550 to 800nm. Similar experiments were then carried out in the presence of benzene at various concentrations.

Figure 2 contains representative results showing the effect of benzene concentration upon the electrochemiluminescence spectrum of $Ru(bipyr)_3$ in water. This figure compares ECL in the absence of benzene and then in the presence of 24 and 48 ppb. The results contained in Figure 2 are interesting and very promising for a number of reasons. First, the observed $Ru(bipyr)_3$ ECL increases significantly as the benzene concentration goes up. In principle, this will allow for the quantitative as well as qualitative determination of a pollutant's presence. Of equal importance is the fact that the change in intensity with benzene is not linear over the whole spectral range. This is important since it should be possible to simultaneously monitor intensity changes at different wavelengths and thereby increase both the sensitivity and signal to noise ratio of the detecting device.

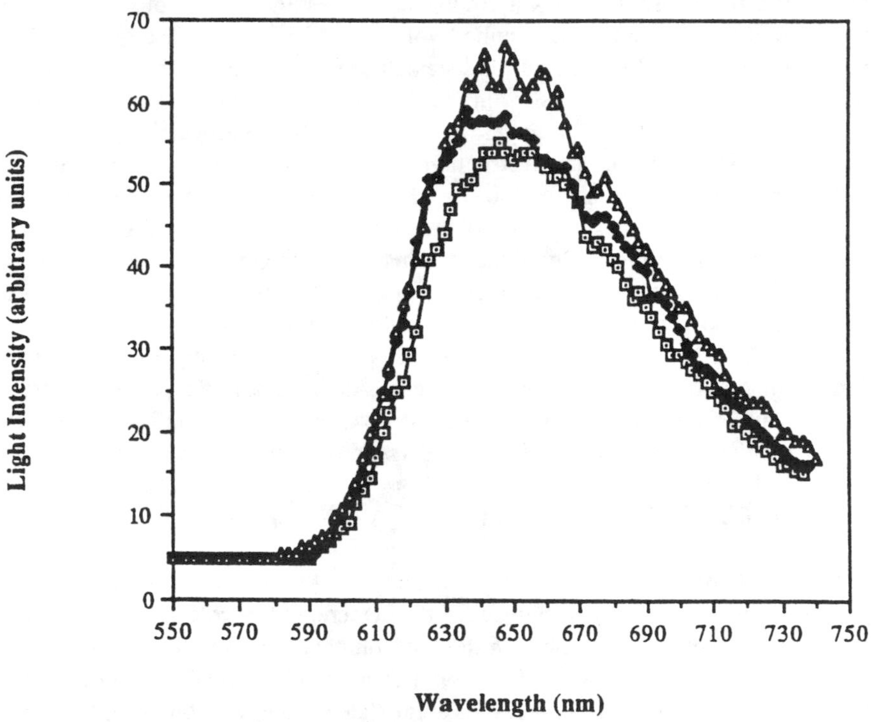

Conditions: : In deionized water; $[Ru(bipyr)_3] = 1 \times 10^{-3}M$; $[Na_2C_2O_4] = 0.2M$; $[NaCl] = 0.1M$; applied voltage = 1.3V.

White squares = without benzene
Black diamonds = with 24 parts per billion benzene
Triangles = with 48 parts per billion of benzene

Figure 2. Effect of benzene concentration upon Ru(bipyr) ECL.

Solid State ECL Studies

The above described homogeneous phase work was then extended to the solid state. Bard and Anson have shown that it is possible to immobilize ruthenium bipyridyl complexes within polymer matrices and carry out solid state ECL.[18-21] In the current work, parallel experiments were carried out wherein the platinum working electrode was first coated with a thin layer of Nafion®, a perfluorinated polymer possessing pendant sulfonic acid groups. This polymer modified electrode was further modified by immersion in a 5×10^{-3}M solution of $Ru(bipyr)_3$ in 0.1M H_2SO_4. This process immobilizes the $Ru(bipyr)_3$ within the polymer via bonding to the sulfonic acid groups.

The modified electrode was placed into the cell so that its flat surface faced toward the end of the fiber optic cable. ECL experiments were then carried out as described above for the homogeneous work. Figure 3 is the solid state counterpart to Figure 1 and includes plots 3a and 3b. Figure 3a shows the results of these solid state scans in the absence of benzene. A comparison of these results with those of Figure 1a yields a couple of interesting conclusions. First, the response of the electrode to the applied voltage, which is immediate in the solution experiments, is slower in the solid state. The decay of the luminescence upon removal of the voltage is also slow in the latter state. These results are not unexpected. Since the Nafion-$Ru(bipyr)_3$ composite is not an especially good electrical or ionic conductor, its electrochemical response is expected to be much slower than the solution phase case. In addition, the restriction of the mobility of the ruthenium complex in the solid state means that the various light quenching mechanisms are drastically reduced. This allows the excited state $Ru(bipyr)_3$ species to have significantly longer lifetimes and the observed luminescence will decay more slowly upon removal of the applied voltage, as is observed.

Figure 3b is the result of a luminescence scan run immediately after the lower run but in the presence of 25 ppb of benzene. As in the solution phase instance, a significant increase in the total luminescence was observed. This result has now been extended to parts per trillion concentrations of benzene as well.

Portable (Field) Device Design

The results of the research to date indicate that it will be feasible to use electrochemiluminescence as a technique for the detection of petroleum contaminants at part per billion concentrations in the environment. The key to the success of this technique, however, is the design and construction of a portable device for field use. As currently envisioned, this solid state device will consist of two sections as detailed in Figure 4. A small PVC pipe containing the optrode sensor and light detector electronics can be lowered underground into a test well, if desired. A larger instrumentation package containing the control electronics for

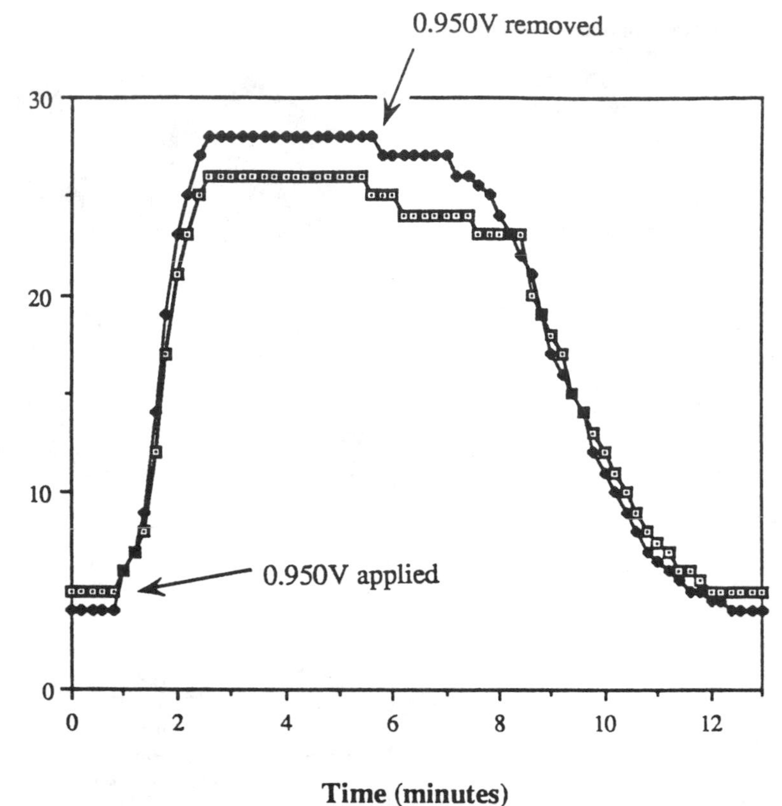

Conditions: [Ru(bipyr)$_3$Cl$_2$] immobilized by ion exchange onto the Nafion® coated, platinum working electrode. H_2O solution of [sodium oxalate] = 0.025M prepared and added to the quartz cell; all solutions prepared in deionized water. Benzene was added from a stock solution. Applied voltage of 0.950V at ~1 minute in both cases.

White squares = no benzene
Solid black diamonds = with 25 parts per billion benzene added

Figure 3. Solid state electrogenerated chemiluminescence of the Ru(bipyr)$_3$Cl$_2$ in Nafion™ with and without benzene in water.

the system will sit on the surface. The two modules will be electrically joined together by a multiconductor shielded cable. The cable will have a connector on the sensor package end to allow for easy replacement of the optrode if it should fail. The PVC pipe containing the sensor will have a number of holes drilled in it to allow groundwater to freely pass over the optrode. The optrode itself will consist of a bundle of glass optical fibers coated with indium-tin oxide as the

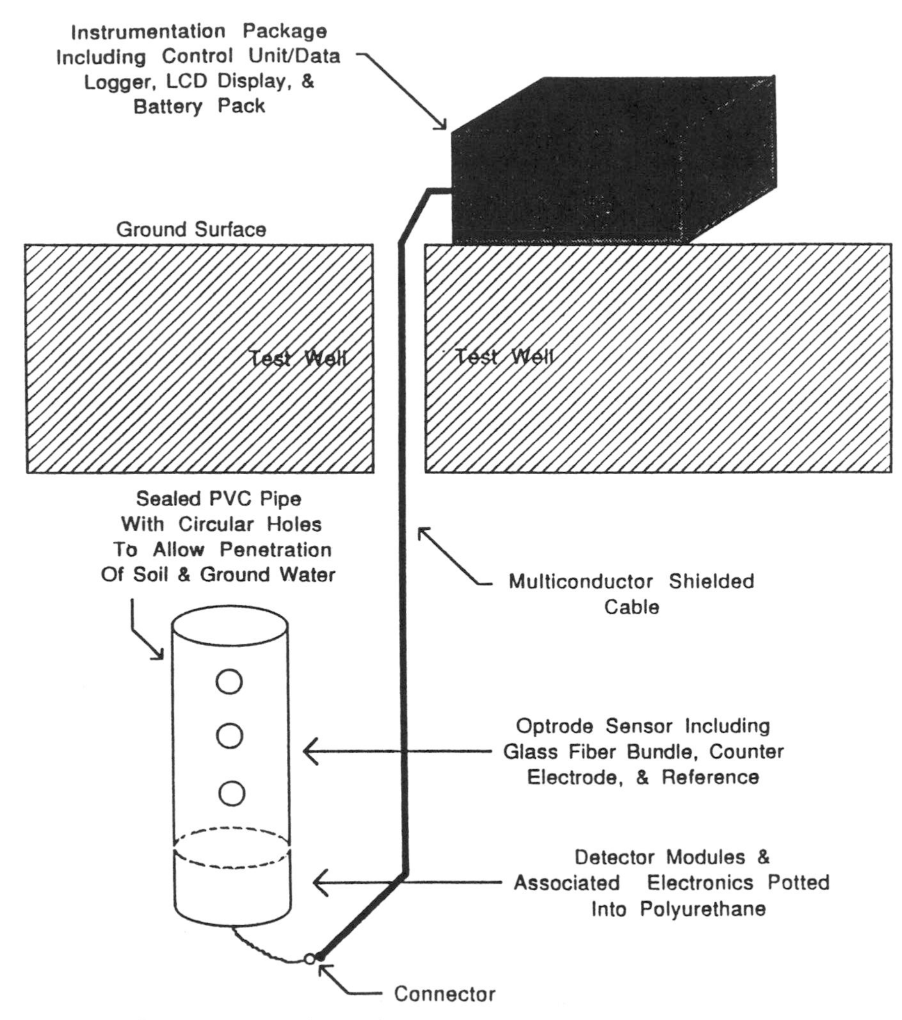

Figure 4. Physical layout of optrode system.

working electrode and the Ru(bipyr)$_3$ compound. The indium-tin oxide electrode is optically transparent, thus excellent light transmission characteristics are to be anticipated. Although it is not required to immobilize the ruthenium complex upon the fiber optic cable, it is attractive to do so since the maximum light sensitivity can be achieved using this arrangement. The electrochemical cell will be completed by a length of ruthenized titanium wire which will serve as the counter electrode, and a piece of silver silver-chloride wire which will be used as the reference electrode. The light generated by the optrode will be optically coupled to detector modules, which will convert the light into an electrical voltage. The detector modules and their associated support electronics will be potted into polyurethane at the end of the PVC pipe. The output voltage signal produced by the detectors will be sent to the instrumentation package on the surface.

The heart of the portable solid state device will be a single board computer. This device will serve as the control unit and data logger for the system. A block diagram of the complete system is shown in Figure 5. The chosen computer is

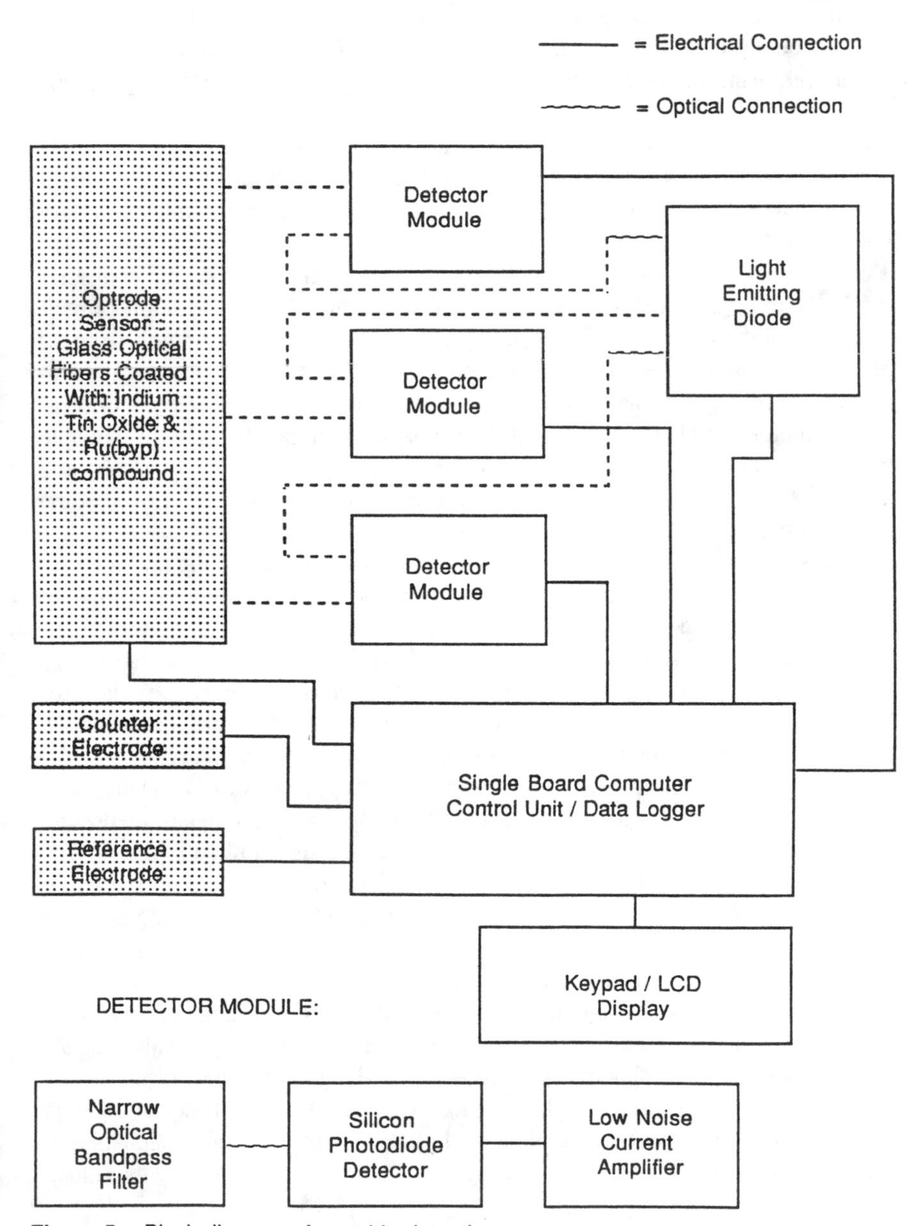

Figure 5. Block diagram of portable detection system.

an extremely versatile device that squeezes a large number of functions into a 2.25″ by 3.725″ by 0.75″ space. The computer is programmable in a hybrid version of BASIC or for greater speed with a compiled version of BASIC. It is also possible to directly program the 8 bit Hitachi 6303Y CPU with assembly language code for minimum program space and maximum speed. Up to 28 kbytes is normally available for program and data storage and an additional add-on memory board allows expansion to 512 kbytes. The computer draws very little power, requiring only 120 mW for normal operation and 24 mW in a standby mode. This will allow the portable device to operate for several months at a time with only a medium sized 12 volt sealed lead acid battery. A solar panel could be used to trickle charge the battery to allow the system to run for much longer periods. The computer also includes a backup lithium battery mounted on board to preserve the program memory and data if the main power should fail. Other features of the computer include a 10 bit, 11 channel analog to digital converter and 16 digital input/output lines for control functions. Also, it generates precision time intervals to allow for accurate sequencing of events ranging from 10 ms to several months. Finally, the computer includes a UART (universal asynchronous receiver and transmitter) to allow for data transfer to a desktop or portable computer at 300, 1200, 9600, or 76800 baud. If telephone lines are available near testing areas, data from a number of test devices can be retrieved and processed by a single central computer located near the testing site or at some other more remote location.

The computer will collect and process data supplied to it from the optrode sensor. To avoid drift problems associated with absolute measurements, a relative measurement system will be used. At regular time intervals, the computer board will pulse a light emitting diode (LED) on and off. The LED will generate bursts of light which will be optically coupled to the detector modules. The detector modules will convert the light into a voltage signal, which in turn will establish the zero level or baseline to which all other measurements will be relative. During alternate time intervals, the computer board will pulse the LED and also supply pulses of power to the optrode sensor. If hydrocarbon contaminants are present in the area, the sensor will generate light when power is supplied to the sensor. The light produced by the optrode will add to the light generated by the LED, causing the output voltage of the detector modules to be greater during the time interval when the sensor and LED is on, as compared to the output voltage produced during the interval when only the LED is on. The difference in voltage levels during these two separate time intervals will indicate the presence and level of concentration of contaminants. This measurement technique will also reduce noise, in addition to eliminating the drift associated with absolute measurements. The computer will synchronize the pulsing of the LED and optrode sensor with the measurements made by its analog to digital converter. Voltage signals outside this synchronization period will be rejected as noise, thus increasing the signal-to-noise ratio of the measurements.

To further increase the accuracy of the testing, three separate detector modules will be used in the system. Each of these integrated optical detectors will look at a different portion of the optical spectrum to verify the effect caused by the presence of contaminants. The optical transducer incorporates three separate components—a narrow optical bandpass filter, a silicon photodiode detector, and a low noise current amplifier into a single miniature package. In addition, the metal case is hermetically sealed for operation in harsh conditions. Incorporating these three elements into a single shielded metal package reduces external noise pickup and reduces the number of optical and electrical connections. The heart of the detector module is a silicon photodiode that converts photons which strike the device into an equivalent electrical current. The photodiode used in this module has a high sensitivity down to 10^{-11} watts of incident optical power, and has an extremely linear response. The useful optical bandwidth of the photodiode extends from 400 to 1100 nm. This is ideal, since the ruthenium complex's ECL emission is from ~500 to 700nm, with a peak at 607nm. Current produced by the photodiode is amplified and converted into a voltage by a high gain, high input impedance operational amplifier. The overall response of the detector module is determined by a narrow optical bandpass filter which is placed over the silicon photodiode. The filter has a typical center wavelength accuracy of $+/-1.5$ nm and 60 dB rejection of signals outside the filter bandwidth.

The complete test data from the optrode sensor will be stored in the computer's memory for later retrieval and analysis. The memory capacity will allow the device to go several months at a time without having to retrieve data.

CONCLUSIONS

To date this ongoing research project has established the overall feasibility of using electrogenerated chemiluminescence as an analytical technique for detecting trace amounts of aromatic hydrocarbons in water. More specifically, the results obtained from this study have led to a number of conclusions as follows:

- Electrochemiluminescence is a sensitive technique for detecting hydrocarbons in water at part per billion concentrations and lower.
- The electrochemiluminescence can be carried out in the solid state as well in aqueous solution.
- ECL can be used to quantitatively measure the concentrations of trace pollutants.
- It will be possible to translate the results obtained in the laboratory to the design of a portable and durable device of reasonably low cost.

FUTURE DIRECTIONS

Research continues on the design and construction of a practical field device. Current efforts seek to determine the scope of the technology's applicability by studying a number of significant factors relating to the use of ECL for pollutants detection in water. The work includes both laboratory and field work and consists of a number of different segments as follows:

- establish the selectivity and ultimate sensitivity of the ECL system to different kinds of compounds, both alone and in mixtures.
- determine the sensitivity of the system to impurities and undesirable quenching mechanisms.
- establish techniques to prevent undesirable quenching mechanisms.
- construct and evaluate a portable device in the laboratory and then perform field tests.
- establish the long-term performance of the engineered device.

A two year program is currently anticipated, at the end of which a low-cost field device will be available for commercial application.

ACKNOWLEDGMENT

The financial support for this research, by the Headquarters Air Force Engineering and Sciences Center, Environics Division, Tyndall Air Force Base, Florida, is gratefully acknowledged.

REFERENCES

1. Peterson, J. L., and G. G. Vurek. *Science* 13:123 (1984).
2. Chudwyk, W. A. et al. *Anal. Chem.* 57:1237 (1985).
3. Wolfbeis, O. S. et al. *Anal. Chem.* 60: 2028 (1988).
4. Seitz, W. R. *Anal. Chem.* 56: 16A (1984).
5. Kenny, J. E. et al. in *Luminescence Applications*, M. C. Goldberg, Ed. ACS Symposium Series No. 383, The American Chemical Society, Washington DC, 1989.
6. Kamlet, M. J., et al. *Environ. Sci. & Tech.* 20: 690 (1986).
7. Schultz, J. S., et al. *Talanta* 35: 145 (1988).
8. De Luca, M. A., and W. D. McElroy. *Bioluminescence and Chemiluminescence* (New York, NY: Academic Press, 1981).
9. Harvey, E. N. *Bioluminescence* (New York, NY: Academic Press, 1952).
10. Herring, P. J. *Bioluminescence in Action* (New York, NY: Academic Press, 1978).
11. Adam, W., and G. Cilento. *Chemical and Biological Generation of Excited States* (New York, NY: Academic Press, 1982).
12. Cormier, M. J., D. M. Hercules, and J. Lee. *Chemiluminescence and Bioluminescence* (New York, NY: Plenum Press, 1973).
13. McCapra, F. *Proc. Org. Chem.* 8: 231 (1973).
14. Bard, A. J. et al. *J. Amer. Chem. Soc.* 103: 512 (1981).
15. Bard, A. J. et al. *J. Amer. Chem. Soc.* 95: 6582 (1973).
16. Bard, A. J. et al. *J. Amer. Chem. Soc.* 104: 6891 (1982).
17. Bard, A.J. et al. *Anal. Chem.* 55: 1580 (1983).
18. Buttry, D. A., and F. C.Anson. *J. Amer. Chem. Soc.* 104: 4824 (1982).
19. Rubenstein, I., and A. J. Bard. *J. Amer. Chem. Soc.* 103: 5007 (1981).
20. Rubenstein, I., and A. J. Bard. *J. Amer. Chem. Soc.* 102: 6642 (1980).
21. Krishnan, M., X. Zhang, and A. J. Bard. *J. Amer. Chem. Soc.* 106: 7371 (1984).

CHAPTER 11

Simplified Soil Gas Sensing Techniques for Plume Mapping and Remediation Monitoring

D. H. Kampbell and **J. T. Wilson,** U.S. Environmental Protection Agency, Ada, Oklahoma
D. W. Ostendorf, Department of Civil Engineering, University of Massachusetts, Amherst

Detection of subsurface zone volatile organics was first used over 60 years ago for oil exploration.[1] Since that time, soil gas surveys have expanded technically to include gasoline contaminated ground water plumes.[2] Many underground volatile organic contaminants have been successfully assayed by grab sampling to identify the extent of plumes.[3]

A spill estimated to be near 100,000 kilograms of aviation gasoline resulted from a transfer pipe failure in 1969 at a U.S. Coast Guard Air Station (Figure 1) and spread laterally, giving rise to a residually contaminated soil mass[4] in the capillary fringe. The bound fuel in turn created a narrow dissolved hydrocarbon plume extending about 1200 meters to a bay of Lake Michigan. In 1980, residents downgradient from the air station reported odors in water from their domestic wells. The U.S. Geological Survey conducted a water quality survey soon afterward. Interdiction wells were installed in 1984 to restrict further dissolved contaminant flow into the residential area.

The captured plume water is currently being renovated by air stripping and activated carbon adsorption. Much of the spilled hydrocarbons remain upgradient from the interdiction field as residual fuel bound to the sandy aquifer material which is a continuing source of contamination.[5]

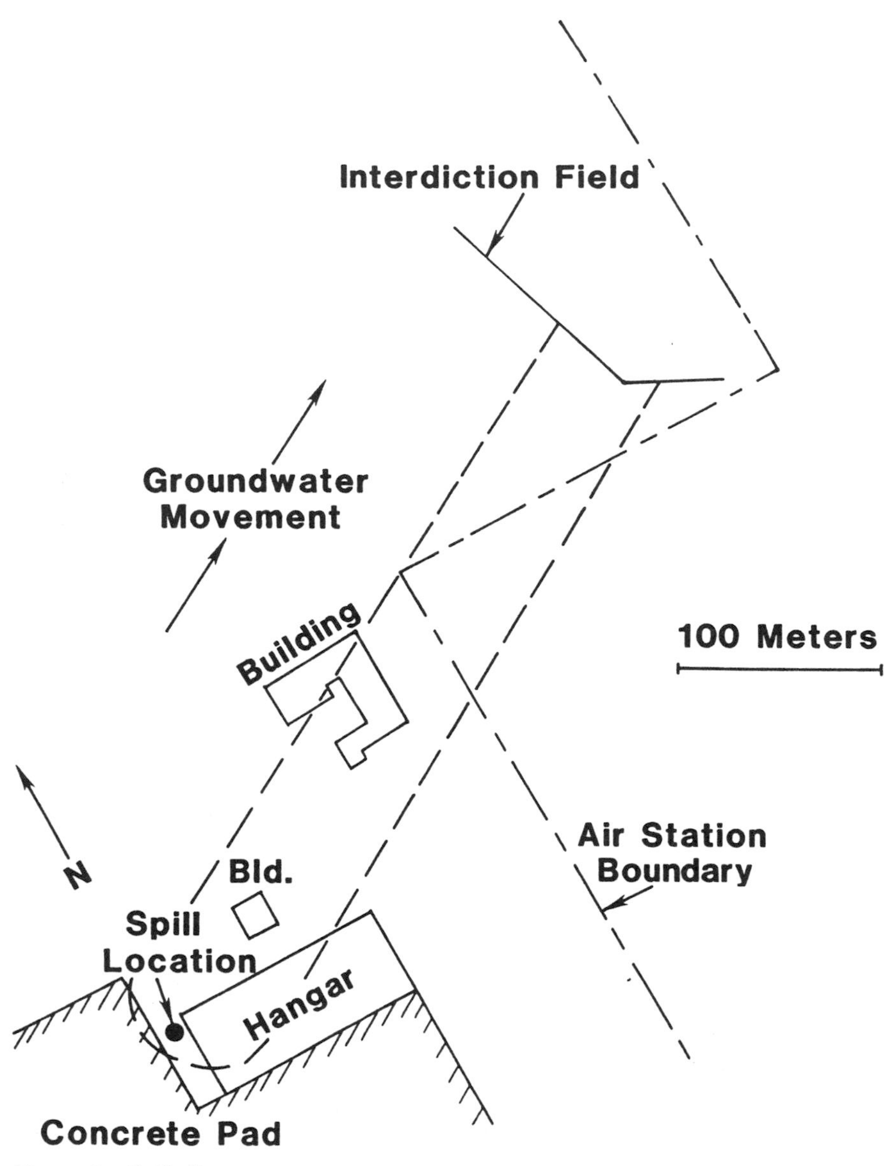

Figure 1. Spill site.

Many factors that complicate interpretation of soil gas measurements did not predominate at the site, since it was characterized by a uniform sand matrix throughout the vertical profile. The objectives of the work were to map the extent of the plume source area and measure the vertical distribution of gasoline vapors, oxygen, and carbon dioxide in the unsaturated zone above the separate phase product. Vertical profile data were applied to derivation of a biodegradation model.

MEASUREMENT METHODOLOGY

Sample Probe

The probe used for plume mapping consisted of a stainless steel tube with dimensions of 1.3 meter length, 1.26 cm diameter, and 0.48 cm internal diameter. The tube with a loose carriage bolt in the bottom end was hand-driven to a depth near 1.1 meter. The tube was then lifted 0.1 meter to remove the carriage bolt. Soil was tightly packed around the tube by striking the surrounding surface with a heavy hammer. A short Teflon® transfer line from the analytical apparatus was attached to the exposed end of the tube. After completion of measurements, the tube was pulled from the ground, rinsed with water, and reinserted at the next sampling location.

Permanently placed tubing clusters were used for measuring the vertical distribution of the soil gas components. Four stainless steel tubes were installed in augered holes at depths near 1, 2, 3, and 4 meters below the soil surface. Drilled material was replaced in the same order of removal. Protective screw caps covered the protruding ends of the tubes. Buried ends of each tube were covered by a porous fine mesh stainless steel nipple. Initial sampling was done several days after placement of the clusters. Each tube was purged of at least one internal volume before a soil gas measurement.

ANALYTICAL APPARATUS

Gasoline Vapors

A portable combustible gas sniffer calibrated to weathered aviation gasoline vapor was used to measure relative levels of volatile fuel hydrocarbons. A meter reading of 195 ppm resulted from 117 μl/L vapor. The resulting conversion factor from direct meter reading to ppm hydrocarbons (v/v) was 0.60. The meter had a catalytical combustion detector, battery power, an integral pump with a flow rate of 2 liter/minute, and a workable range of 10 to 10,000 ppm (hexane). A maximum meter reading after 15 seconds measure time from each sample probe was recorded as the soil gas concentration.

Oxygen

A portable oxygen indicator calibrated to 21.0% oxygen in air was used to measure soil gas oxygen. The meter had an electrochemical oxygen sensor, battery power, an integral pump, and a workable range of 0.2 to 25% oxygen. A 30 second measure time was usually needed to obtain a stabilized reading.

Carbon Dioxide

Draeger detector tubes calibrated for 0.5% to 10% carbon dioxide based on 100 mL sample volume were placed into an exposure device shown in Figure 2. A hand-operated vacuum pump drawing 15 mL per stroke was used to carry out the suction process and measure volume simultaneously. Initial soil gas of at least one probe volume was bypassed using the flow direction control valve. Then flow was diverted to pass through the detector tube. Soil gas aliquots of 15 mL volume were pulled through the Draeger tube until a readable carbon dioxide level based on color indicator change was obtained. Since the detector tubes were calibrated for 100 mL, concentrations were adjusted by calculation for sample volumes other than 100 mL.

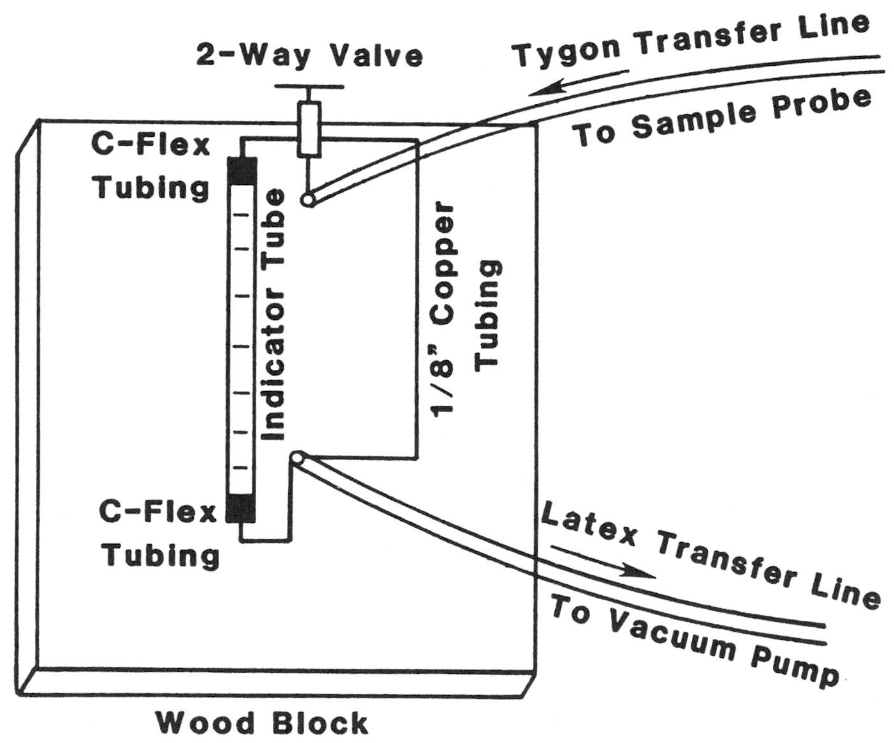

Figure 2. Indicator tube holder.

Data Evaluation

Performance of the analytical apparatus used was evaluated using a calibration gas containing 1000 μl/L each of methane, oxygen, and carbon dioxide. Replicate readings by the combustible gas meter had a coefficient of variance equal to 2.7%. Our calculated conversion factor for hexane to methane was 1.47 compared to a value of 1.58 specified by the manufacturer. Meter readings for eight different samples in the field, ranging from 76 to 3320 ppm combustible gas, had a correlation coefficient of 0.901 when compared to total peak area on gas chromatograms obtained with duplicate samples collected by a Tenax-cartridge trap technique. The oxygen meter set at 21.0% oxygen for air gave a reading of 0.3% after a 30-second measure time using the 0.1% calibration gas. Averaged variation for duplicate readings at five sample points in the field was 2.4% where the oxygen levels ranged from 1.3% to 11.3%. A coefficient of variance of 13.7% was obtained for 14 replicate analyses of the 0.1% carbon dioxide calibration gas with the 0.5% to 10% indicator tubes. Replicate analyses at a field sample point having 9.9% carbon dioxide showed a coefficient of variance of 3.5%. Background level readings at a one-meter depth in March at a location 130 meters from the plume were 40 ppm for combustible gas and 0.33% for carbon dioxide. The carbon dioxide concentration was similar to that of 0.3% in the C horizon of a forest soil at the start of the growing season as reported by Fernandez and Kosian.[6] Additional measurements taken during August at locations Bkg1 and Bkg2 on Figure 3 at one and two meter depths ranged from 22 to 35 ppm hydrocarbons, 20.6% to 20.8% oxygen, and 0.003% to 0.02% carbon dioxide.

APPLICATION OF MEASUREMENTS

Plume Mapping

Transects for gasoline vapor detection are shown in Figure 3. The one-meter probe measurements indicated that the major portion of the plume source area was about 80 meters wide and 360 meters in length. Relative responses by the combustible gas sniffer are shown for the transects in Figure 4. A hot spot was located about 260 meters downgradient from the original spill location. Meter readings in the hot spot area were about 40 times greater than those upgradient in the plume. A past history of the upgradient plume area A, B, and C and transect d-c of excessive watering and fertilization of the overlying turf suggests flushing and enhanced biodegradation as one explanation of lowered gasoline vapor concentrations near the surface, compared to the unfertilized area at cluster D and transect e-h, e-f, e-g, and k-i. Relative levels of vapor hydrocarbons were

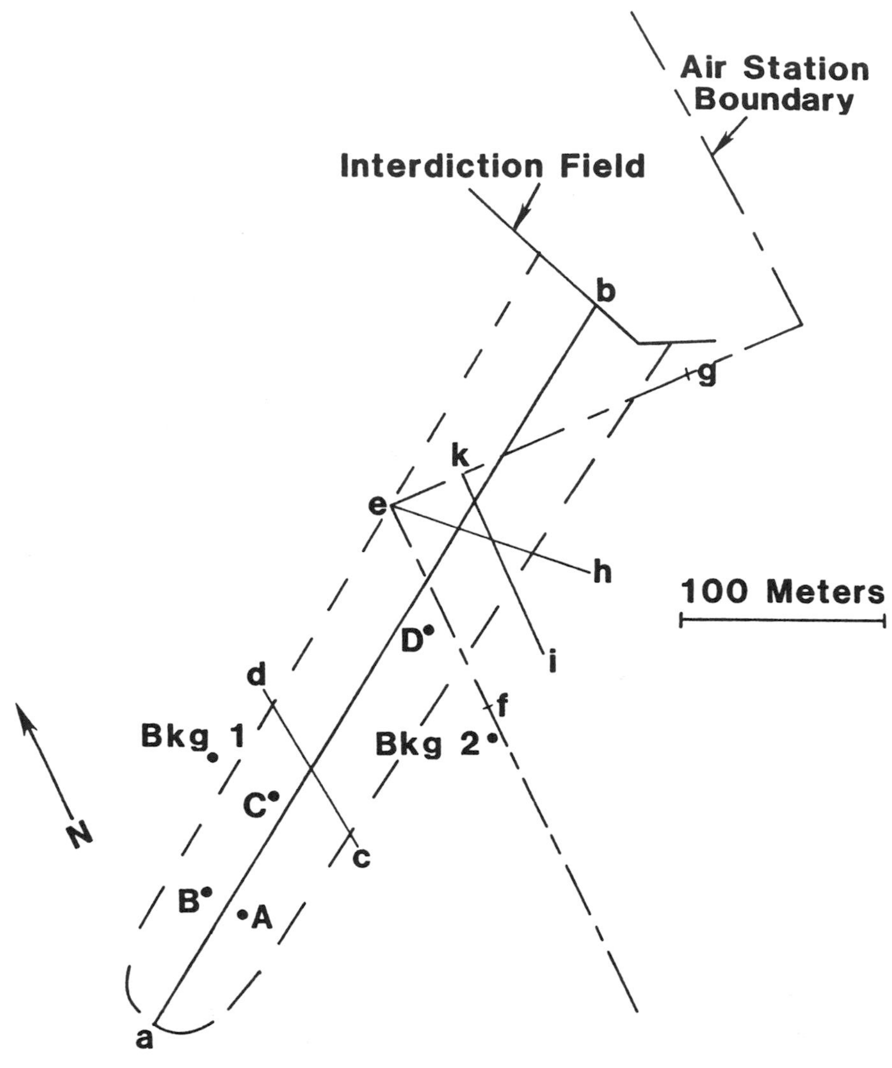

Figure 3. Plume mapping transects and location of tubing clusters A, B, C, and D.

in agreement with solid core sampling.[5] A data distribution contour line computer program was used to generate Figure 5 map.

Vertical Profiles

Soil gas constituents at the sampling points of permanently placed tubing clusters are listed in Table 1. The water table was near a five-meter depth. Maximum gasoline vapors were at the lowest depth and declined near the surface. Oxygen levels decreased with greater depth but varied in magnitude at different sampling

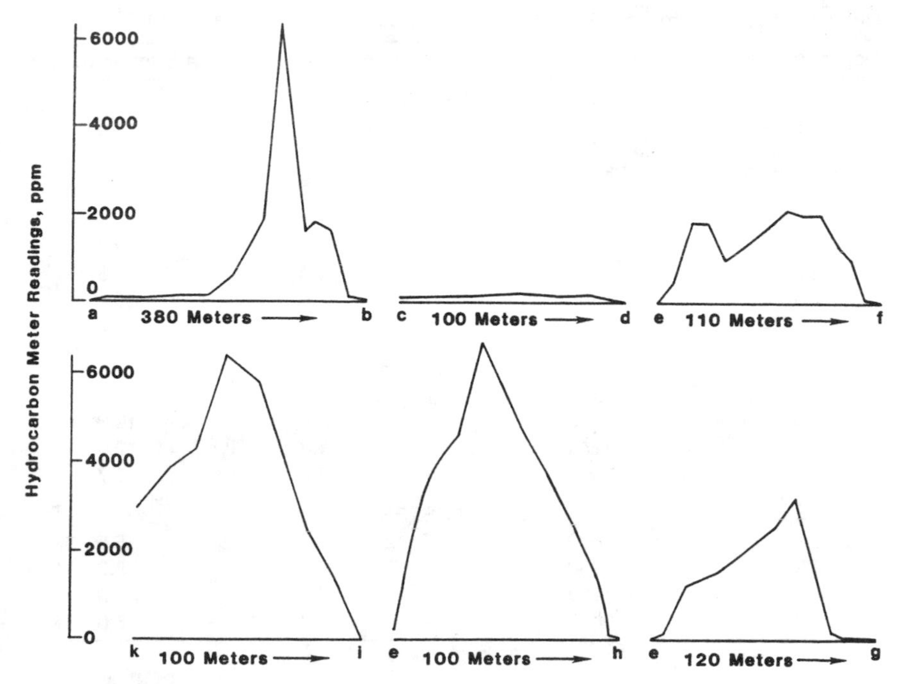

Figure 4. Gasoline vapors detected at one meter depth of transects.

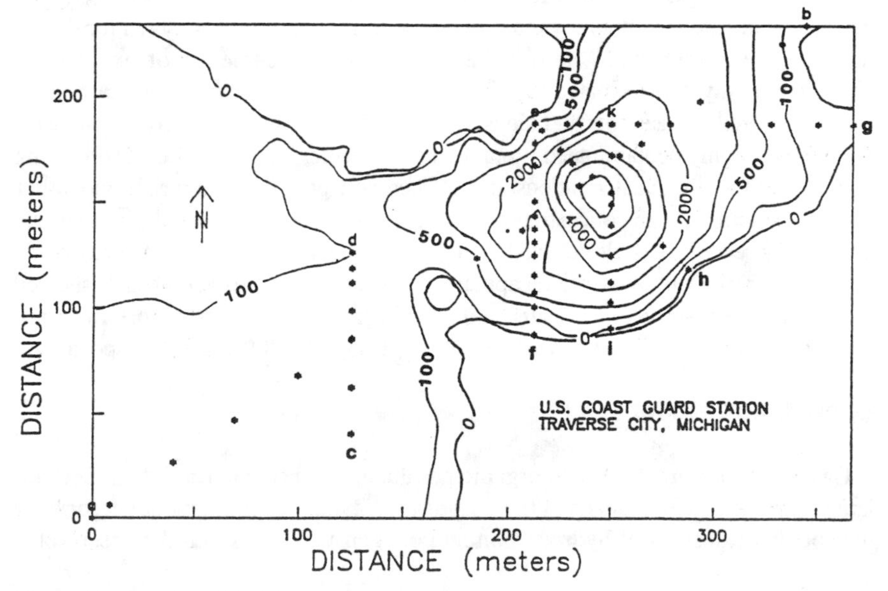

Figure 5. Isoconcentration contour map.

Table 1. Vertical Profile Measurements as Direct Readings—March 1989

Depth, meters	HC, ppm	Oxygen, %	Carbon dioxide, %
		CLUSTER A	
1.8	300	9.1	7.8
3.0	2700	5.4	9.7
4.3	6200	0.9	11.7
		CLUSTER B	
1.0	200	6.0	9.3
2.0	220	4.8	11.3
3.0	1300	3.5	10.3
4.0	2800	1.0	12.2
		CLUSTER C	
0.9	110	5.5	9.4
1.8	160	3.4	10.3
2.7	1200	1.3	12.0
3.6	10000	0.8	11.7
		CLUSTER D	
0.9	2300	11.5	5.6
1.9	4300	8.8	8.0
2.9	7200	5.1	11.0
3.9	>10000	1.4	15.6

locations. The general trend of carbon dioxide concentrations were related to levels of gasoline hydrocarbons. The data suggest that elevated carbon dioxide represented biogenic gases formed from viable degradation processes. Steep oxygen gradients indicated that there was a downward limit of sufficient oxygen for degradation of hydrocarbons by aerobic bacteria. A severe depletion of oxygen just above the highly contaminated capillary fringe mirrored an increase of carbon dioxide. Elevated soil gas concentrations of carbon dioxide over 20% have been measured above highly contaminated groundwater.[7] Further studies at the site are being pursued on the chemical relationship between soil gas, core material, and water quality. Generally, high concentrations of fuel hydrocarbons in the soil gas at the site are directly related to levels detected in groundwater and capillary fringe core material. Upper level groundwater and capillary fringe core analyses at cluster C showed 1,610 μg hydrocarbons/liter water and 11,100 mg fuel carbon/kg core material. Similar analyses at cluster D were 2,840 μg/L and 10,200 mg/kg, respectively.

Degradation Model

Gaseous transport of volatile organic pollutants has been an important mechanism in mathematical analysis of the unsaturated zone. A simple model describing the coupled transport of hydrocarbon and oxygen vapor through the unsaturated zone was derived and tested using the following logic: The hydrocarbon H and oxygen O concentrations are governed by a steady balance of diffusion and biological degradation R

$$nD \frac{d^2H}{dz^2} = nR \quad (1a)$$

$$nD \frac{d^2O}{dz^2} = n\gamma R \quad (1b)$$

$$\gamma = 3.51 \quad (1c)$$

with air filled porosity n and gaseous diffusivity D taken as constant in the unsaturated zone bound by the top of the capillary fringe at $z=0$ and the ground surface at $z=\zeta$. The reaction ratio γ value was based on the stoichiometry of the biodegradation process for weathered aviation gasoline.

Following Yates and Enfield,[8] we defined the combined constituent variable χ and subtract Equation 1b from 1a with the result

$$\chi = H - \frac{O}{\gamma} \quad (2a)$$

$$\frac{d^2\chi}{dz^2} = 0 \quad (2b)$$

$$\chi = H_O \qquad (z=0) \quad (2c)$$

$$\chi = -\frac{O_\zeta}{\gamma} \qquad (z=\zeta) \quad (2d)$$

The boundary conditions 2c and 2d reflect an assumption of an active aerobic biomass in the vadose zone, which depletes hydrocarbon and oxygen vapors as they diffuse away from their sources. Thus, χ is comprised solely of oxygen (O_ζ) and hydrocarbon (H_O) at the ground and contaminated capillary fringe surfaces, respectively. Equation 2 yielded a simple straight line relationship for the combined constituent variable

$$\chi = H_O \left(1 - \frac{z}{\zeta}\right) - \frac{O_\zeta z}{\gamma \zeta} \quad (3)$$

It is important to note that this profile is independent of the reactive term; our χ profile reflects reaction stoichiometry rather than reaction kinetics. The model validity rests upon biodegradation as the sole sink for both oxygen and hydrocarbons in the unsaturated zone, along with the constancy of D and n in this region.

Figure 6 and Table 2 summarize a two parameter regression for the combined constituent model that yielded calibrated values for the hydrocarbon and oxygen intercepts at the ground and contaminated capillary fringe surfaces for stations A, B, C, and D. We used statistics of error δ_χ to calibrate the model[9]

$$\delta_\chi = \frac{\chi\ (\text{measured}) - \chi\ (\text{predicted})}{H_O} \tag{4a}$$

$$\bar{\delta} = \frac{1}{j}\, \Sigma\ (\delta) \tag{4b}$$

$$\sigma = [\frac{1}{j}\, \Sigma(\delta^2) - \bar{\delta}^2]^{1/2} \tag{4c}$$

The H_O value was used to zero the mean error predicted by Equation 3, while the O_ζ intercept minimized the error standard deviation σ_χ. The latter parameter ranged from 3% to 29% for the four stations, which indicated substantially reasonable degree of calibration accuracy. The O_ζ values were substantially less than the atmospheric content of 0.287 kg/m^3, which implied consumption of oxygen by surface vegetation prior to entry into the unsaturated zone. The H_O intercepts were likewise considerably less than the saturated value of 0.14 kg/m^3 previously estimated by us, which suggested mass transfer limitations[10] within the capillary fringe itself.

We specified the reaction R in Equation 1 to proceed with the description of separate hydrocarbon and oxygen concentration profiles. The term was presumed to follow Monod kinetics[11] with hydrocarbon as the limiting constituent in the (aerobic) unsaturated zone

$$R = \frac{VH}{K+H} \tag{5}$$

with maximum reaction rate V and half saturation constant K for hydrocarbons.

Our model was predicated on the abundance of oxygen, which required concentrations in excess of 0.003 kg/m^3, based on an aqueous oxygen half saturation constant[12] of 10^{-4} kg/m^3, and an air/water oxygen partition coefficient of 26.

We also derived a nonlinear solution to Equations 1a and 5, making use of the hydrocarbon flux as an intermediate variable, as suggested by Suidan and Wang.[13] An implicit dimensionless hydrocarbon profile followed

$$z = (\frac{DK}{2V})^{1/2} I\,(H^*) \tag{6a}$$

$$O = \frac{O_\zeta z}{\zeta} + \gamma\,[H - H_O\,(1 - \frac{z}{\zeta})] \tag{6b}$$

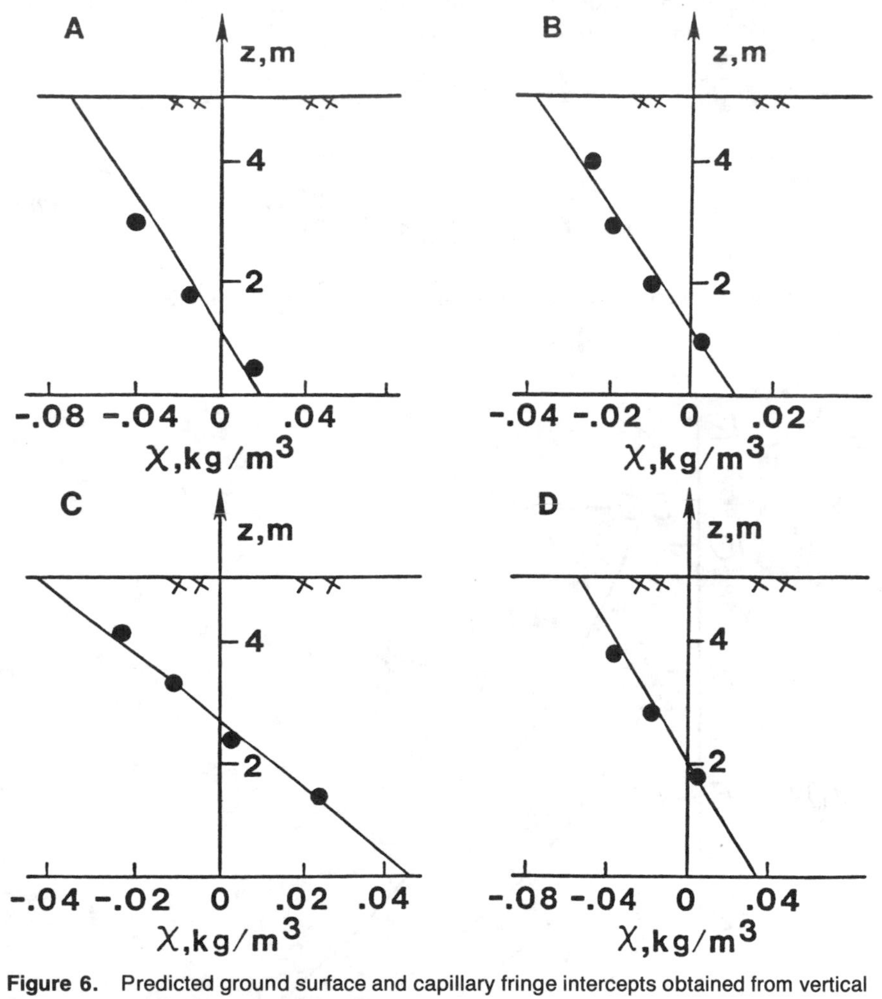

Figure 6. Predicted ground surface and capillary fringe intercepts obtained from vertical profile measurements.

Table 2. Coupled Constituent Profile Parameters

Station	ζ m	O_ζ kg/m³	H_O kg/m³	σ_χ %	V Kg/m³ – s	σ_V %
A	4.70	0.234	0.0191	8	7.25×10^{-9}	6
B	4.82	0.129	0.0115	23	6.83×10^{-9}	6
C	4.97	0.149	0.0455	11	1.39×10^{-8}	10
D	4.54	0.201	0.0320	5	4.22×10^{-9}	4

Equations 3 and 6a yielded the coupled oxygen profile 6b. Our dimensionless variable and integral function were defined by

$$H^* = \frac{H}{K} \tag{7a}$$

$$I(H^*) = \int_{H^*}^{H_O^*} \frac{dx}{[x - \ln(1+x)]^{1/2}} \tag{7b}$$

The integral function was sketched in Figure 7 to aid in model usage.

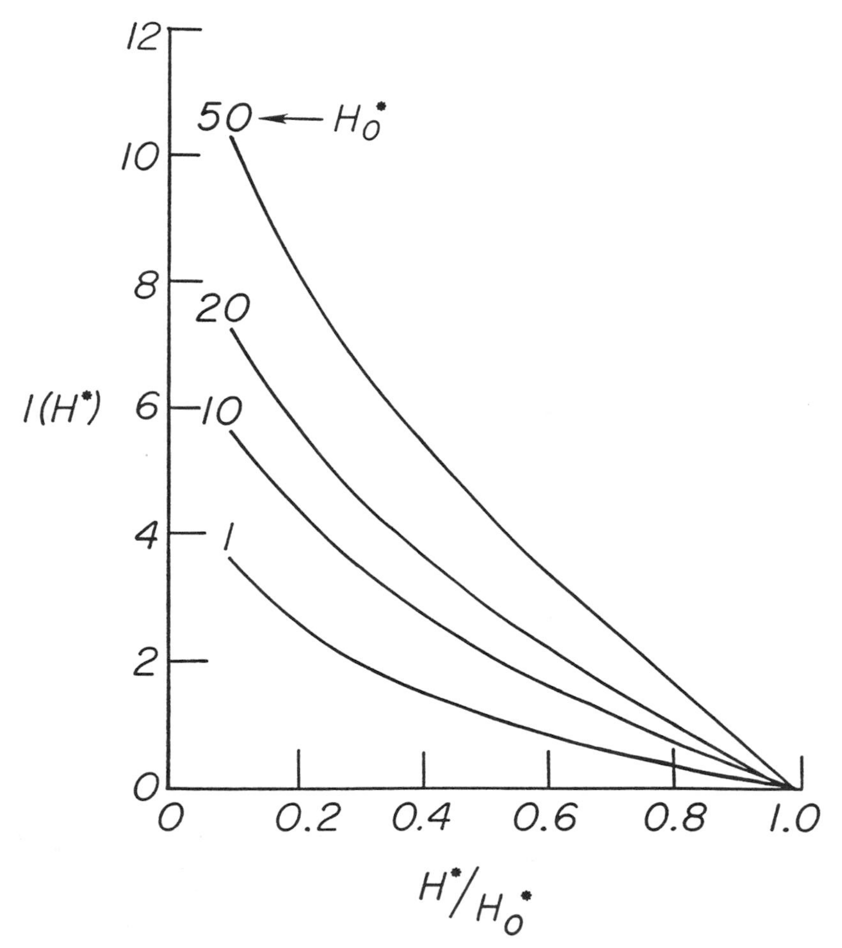

Figure 7. Integral function 1 (H*) for various values of the dimensionless hydrocarbon concentration at the capillary fringe boundary H_O^*.

Table 3. Measured and Predicted Vapor Concentrations.

z m	H (meas.)[a] kg/m³	H (pred.)[b] kg/m³	δ_H %	O (meas.)[a] kg/m³	O (pred.)[b] kg/m³	δ_O %
			Station A			
0.40	0.0175	0.0157	10	0.0123	0.0136	−1
1.70	0.00760	0.00737	1	0.0737	0.0677	3
2.90	0.000844	0.00287	−10	0.124	0.129	−2
			Station B			
0.82	0.00788	0.00700	8	0.0136	0.0130	0
1.82	0.00366	0.00350	1	0.0478	0.0359	9
2.82	0.000619	0.00153	−8	0.0655	0.0641	1
3.82	0.000563	0.000594	0	0.0819	0.0959	−11
			Station C			
1.37	0.0281	0.0212	15	0.0109	0.0001	7
2.27	0.00338	0.0113	−17	0.0177	0.0209	−2
3.17	0.000450	0.00488	−10	0.0463	0.0543	−5
4.07	0.000310	0.00175	−3	0.0751	0.0992	−16
			Station D			
1.64	0.0203	0.0185	5	0.0696	0.0660	2
2.64	0.0121	0.0125	−1	0.120	0.114	3
3.64	0.00647	0.00785	−4	0.157	0.166	−5

[a]Converted with ideal gas law at 12°C, using molar mass of 0.110 kg for hydrocarbon vapor and 0.032 kg for oxygen.
[b]Predicted using air filled porosity of 0.258 and gaseous diffusivity of 2.3×10^{-6} m²/s.

The individual oxygen and hydrocarbon data of Table 3 was used to calibrate the kinetic parameters of Equation 6, based on statistics of the profile errors δ_H and δ_O

$$\delta_H = \frac{H(\text{measured}) - H(\text{predicted})}{H_O} \tag{8a}$$

$$\delta_O = \frac{O(\text{measured}) - O(\text{predicted})}{O_\zeta} \tag{8b}$$

$$\bar{\delta}_V = \frac{1}{2j} \Sigma (\delta_H + \delta_O) \tag{8c}$$

A literature value[12,14] of K equal to 0.003 kg/m³ was adopted for hydrocarbon vapors. The maximum reaction rate was used to zero the mean profile error. The results are summarized in Table 3 and Figure 8. Our error standard deviations σ_V were much less than 10% in magnitude, which indicated a good fit of the non-linear profile theory with the separate constituent profile data. The maximum reaction rates ranged from 2.7×10^{-9} to 1.3×10^{-8} kg/m³−s in magnitude for the March 1989 profile data. Basically the intercepts as shown in Figure 8 predict that biodegradation would prevent both oxygen from reaching the water table and escape of appreciable hydrocarbon vapor to the atmosphere at locations A, B, and C.

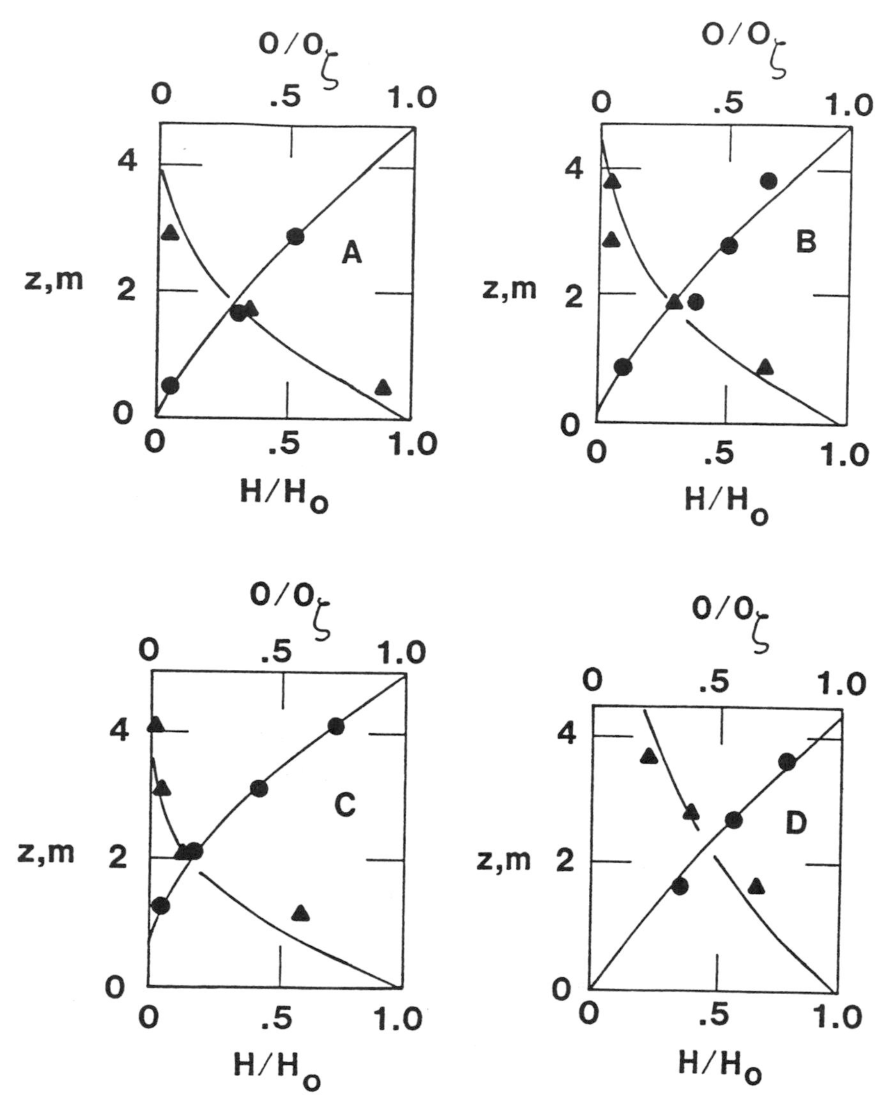

Figure 8. Predicted profile value curves of observed, oxygen (circles) and hydrocarbon (triangles) vapor concentrations.

CONCLUSIONS

The soil gas sampling and analysis strategy provided required information for mapping the plume and vertical profile measurements with a minimal expenditure of resources and work time. Analysis of a calibration gas and replicate sampling showed that the apparatus used gave reasonably correct soil gas constituent measurements. Boundaries of the plume were defined and a hot spot was located downgradient from the original spill location. Elevated carbon dioxide above the contaminated capillary fringe indicated microbial respiration activity. A biodegradation model developed from the vertical profile data predicted very low oxygen at the water table and little or no loss of hydrocarbon emissions to the atmosphere.

REFERENCES

1. Horowitz, L. "Geochemical Exploration for Petroleum," *Science,* 229(4716):821–827 (1985).
2. Marrin, D. L. "Delineation of Gasoline Hydrocarbons in Groundwater by Soil Gas Analysis," in *Proceeding of the Hazardous Materials Management Conference* (Wheaton, IL: Touer Conference Management Company, 1985), pp. 112–119.
3. Marrin, D. L., and H. B. Kerfoot. "Soil-Gas Surveying Techniques." *Environ. Sci. Technol.*, 22(7): (1988).
4. Ostendorf, D. W., D. H. Kampbell, J. T. Wilson, and J. H. Sammons. "Mobilization of Aviation Gasoline from a Residual Source." *Research Journal Water Pollution Control Federation,* in press, 1989.
5. Ostendorf, D. W. "Long Term Fate and Transport of Immiscible Aviation Gasoline in the Subsurface Environment." Submitted to *Water Science and Technology,* 1989.
6. Fernandez, I. J., and P. A. Kosian. "Soil Air Carbon Dioxide Concentrations in a New England Spruce-Fir Forest." *Soil Sci. Soc. Am. J.* 51: 261–263 (1987).
7. Bishop, W. D., R. C. Carter, and H. F. Ludwig. "Water Pollution Hazards from Refuse-Produced Carbon Dioxide," Third International Conference on Water Pollution Research, Section 1, Paper No. 10, Water Pollution Control Federation, Washington DC (1966).
8. Yates, S. R. and C. G. Enfield, C. G. "Decay of Dissolved Substances by Second Order Reaction," *J. Environ. Sci. Health* 23:59–84 (1988).
9. Benjamin, J. R. and C. A. Cornell. *Probability, Statistics, and Decision for Civil Engineers* (New York: McGraw-Hill Book Company, 1970), pp. 684.
10. Pfannkuch, H. O. "Determination of the Contaminant Source Strength from Mass Exchange Processes at the Petroleum Groundwater Interface in Shallow Aquifer Systems," in *Proceedings Petroleum Hydrocarbons and Organic Chemicals in Groundwater,* National Water Well Association/American Petroleum Institute, Houston, TX, 1984, pp. 111–129.
11. McCarty, P. L., B. E. Rittman, and E. J. Bouwer. "Microbial Processes Affecting Chemical Transformations in Groundwater," *Groundwater Pollution Microbiology* (New York: Wiley-Interscience, 1984), pp. 89–115.
12. Borden, R. C., and Bedient, P. B. "Transport of Dissolved Hydrocarbons Influenced by Oxygen Limited Biodegradation 1. Theoretical Development," *Water Resour. Res.*, 22:1973–1982 (1986).
13. Suidan, M. T. and Y. T. Wang. "Unified Analysis of Biofilm Kinetics," *J. Environ. Eng.*, 111:634–646 (1985).
14. Molz, F. J., M. A. Widdowson, and L. D. Benefield. "Simulation of Microbial Growth Dynamics Coupled to Nutrient and Oxygen Transport in Porous Media," *Water Resour. Res.*, 22:1207–1216 (1986).

CHAPTER 12

Differentiation of Crude Oil and Refined Petroleum Products in Soil

Ann L. Baugh, Unocal Corporation, Los Angeles, California
Jon R. Lovegreen, Applied Geosciences Inc., Tustin, California

As part of the societal recycling of depleted oil fields, assessing whether a petroleum hydrocarbon detected in soil samples is a crude oil or a refined petroleum product has increased in importance. Neither federal nor California regulations list soils containing crude oil to be hazardous wastes.[1,2] Soils containing refined petroleum products typically require mitigation even though regulations do not define these soils as a hazardous waste, especially in the case of leaking underground fuel tanks.[3] The differentiation of crude oil from refined petroleum products has important economic impacts. Crude oil can be left in place provided it can be demonstrated that it is not hazardous[2] and that there is a low likelihood of migration to the water table.[3] However, for refined petroleum products, gasoline may be required to be mitigated to levels as low as 10 parts per million (ppm) and diesel fuel to 100 ppm.[3] Benzene and toluene may be required to be mitigated to 0.3 ppm, and total xylenes and ethylbenzene to 1 ppm.[3]

CHEMICAL ANALYSES

Time and economic constraints of a typical site investigation call for use of data provided by standard analyses available from state-certified and Environmental Protection Agency (EPA) contract laboratories. The approach for differentiating between crude oil and refined petroleum products reported here emerged in the course of site investigations[4-8] associated with the redevelopment of depleted oil field properties in southern California. Use was made of analyses routinely performed in the course of the site investigation. In Table 1, a list is given of the analyses that have been found to be helpful in making judgments as to the likelihood that the origin of a petroleum hydrocarbon is from crude oil or refined petroleum product.

Table 1. Chemical Analyses Used for Differentiation of Crude Oil and Refined Petroleum Products in Soil

1. Total Recoverable Petroleum Hydrocarbons (TRPH), EPA Method No. 418.1[9]
2. Total Petroleum Hydrocarbons (TPH), EPA Method No. 8015 (Modified)[3]
3. Benzene, Toluene, Total Xylene and Ethylbenzene (BTXE), EPA Method No. 8020[10]
4. Volatile Organic Compounds (VOCs), EPA Method No. 8240[10]
5. Semi-Volatile Organic Compounds (Semi-VOCs), EPA Method No. 8270[10]
6. Organic Lead, Inductively Coupled Plasma/Mass Spectrometry (ICP/MS)[11]

The total recoverable petroleum hydrocarbon (TRPH) analysis, EPA Method No. 418.1, is an infrared spectrophotometric procedure[9] in which an extract using fluorocarbon-113 (1,1,2-trichloro-trifluoroethane) is quantified using the C-H stretch band at about 2930 cm-1 against a standard containing n-hexadecane ($C_{16}H_{34}$), a straight chain hydrocarbon; isooctane (C_8H_{18}), a branched hydrocarbon; and chlorobenzene (C_6H_5Cl), an aromatic hydrocarbon. These compounds are representative of those found in a mineral oil or a light fuel,[9] but do not include the heavier fractions (C_{20}+) found in crude oil. Hence, the TRPH analysis provides only an estimate of the concentration of a crude oil in soil. It is described as a useful survey tool in the Leaking Underground Fuel Tank (LUFT) Field Manual[3] because of the low cost. Also, the loss of about half of any gasoline present during the extraction process is reported in the analytical procedure.[9] Nyer and Skladany[12] report 25% or more variability. The result of a TRPH analysis is a single concentration, reported generally in ppm.

The total petroleum hydrocarbon (TPH) analysis, which is routinely referred to in California as EPA Method No. 8015 (modified), in reality is a method described in the LUFT Field Manual[3] and developed by the California Department of Health Services (DOHS). This method emerged in the development of analyses for gasoline and diesel fuel from leaking underground fuel tanks. It resembles the EPA method in that it uses a temperature programmable gas chromatograph (GC) with a flame ionization detector (FID) and a similar column. The DOHS method[3] describes the GC operating conditions as follows:

"Column temperature is set at 40 degrees Celsius at the time of injection, held for four minutes, and programmed at 10 degrees Celsius per minute to a final temperature of 265 degrees Celsius for 10 minutes."

In California, the method used from laboratory to laboratory when requesting EPA Method No. 8015 (modified) is not standardized, and can vary to some degree, especially in the temperature programming. DOHS is attempting to standardize this and recommends that laboratories be given the procedure as described in the LUFT Field Manual.[3]

When TPH analyses are being performed on soils for the purpose of differentiating crude oil and refined petroleum products, the extraction procedure is used. The solvent suggested in the DOHS procedure is carbon disulfide; although other solvents such as ethyl acetate or methylene chloride may be used, provided the solvent can extract the petroleum hydrocarbons, and does not interfere with the resulting gas chromatogram.[3] When the extraction method is used, the detection limit for TPH is normally 10 ppm for both gasoline and diesel.[3] The results are reported as ppm for generally no more than two of the four refined petroleum products in this boiling range, i.e., gasoline, mineral spirits, kerosene and diesel fuel. Peaks above the diesel range are reported as $C_{20}-C_{30}$ hydrocarbons. Some laboratories report only one number as TPH in ppm, which is calculated using only one standard, generally gasoline or diesel. When the analysis is being performed for the purpose of a leaking underground fuel tank, often the contents of the tank are known and this method is appropriate. However, for our purposes in differentiating crude oil and refined petroleum products, the use of two standards with a report of gasoline, diesel fuel, or $C_{20}-C_{30}$ segments is preferred. The $C_{20}-C_{30}$ segment is usually calculated using the diesel standard for comparison. This practice is likely to produce slightly low results for the $C_{20}-C_{30}$ fraction because the instrument response per unit concentration of a heavier hydrocarbon is probably lower than that of diesel fuel.

The EPA Method No. 8020 analysis is for aromatic volatile organic compounds (VOCs) and provides data on chlorobenzene and dichlorobenzenes as well as benzene, toluene, xylene and ethylbenzene (BTXE). A purge-and-tap method is used with soils for sample injection into the GC. A temperature programmable GC is used with a photoionization detector (PID). Due to the fact that samples can be contaminated by the diffusion of volatile organics, especially chlorofluorocarbons and methylene chloride, through the sample container septum during shipment and storage, a field sample blank prepared from reagent water and carried through sampling and subsequent storage and handling is recommended to serve as a check.[10] With soil samples, the detection limit for BTXE can be as low as 1 part per billion (ppb). This will vary depending on actual concentration in the sample and interference from other compounds and can reach 250 ppb.[10] The results are reported in ppm or ppb, as appropriate.

The EPA Method No. 8240 is used to determine VOCs in a variety of solid waste matrices, including soil. The method is applicable to most VOCs that have a boiling point below 120°C and vapor pressures of a few millimeters of mercury at 25°C and that are insoluble or slightly soluble in water. More than 50 compounds are reported to be analyzed by this method,[10] including BTXE. When this analysis is performed on a sample, EPA Method No. 8020 can usually be omitted. The method consists of a purge-and-trap process for injection of the volatiles into the GC; a temperature programmable GC; a scanning, electron impact mass spectrometer (MS); and a computerized data system.

The detection limits for BTXE with this method is generally 5 ppb for soils with low concentrations, and increases to over 600 ppb for high concentrations. Samples require the same precautions for contamination by other VOCs as discussed above for EPA Method No. 8020.[10] The VOCs other than BTXE detected by this method are for the most part not petroleum hydrocarbons. However, some laboratories report mixtures of compounds such as C_5-C_{11} aliphatic and alicyclic hydrocarbons and C_9-C_{10} alkylbenzenes, which are constituents of gasoline and the gasoline fraction of crude oil. The results of an EPA Method No. 8240 analysis are reported in a computer printout from the mass spectrometer data system with chemical abstract number, compound name, concentration in either ppb or ppm, as appropriate, and detection limit. All chemicals routinely analyzed are listed, whether present or not; so typically, a data sheet will contain a large number of NDs (not detected).

The EPA Method No. 8270 is similar to 8240 in that it is a gas chromatography/mass spectrometry (GC/MS) method. It is used to determine the concentration of semivolatile organic compounds (semi-VOCs) in extracts prepared from all types of solid waste matrices, soils, and groundwater. The method is applicable to most neutral, acidic, and basic organic compounds that are soluble in methylene chloride and are capable of being eluted without derivatization as sharp peaks from a GC fused-silica capillary column with a slightly polar silicone. Such compounds include polynuclear aromatic hydrocarbons, chlorinated hydrocarbons, pesticides, esters, aldehydes, ethers, ketones, anilines, pyridines, quinolines, aromatic nitro compounds, and phenols. The EPA[10] lists more than 100 compounds as routinely detected by this method, which includes semi-VOCs found in crude oil. As with EPA Method No. 8240, the results of an EPA Method No. 8270 analysis are reported in computer printout from the mass spectrometer data system with chemical abstract number, compound name, concentration in either ppb or ppm, as appropriate, and detection limit. Also, it includes typically more than 50 standard compounds with ND (not detected) reported for most compounds.

The analysis for organic lead does not appear to be standardized. The LUFT Field Manual[3] describes a DOHS procedure using a xylene extraction followed by a flame atomic absorption spectroscopic (AA) method. A similar method using inductively coupled plasma/mass spectrometry (ICP/MS) has been developed[11] which reports total organic lead. The ICP/MS method was used for the analyses reported in the case studies.

CRITERIA USED FOR THE DIFFERENTIATION

The criteria used for the differentiation of crude oil and refined petroleum products in soil samples are listed in Table 2. The primary criterion is the interpretation of gas chromatograms obtained from the TPH analysis. In this analysis, the GC scans eluted compounds in the temperature range from 50°C to 300°C at a heating rate of 10° per minute and held at 300°C for five minutes, resulting in a total run time of 30 minutes. The chromatograms are visually

Table 2. Criteria Used for the Differentiation of Crude Oil and Refined Petroleum Products in Soil

1. Appearance of TPH gas chromatogram
2. Comparison of TRPH and TPH
3. BTXE concentrations
4. Other VOCs present
5. Semi-VOCs present
6. Presence of organic lead

compared to standards, then calculated and reported in generally no more than two of four refined petroleum products in this boiling range, i.e., gasoline, mineral spirits, kerosene and/or diesel fuel. Peaks above the diesel range are generally reported as $C_{20}-C_{30}$ hydrocarbons. The peaks that occur in the last five minutes of the scans are from the higher boiling point tars and asphaltenes in the sample. Their occurrence is a primary indication of a crude oil. In Figures 1, 2, and 3, the TPH gas chromatograms of the standards for gasoline, kerosene, and diesel fuels are given, respectively. The number printed above the peaks are retention times in minutes. The occurrence of peaks for each standard is confined to a particular section of the chromatogram representing the boiling fraction of the crude from which it was refined. Gasoline elutes from the beginning of the scan to approximately 10 minutes, kerosene from approximately 5 to 16 minutes, and diesel fuel from approximately 7 to 22 minutes. Also, the chromatograms of the refined products have a bell-shaped appearance. In Figure 4, a representative chromatogram is given of crude oil collected from operating wells in the oil field in which all of the case studies are located.

The laboratory results for the TRPH, TPH, and BTXE analysis of the four crude oil samples are given in Table 3. The occurrence of peaks for the crude oil is relatively uniform throughout the chromatogram from 2 to 29 minutes. This uniform occurrence of peaks throughout the chromatogram is also a primary indication of crude oil, despite the fact that the laboratory analysis for TPH will report the different segments as gasoline, kerosene, or diesel fuel fractions. In summary, the uniform occurrence of peaks throughout the 30-minute scan, plus the occurrence of peaks in the last five minutes, are primary indications of a crude oil, rather than a refined product, whereas the occurrence of segmented peaks, especially with a bell-shaped appearance, is an indication of a refined petroleum product.

A second criterion used to assess whether petroleum hydrocarbons detected in soil samples are refined products or crude oil is the comparison of the TPH results with the results of analyses for TRPH. The TPH analysis measures hydrocarbon content to a maximum of about C_{30} with the ability to quantify using standards terminating with the carbon range of diesel fuel, which is generally up to approximately C_{23}.[3] The TRPH analysis is reported[9] to lose approximately half of the gasoline present during the extraction process, and extracts the higher molecular weight fuels and oils. Hence, the TPH analysis is biased to the lower

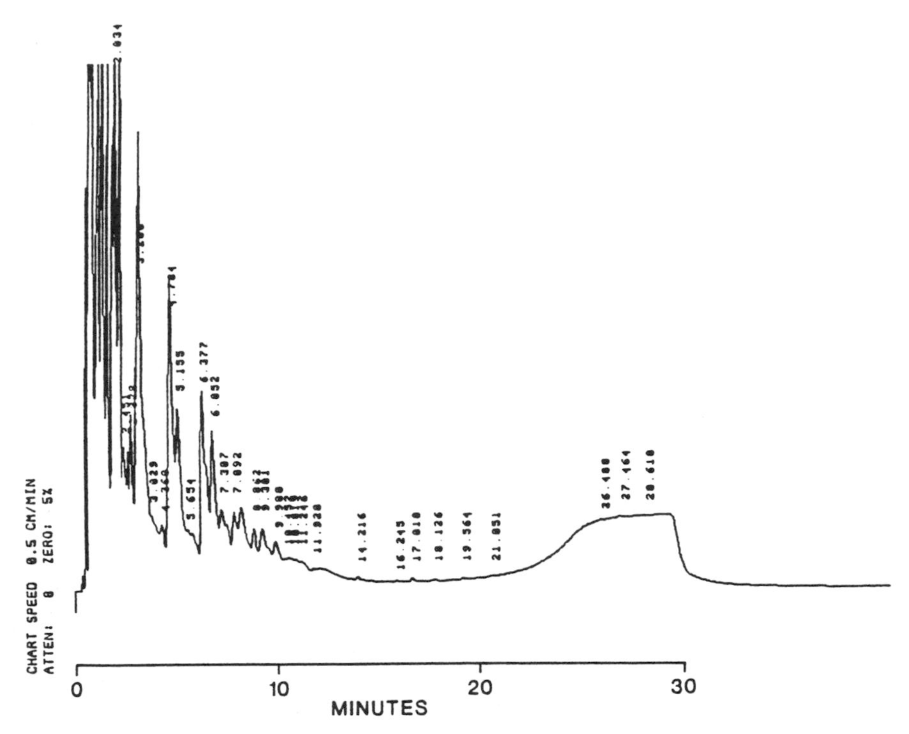

Figure 1. Standard TPH gas chromatogram for gasoline.

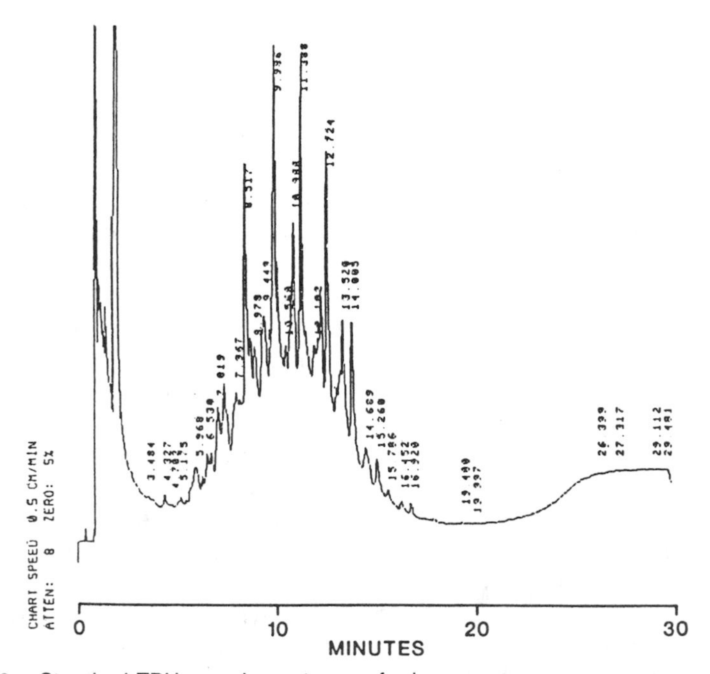

Figure 2. Standard TPH gas chromatogram for kerosene.

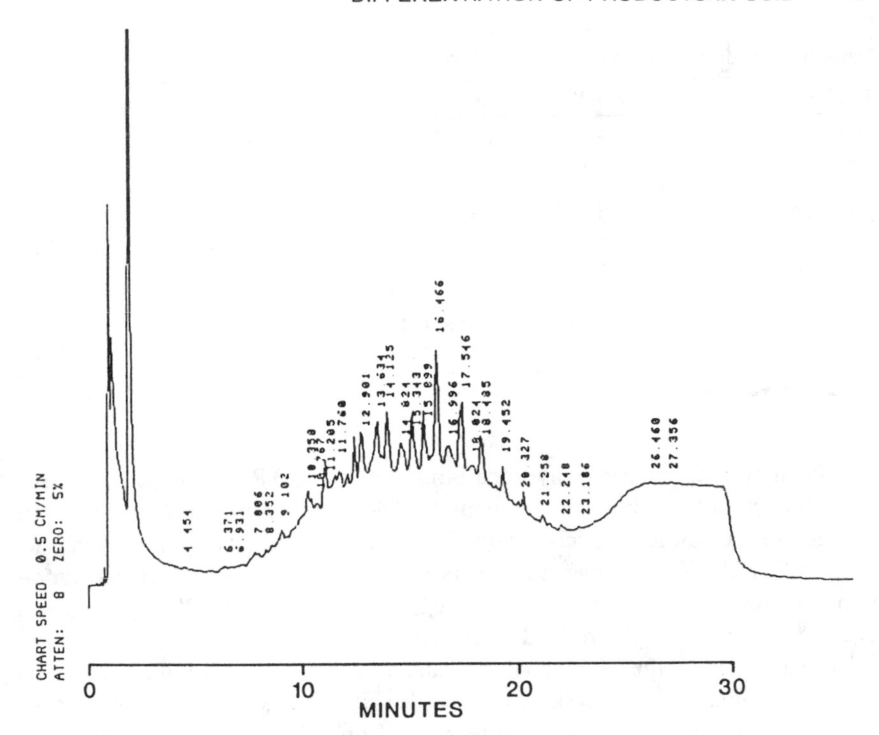

Figure 3. Standard TPH gas chromatogram for diesel fuel.

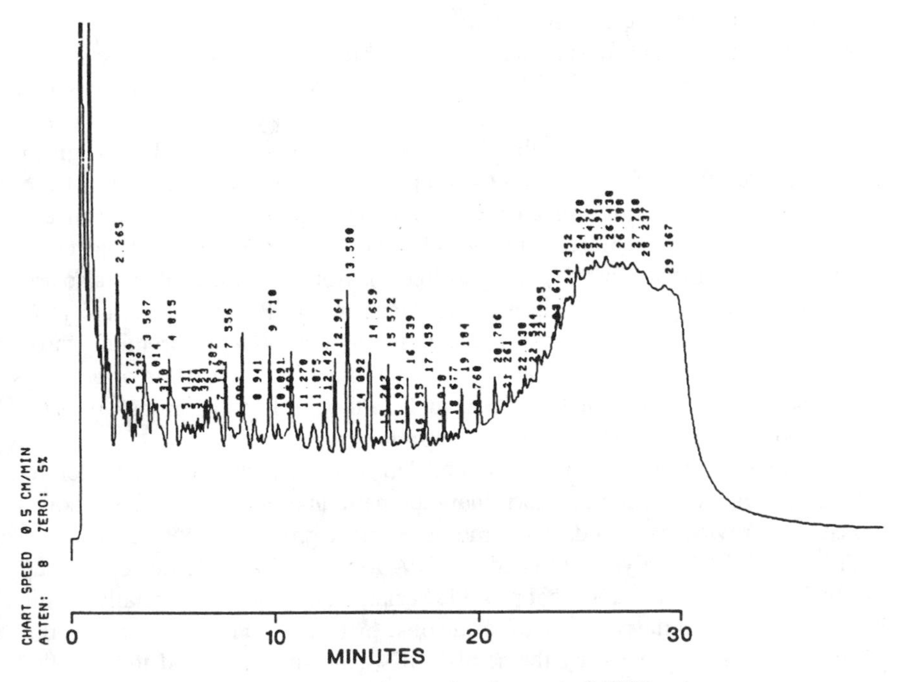

Figure 4. TPH gas chromatogram for crude oil sample 0-639B.

Table 3. Crude Oil Sample Analysis (Percent)

Analysis	O-620N	O-634D	O-639B	O-645C
TRPH	99	83	47	87
TPH	48.1	45.5	44.8	45.4
Gasoline	28.8	24.8	24.7	23.6
Diesel	10.4	9.1	10.6	14.2
$C_{20}-C_{30}$	14.9	11.6	9.5	7.6
Benzene	0.24	0.14	NA[a]	NA
Toluene	0.54	0.33	NA	NA
Total xylenes	0.16	0.10	NA	NA
Ethylbenzene	0.83	0.59	NA	NA

[a]Indicates not analyzed.

molecular weight petroleum hydrocarbons, while the TRPH analysis is biased to the higher molecular weight compounds. The comparison of TRPH with TPH concentrations indicates the relative distribution of low and high hydrocarbon fractions. When the TRPH concentration is much larger than the TPH concentration, the presence of a crude oil is indicated, whereas a higher TPH concentration is a clear indication of a refined petroleum product.

The variability of the TRPH analysis can be anticipated to limit the application of this criterion. In the analysis of four crude soil samples collected from operating wells in Table 3, the TRPH results range from 47% to 99%, while the ratio of the TRPH and TPH concentrations for three samples is approximately 2:1 and greater than 1:1 for the fourth sample.

A third criterion that can be used to assess the presence of gasoline is the results of analyses for BTXE, as provided in either EPA Method No. 8020 or 8240. BTXEs are natural constituents of crude oil, and typically are found in concentrations of a few percent.[13] In Table 3, the percent by volume of BTXE is given for two crude oil samples collected from operating oil wells on the site in Case Study I. The concentrations for BTXE in Table 3 are quite similar to the concentrations reported for a representative petroleum of the API Research Project 6.[13]

BTXEs are primarily found in the gasoline fraction of a crude oil and also are added to gasoline in the refining process. Historically, BTXEs constitute 15% to 25% of regular or unleaded gasolines, and as high as 40% for premium gasolines.[14] Since BTXEs are in the boiling range of the gasoline fraction of petroleum, they tend not to be found in diesel fuels, although higher boiling aromatic compounds may be found in low concentrations.

The VOCs analysis using EPA Method No. 8240 can detect, in addition to BTXEs, other VOCs such as short chain aliphatic and alicyclic hydrocarbons, which can provide backup data for the assessment process.

The semi-VOCs analysis provided by EPA Method No. 8270 detects higher molecular weight compounds that provide backup data which is especially useful for the indication of the presence of crude oils. In Table 4, a list is given of some of the compounds detected by the semi-VOCs analysis that would indicate the presence of a crude oil and not a gasoline or diesel fuel.

Table 4. Semivolatile Organic Compounds in Crude Oil[a]

Benz(a)anthracene
Chrysene
Dimethylnaphthalene
1-Methylnaphthalene
Fluoranthene
2-Methylnapthalene
Napthalene
Phenanthrene
Pyrene

[a]Kirk and Othmer.[15]

An additional indicator of gasoline as a refined product is the presence of organic tetraethyl or tetramethyl lead. Organic lead has, historically, been added to gasoline at concentrations up to 800 ppm. It can be analyzed for using ICP/MS.[11] The LUFT Field Manual[3] describes a DOHS method for organic lead using flame AA spectroscopy.

CASE STUDY I

One of the first sites[6] on which the differentiation criteria were used was located in the north-central portion of a large southern California oil field. The 20-acre site was the location for a proposed business park. The site was bound on the north, west and, south by major city streets, and on the west by railroad tracks. At the time of the site investigation, there were 18 active/idle and 19 abandoned oil wells within the site boundaries. Numerous oil-field-related pipelines traversed the site.

In the immediate site vicinity (the area within approximately one mile of the site), past and present primary land use included oil field development and chemical manufacturing/refining. Commercial development had been taking place since the early 1980s, resulting in the conversion of land that had been used historically for oil field and industrial purposes to office, warehouse, and light manufacturing facilities.

Oil-field-related potential source areas onsite that were judged to warrant investigation for the presence of crude oil or other potentially hazardous waste included former sump areas, former aboveground storage tank areas, and the abandoned, active and idle oil wells. A site schematic is given in Figure 5 that indicates the location of these potential source areas, along with the approximate location of trenches and borings used for the characterization of the potential source areas. Sumps and aboveground storage tanks (ABT) are designated with letters A through T as follows:

Sumps: A, B, C, D, H, K, L, N, Q, R, S, T (12 total).

Aboveground storage tanks: E, F, G, I, J, M, O, P (8 total).

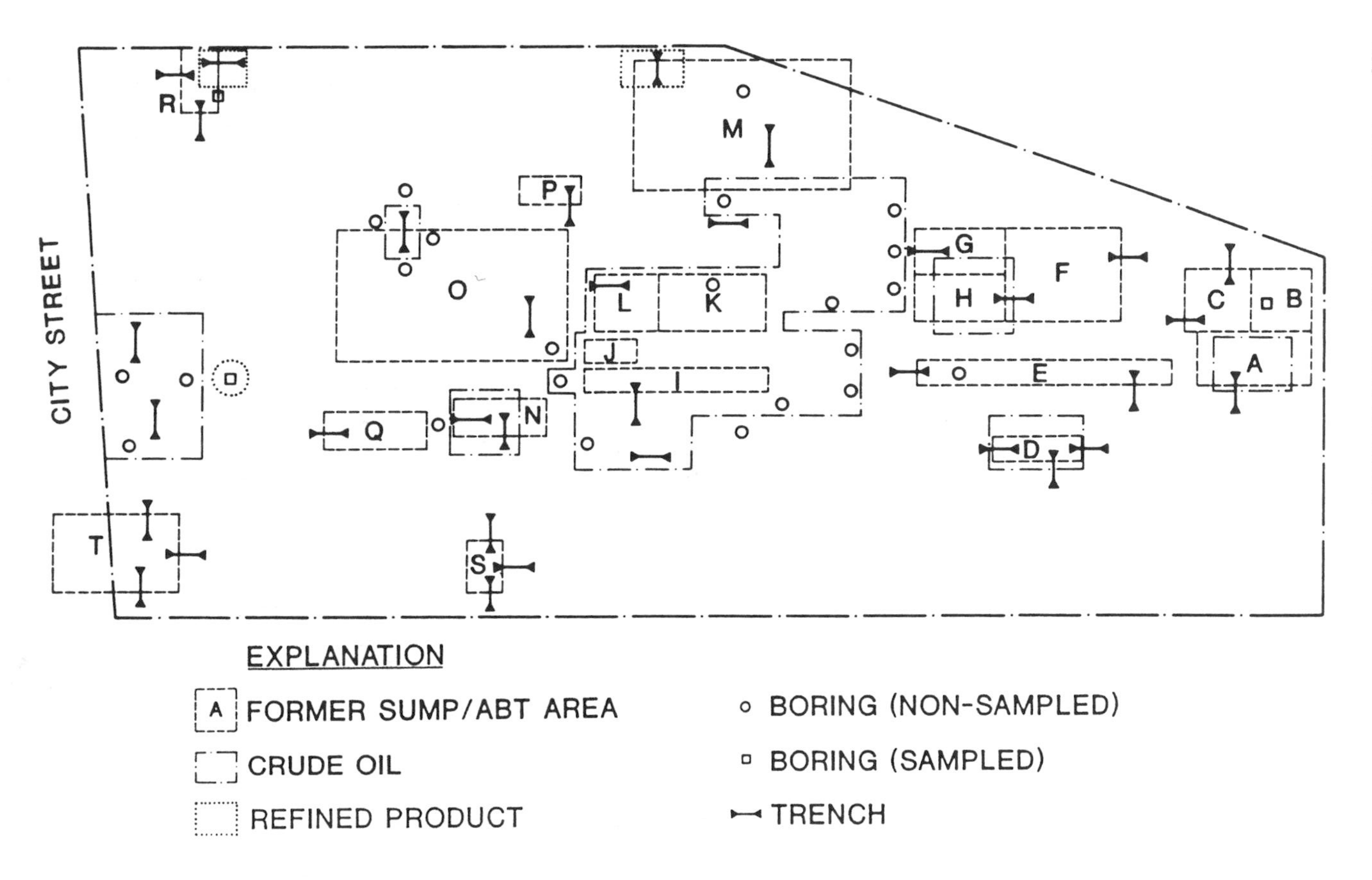

Figure 5. Site I schematic with petroleum hydrocarbon degraded areas.

The areas that were judged to be degraded with crude oil and refined petroleum product as a result of the site investigation are also shown in Figure 5. Most of the petroleum hydrocarbons detected at this site were assessed to be crude oil, with a number of areas exceeding 10,000 ppm TRPH for which remediation may be required. Three areas were assessed to contain refined petroleum product: Sump R, as primarily diesel fuel; Tank M, as primarily gasoline; and a background area, as a mixture of gasoline and diesel. Representative laboratory results for source areas assessed as crude oil and refined petroleum products are provided in Table 5. In Table 6, the evaluation of these data in terms of the criteria described above are given to exemplify the application of this approach. The assessment of crude oil for the majority of the samples is based on the appearance of the chromatograms and on the comparison of the TRPH and TPH results. The chromatograms for Sump H and Tank O, which were judged to be crude oil, are provided in Figure 6. These chromatograms are typical of those that are judged to be crude oil. They have a uniform occurrence of peaks throughout the scan, including peaks in the last five minutes of the run. While the intensity of the peaks in the diesel range is greater than the other areas, it is not interpreted to represent diesel fuel. In aged crude oil soil samples, the gasoline fraction is lost through volatilization and biodegradation. The C_{20} to C_{30} fraction of crude oil has a lower concentration than the diesel fraction and the instrument response to this fraction is also reduced. The diesel fuel standard in Figure 3 can be seen to have a much more dramatic bell-shape than in these chromatograms. The interpretation of these chromatograms as a crude oil was made with confidence.

The chromatograms for Sump R and Tank M are given in Figure 7. Each of these chromatograms was interpreted to be a refined petroleum product with a background of crude oil. For each of the soil samples, the TPH result was larger than the TRPH concentration as given in Table 5. The chromatogram for Sump R area exhibits an intense segmented, bell-shaped scan from about 2 minutes to 12 minutes, with scattered weak peaks throughout the balance of the run. These weak peaks are believed to indicate a background concentration of crude oil underlying the refined petroleum product. The TPH concentration was 14,700 ppm, primarily in the gasoline range, versus a TRPH concentration of 1,300 ppm. The chromatogram for Tank M area also has a prominent segmented, bell-shaped portion from approximately 7 minutes to 20 minutes, primarily in the diesel range. In addition, the chromatogram has weak peaks throughout the balance of the run which are interpreted to be a background of crude oil. The TPH concentration was 1,730 ppm versus a TRPH concentration of 1,100 ppm.

In this case, the appearance of the gas chromatograms and the comparison of TPH and TRPH provided consistent evidence for the presence or absence of refined petroleum products. The judgments were made with confidence. While reliance on the analyses for BTXE, other VOCs, semi-VOCs, or organic lead was not used, the results for these analyses are consistent with the interpretations.

Table 5. Case Study I: Laboratory Results (ppm)

Analysis	Sump H	Tank O	Sump R	Tank M
TRPH	104,000	12,000	1,300	2,100
TPH	44,000	7,350	14,700	3,620
Gasoline	4,000	370	14,700	ND
Diesel	36,000	6,300	ND	3,500
$C_{20}-C_{30}$	4,000	680	ND	120
Benzene	ND[a]	12	ND	ND
Toluene	ND	29	18	ND
Total xylenes	4	27	130	ND
Ethylbenzene	1	3	19	ND
Organic lead	ND	NA[b]	NA	ND
Other VOCs			NA	ND
C_5-C_{11} Aliphatic and alicyclic hydrocarbons	1,000	900		
C_9-C_{10} Alkylbenzenes	300	200		
Semi-VOCs			NA	
C_8-C_{35} Hydrocarbon matrix	300,000	40,000		10,000
Benzo (B&K) fluoranthenes	ND	1		
Dimethylnaphthalenes	600	90		
Fluorene	19	3		
1-Methylnaphthalene	100	20		
2-Methylnaphthalene	230	35		
Naphthalene	92	12		
Phenanthrene	56	8		
Pyrene	ND	1		

[a]Indicates not detected.
[b]Indicates not analyzed.

Table 6. Case Study I: Assessment of the Origin of Petroleum Hydrocarbons in Soil

Criterion	Sump H	Tank O	Sump R	Tank M
1. Appearance of TPH gas chromatogram				
a. Peaks last 5 min.	YES	YES	YES	YES
b. Segmented, bell-shaped peaks	NO	NO	YES	YES
c. Uniform occurrence of peaks	YES	YES	NO	YES
2. TRPH > > TPH[a]	YES	YES	NO	NO
3. BTXE present	YES	YES	YES	NO
4. Other VOCs present	YES	YES	NA[b]	YES
5. Semi-VOCs present	YES	YES	NA	NO
6. Organic lead present	NO	NA	NA	NO
JUDGMENT	Crude oil	Crude oil	Refined product, crude oil background	Refined product, crude oil background

[a]Indicates concentration of total recoverable petroleum hydrocarbons (TRPH) is much greater than concentration of total petroleum hydrocarbons (TPH) detected by EPA Method Nos. 418.1 and 8015 (modified), respectively.
[b]Indicates not analyzed.

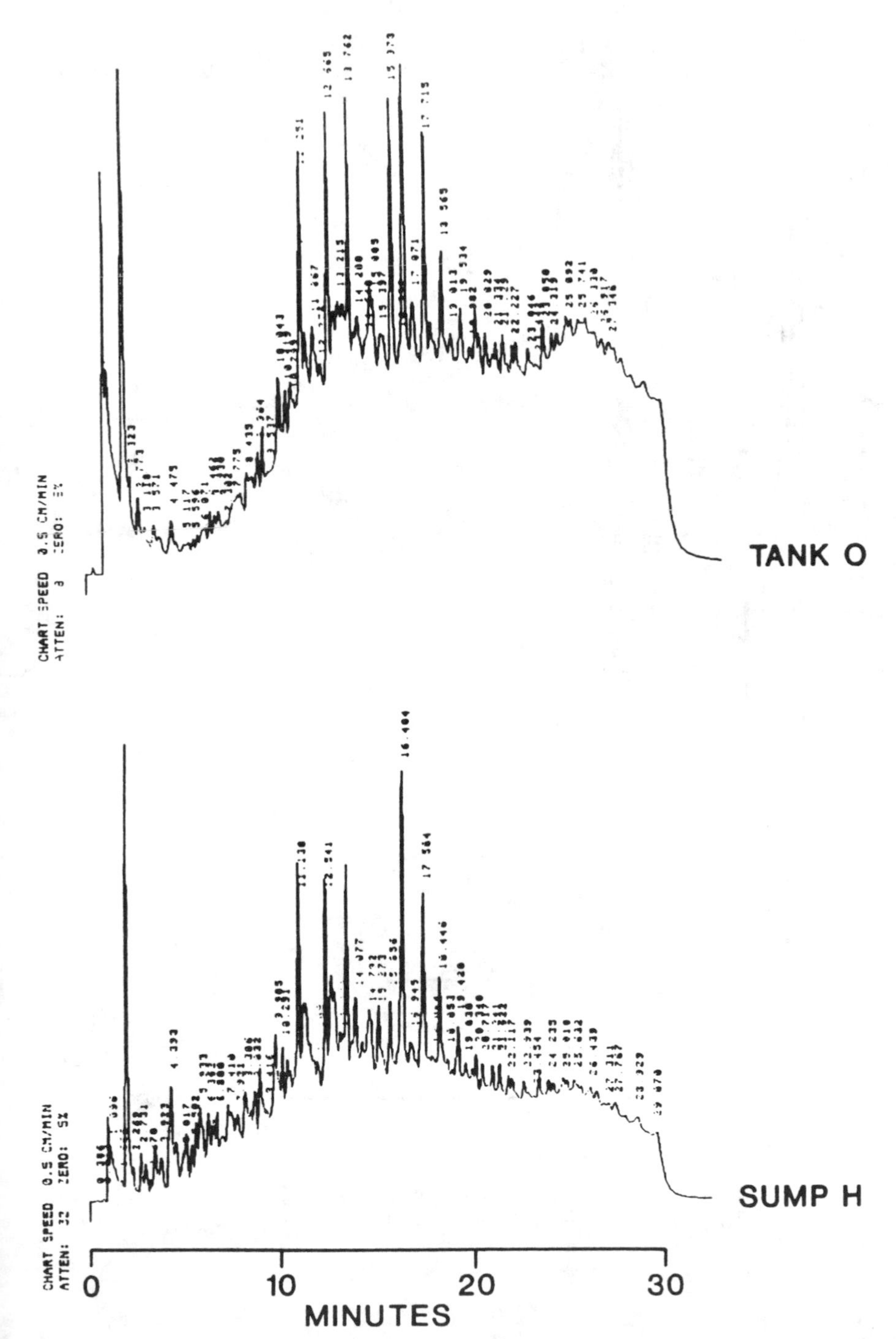

Figure 6. TPH gas chromatogram for Site I, Sump H, and Tank O areas, judged to be crude oil.

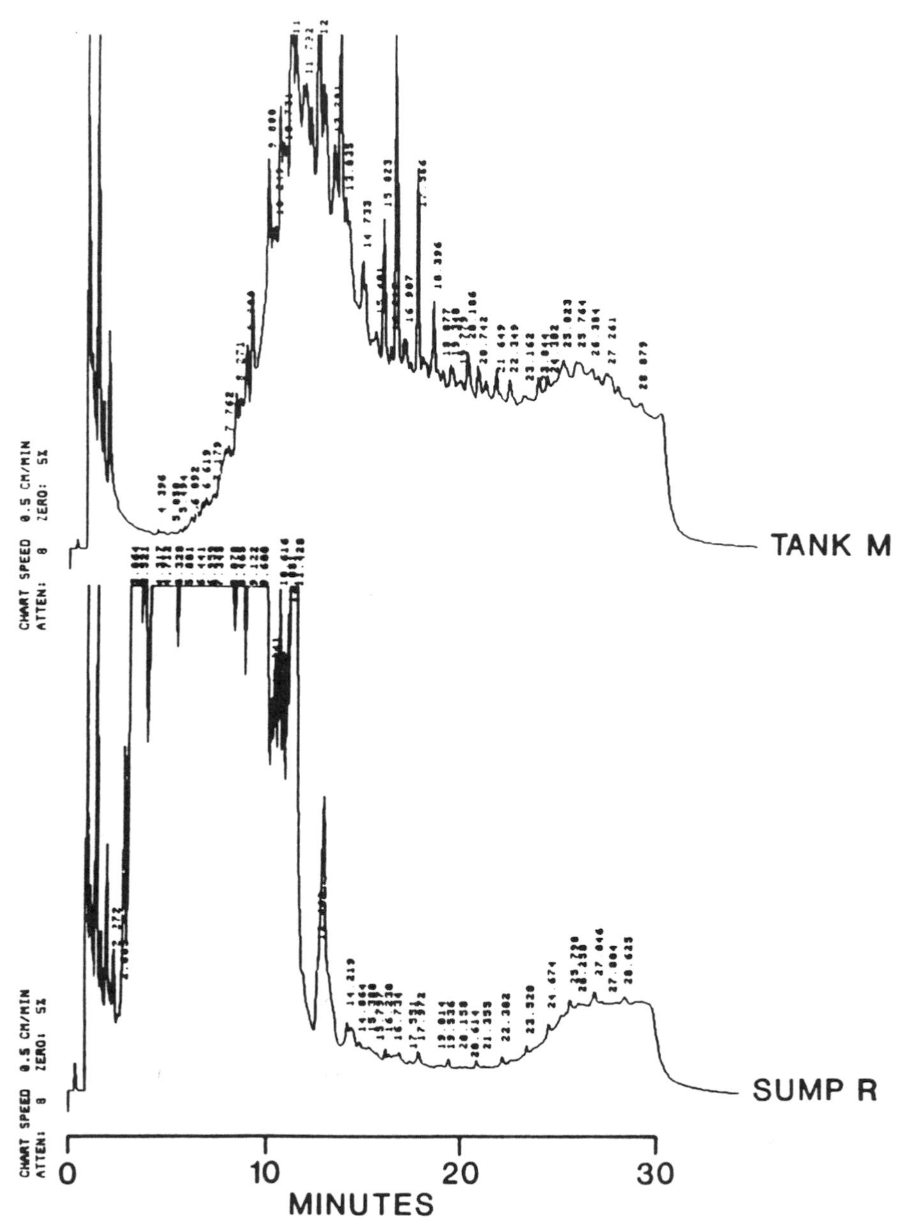

Figure 7. TPH gas chromatogram for Site I, Sump R, and Tank M areas, judged to be refined petroleum product and crude oil background.

CASE STUDY II

Site II was a 2.5-acre parcel located in the north-central portion of a southern California oil field. The site was bound on the south by a city street, on the north by an operating oil field, and on the east and west by light industrial buildings. The immediate site vicinity, the area within approximately a one-mile radius of the site, consisted largely of operating oil fields and some industrial properties. The objective of this investigation was to assess the nature, extent, and potential migration of petroleum hydrocarbons reported to be in the subsurface from a previous investigation.

A schematic of Site II is given in Figure 8 with the location of borings from the previous investigation, as well as from the existing investigation shown. Also, the areas judged to have petroleum degraded soils are indicated in Figure 8. In Table 7, the results of chemical analyses used for the differentiation of crude oil and refined petroleum products are given. The chromatograms of two of the soil samples used for the evaluation of the appearance criteria are provided in Figures 9 and 10. The assessment of the origin of the petroleum hydrocarbons detected, including the judgment reached for each sample, is described in Table 8. The Area I sample was interpreted to be crude oil, while the Area II sample was judged to be refined petroleum product (gasoline). The chromatogram for Area I clearly exhibits a uniform occurrence of peaks throughout the scan, as well as peaks in the last five minutes of the run, for indications of the presence of predominately crude oil. The chromatogram for Area II, which was judged to be gasoline, exhibits peaks in the first 7 to 10 minutes of the run, with a bell-shaped appearance and no peaks in the balance of the scan.

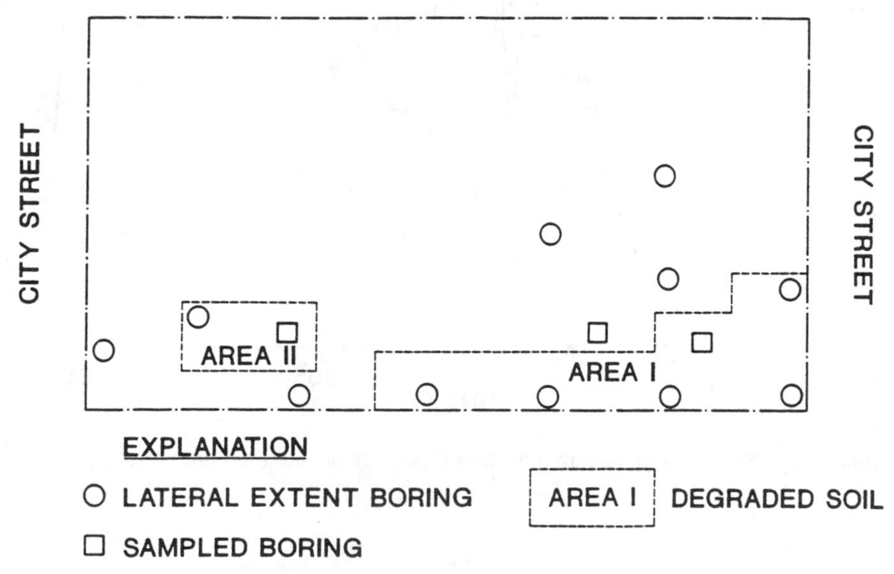

Figure 8. Site II schematic with petroleum hydrocarbon degraded areas.

Table 7. Case Study II: Laboratory Results (ppm)

Analysis	Area I	Area II
TRPH	9,000	4,300
TPH	790	3,500
Gasoline	ND[a]	3,500
Diesel	630	ND
$C_{20} - C_{30}$	160	ND
Benzene	ND	1.4
Toluene	0.3	44
Total xylenes	2.7	93
Ethylbenzene	0.9	30
Organic lead	ND	ND

[a]Indicates not detected.

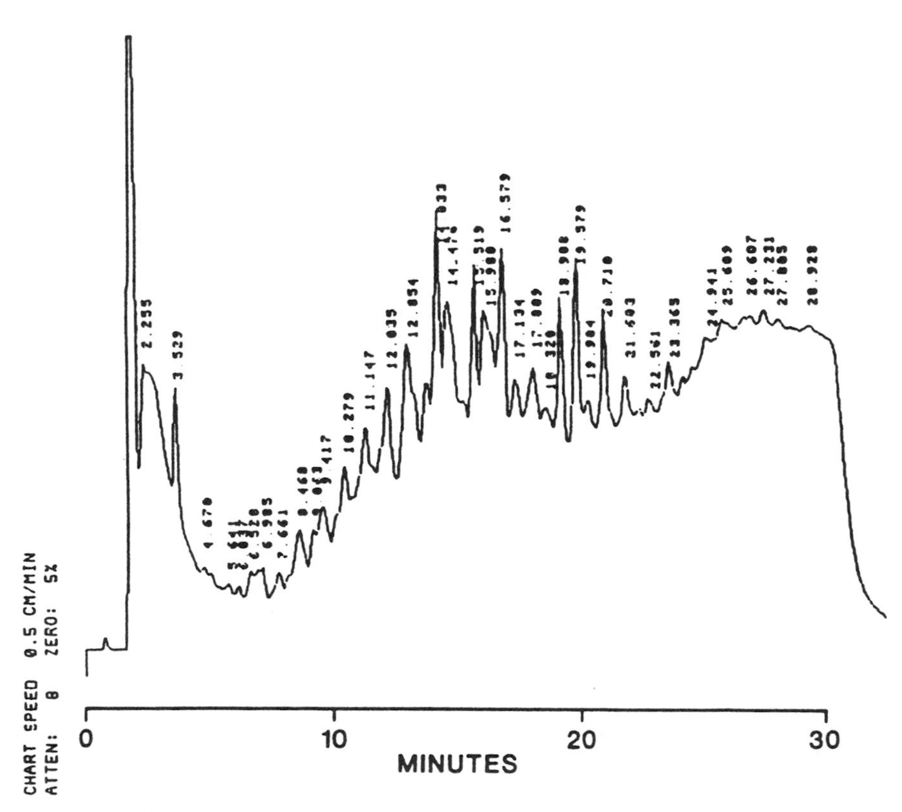

Figure 9. TPH gas chromatogram for Site II, Area I, judged to be crude oil.

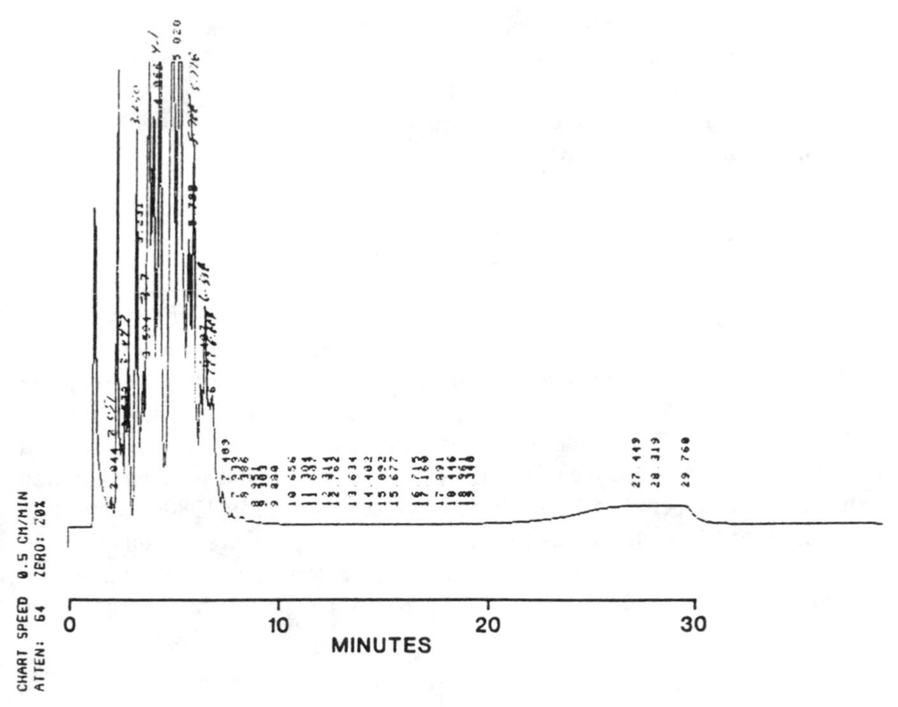

Figure 10. TPH gas chromatogram for Site II, Area II, judged to be refined petroleum product.

Table 8. Case Study II: Assessment of the Origin of Petroleum Hydrocarbons in Soil

Criterion	Area I	Area II
1. Appearance of TPH gas chromatogram		
a. Peaks last 5 min.	YES	NO
b. Segmented, bell-shaped peaks	NO	YES
c. Uniform occurrence of peaks	YES	NO
2. TRPH>>TPH[a]	YES	NO
3. BTXE present	YES	YES
4. Other VOCs present	NA[b]	NA
5. Semi-VOCs present	NA	NA
6. Organic lead present	NO	NO
JUDGMENT	Crude oil	Gasoline

[a]Indicates concentration of total recoverable petroleum hydrocarbons (TRPH) is much greater than concentration of total petroleum hydrocarbons (TPH) detected by EPA Method Nos. 418.1 and 8015 (modified), respectively.
[b]Indicates not analyzed.

Based on these judgments and the results from a previous investigation, it was recommended that the degraded soil in both Areas I and II be excavated and remediated to a depth of 5 feet. Also, the soil in Area 2 that was judged to be degraded with gasoline should be studied for the feasibility of mitigating with an in situ vapor extraction system.

CASE STUDY III

Site III provides an example in which the application of the criteria resulted in conflicting assessments. Site III was a four-acre parcel located in the south-central portion of a large southern California oil field. The site was bound on the east and south by major city streets, on the north by a landscaping business, and on the west by light industry. In the immediate site vicinity, historical and present primary land use included oil field development and refining. Limited industrial development had been taking place since the early 1980s, converting the depleted oil field properties to office, warehouse, and light manufacturing facilities. The objective of the effort was to characterize the extent and degree of soil degradation and to assess the origin of the petroleum hydrocarbons

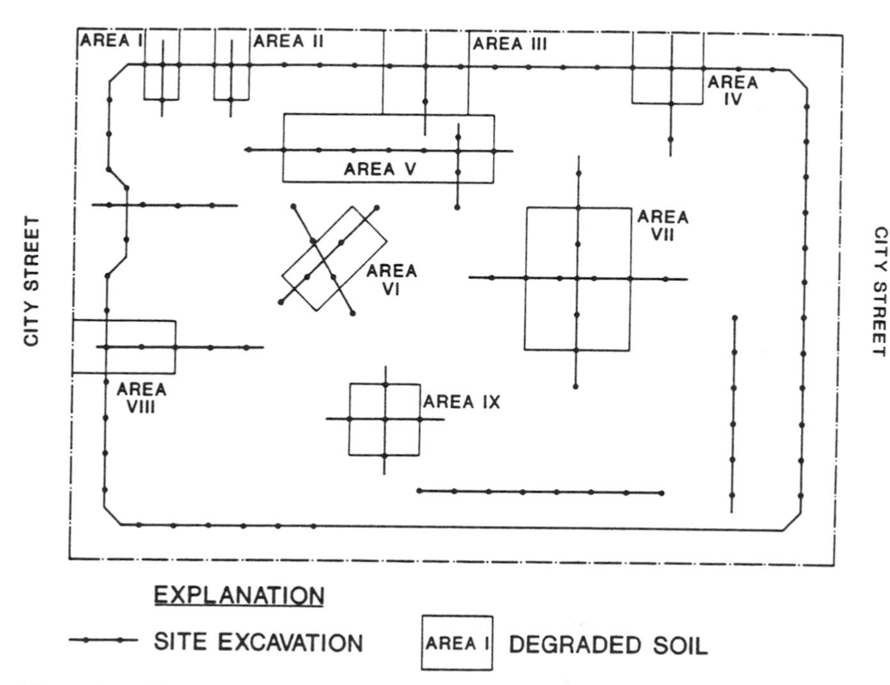

Figure 11. Site III schematic with petroleum hydrocarbon degraded soil areas.

detected in terms of crude oil or refined petroleum products. In Figure 11, a schematic of the site is given showing potential source areas, test pits, and the perimeter excavation used for characterization and sampling. The areas assessed to be degraded are also indicated.

In the investigation, 1 perimeter trench and 18 test pits were excavated. The degraded areas included three pipeline leaks along the northern property boundary, Area III along the northern boundary, and a portion of Area VIII along the western boundary. In Table 9, laboratory results used to assess the origin of petroleum hydrocarbons in terms of crude oil or refined petroleum products are given for three typical soil samples at the site. In Figures 12 and 13, the TPH gas chromatograms are given for the three representative samples. In Table 10, the assessment of the data and the gas chromatograms are given. Each of the chromatograms exhibits peaks in the last five minutes of the run, a uniform occurrence of peaks throughout the 30-minute run, and no segmented, bell-shaped features. All of these characteristics are consistent with an assessment of crude oil. However, only the Area V sample had a TRPH result greater than the TPH analysis and the presence of a refined product cannot be ruled out with confidence. The other criteria provide no assistance, because of no analysis or nondetect. The detection of phenanthrene in the semi-VOC analysis for Area V sample does confirm the assessment of crude oil for this sample, but does not help to rule out a refined product. The BTXE also detected for the Area V sample adds strength to the crude oil assessment because the refined petroleum product, in this case, is judged to be diesel fuel which generally contains little or no BTXE. The final judgment for this site reported the presence of predominantly crude oil, with the possibility of some overlapping diesel fuel. This assessment is likely to require remediation to a much lower level than might otherwise have been needed if an assessment of crude oil only could have been justified on the basis of the appearance of the chromatograms.

Table 9. Case Study III: Laboratory Results (ppm)

Analysis	Area V	Area IX	Area II
TRPH	7,600	3,500	15,000
TPH	6,600	26,800	17,700
Gasoline	1,000	7,400	1,300
Diesel	2,900	13,000	10,000
$C_{20} - C_{30}$	2,700	6,400	6,400
Benzene	ND [a]	0.080	0.081
Toluene	0.029	9.3	12
Total xylenes	0.120	4.7	6.1
Ethylbenzene	ND	4.0	7.2
Organic lead	ND	NA[b]	NA

[a]Indicates not detected.
[b]Indicates not analyzed.

Table 10. Case Study III: Assessment of the Origin of Petroleum Hydrocarbons in Soil

Criterion	Area V	Area IX	Area II
1. Appearance of TPH gas chromatogram			
a. Peaks last 5 min.	YES	YES	YES
b. Segmented, bell-shaped peaks	NO	NO	NO
c. Uniform occurrence of peaks	YES	YES	YES
2. TRPH>>TPH[a]	YES	NO	NO
3. BTXE present	YES	YES	NA
4. Other VOCs present	NA[b]	NA	NA
5. Semi-VOCs present	YES[c]	NA	NA
6. Organic lead present	ND[d]	NA	NA
JUDGMENT	Crude oil	Crude oil and diesel fuel	Crude oil and diesel fuel

[a]Indicates concentration of total recoverable petroleum hydrocarbons (TRPH) is much greater than concentration of total petroleum hydrocarbons (TPH) detected by EPA Method Nos. 418.1 and 8015 (modified), respectively.
[b]Indicates not analyzed.
[c]Phenanthrene detected at 0.640 ppm.
[d]Indicates not detected.

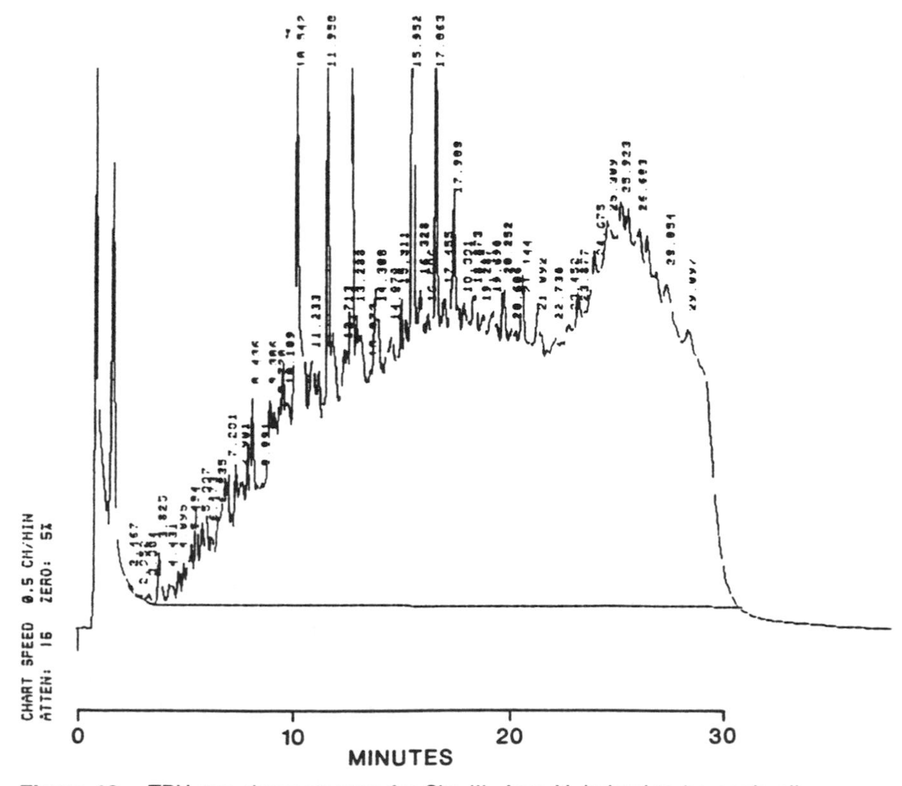

Figure 12. TPH gas chromatogram for Site III, Area V, judged to be crude oil.

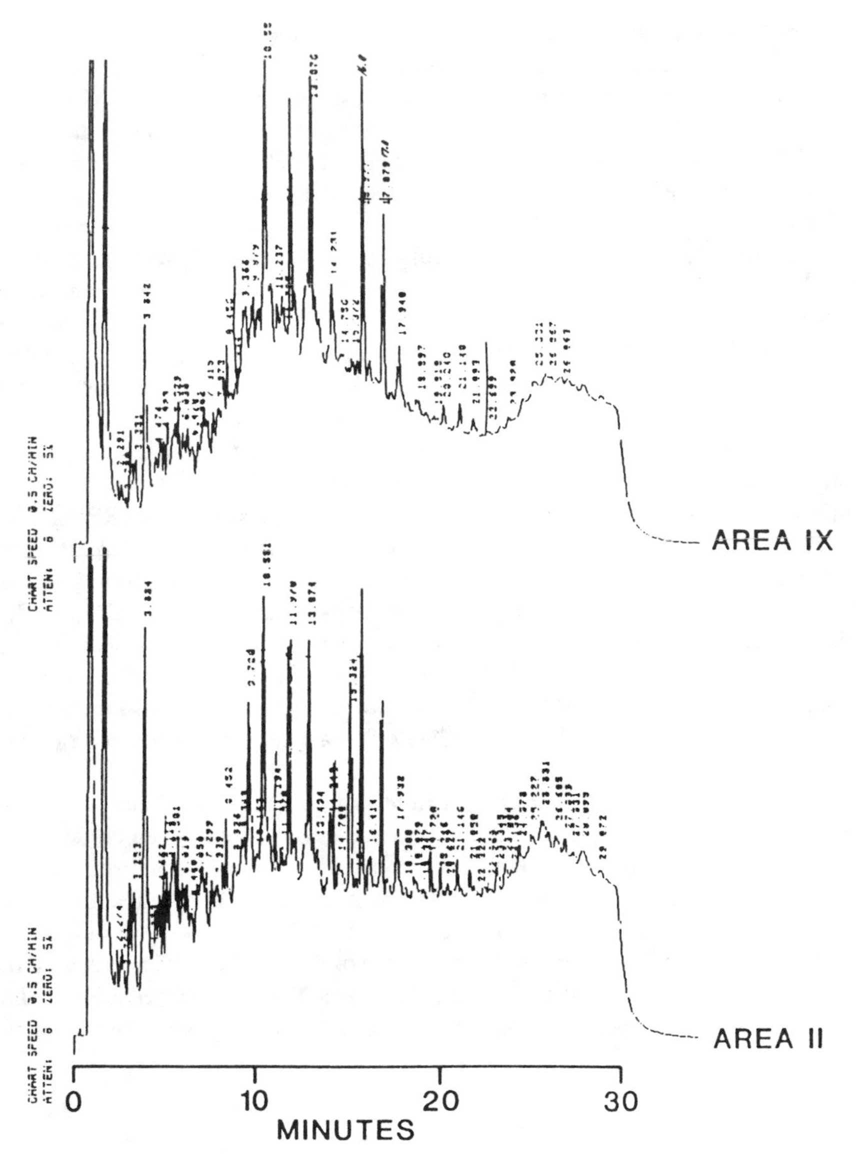

Figure 13. TPH gas chromatogram for Site III, Areas II and IX, judged to be crude oil and refined petroleum product.

LIMITATIONS AND FUTURE WORK

The assessment of the origin of petroleum hydrocarbons using the criteria described above has, in a number of cases, provided regulators with the evidence needed to justify reduced mitigation measures based on the absence of refined petroleum product. To that end, these criteria have served in reducing the cost of remediation. However, the reliance on TRPH is a recognized limitation of this approach because it is more of a survey tool than a reliable site characterization analysis. Future improvements in the analysis for the concentration of crude oil in soils could be made on two bases. First, a better understanding of the ability of EPA Method No. 418.1 to measure crude oil could improve the interpretation of the comparison of this TRPH with the TPH result. Retaining the use of the TRPH analysis has the advantage of being a simple, inexpensive method. Second, more reliable improvement would be to find another analysis, perhaps one that is routine in the petroleum industry, to substitute for the TRPH analysis.

When investigating a site which historically has been used as an oil field, the question of the presence of crude oil is not generally the issue. The issue to resolve is the absence of a refined petroleum product. A further development in the TPH analysis that included standards for crude oil and for mixtures of crude oil and diesel fuel might provide the evidence needed. If reliable standards were found, convincing evidence for the absence of diesel fuels might be possible using the appearance of TPH gas chromatograms without a comparison to TRPH.

Senn and Johnson[16] have discussed the use of capillary column GC to estimate the degree of degradation and relative age of petroleum hydrocarbons in soil and groundwater. Here, use is made of ratios of pristane and phytane, C_{17} and C_{18} alicyclic hydrocarbons, respectively, which biodegrade more slowly than the C_{17} and C_{18} aliphatic hydrocarbons. Therefore, larger C_{17} to pristane and C_{18} to phytane peak ratios indicate less degradation between samples. They also report that the presence of light-end peaks also indicate that a gasoline product in a sample is relatively fresh.

In our case studies, the occurrence of light-end peaks is rare. Hence, our concern centers primarily on assessing the absence of diesel fuel when crude oil is believed to be present. Diesel fuels are reported[3] to have predominate carbon chain lengths of C_{16} and C_{17} in the C_{10} to C_{23} mixture. If this predominance of C_{16} and C_{17} does not occur in the crude, this feature could be very helpful in providing evidence for the presence or absence of diesel fuels. Even with a more expensive analysis to provide this data, the reduced costs in remediation if the absence of a refined product is accepted by a regulatory agency would overshadow the cost of the analysis.

REFERENCES

1. CFR (Code of Federal Regulations), Identification and Listing of Hazardous Waste: 40 CFR 261, Revised 1 July 1987.
2. CCR (California Code of Regulations), Title 22, Division 4, Environmental Health, Chapter 30, Minimum Standards for Management of Hazardous or Extremely Hazardous Waste, 1988.
3. SWRCB (State Water Resources Control Board), *Leaking Underground Fuel Tank Manual: Guidelines for Site Assessment, Cleanup, and Underground Storage Tank*

Closure State of California Leaking Underground Fuel Tank Task Force, Sacramento, California, revised 5 April 1989, 121 pp.

4. Applied Geosciences Inc. "Report on the Phase II Site Characterization of a 4.3-Acre Property," unpublished report prepared for a confidential client, 2 June 1988(a), 33 pp.
5. Applied Geosciences Inc. "Report on the Results of an Assessment of Oily Soil Adjacent to Two Oil Wells," unpublished report prepared for a confidential client, 6 July 1988(b), 22 pp.
6. Applied Geosciences Inc. "Phase II Toxic Hazard Investigation of a Planned Business Park," unpublished report prepared for a confidential client, 16 August 1988(c), 49 pp.
7. Applied Geosciences Inc. "Oily Soil Assessment," unpublished report prepared for a confidential client, 14 December 1988(d), 7 pp.
8. Applied Geosciences Inc. "Phase II Site Characterization Report," unpublished draft report in preparation for a confidential client, in preparation September 1989.
9. U.S. EPA (Environmental Protection Agency), Methods for Chemical Analysis of Water and Wastes, EPA 600/4-79-020, revised March 1983, U.S. EPA Environmental Monitoring Laboratory, Cincinnati, Ohio.
10. U.S. EPA, Test Methods for Evaluating Solid Waste; Physical/Chemical Methods, SW-846, Third Edition, Office of Solid Waste and Emergency Response, U.S. EPA, Washington, D.C., 1986.
11. Northington, J. D., Technical Director, West Coast Analytical Service, Inc., Santa Fe Springs, California, oral communication, 1988.
12. Nyer, E. K., and G. J. Skladany. "Relating the Physical and Chemical Properties of Petroleum Hydrocarbons to Soil and Aquifer Remediation," *Ground Water Monitoring Review* Winter 1989, pp 54–60.
13. Rossini, F. D. "Hydrocarbons in Petroleum," *J. Chem. Ed.* 37:554–561 (1960).
14. Guard, H. E., J. Ng, and R. B. Laughlin, Jr., Characterization of Gasoline, Diesel Fuels and Their Water Soluble Fractions, Naval Biosciences Laboratory, Naval Supply Center, Oakland, California, September 1983.
15. Kirk, R. E., and D. F. Othmer, Eds., *Encyclopedia of Chemical Technology* (New York: John Wiley & Sons, 1983).
16. Senn, R. B., and M. S. Johnson. Interpretation of Gas Chromatography Data as a Tool in Subsurface Hydrocarbon Investigations, in *Proceedings of the NWWA/API Conference on Petroleum Hydrocarbons and Organic Chemicals in Ground Water—Prevention, Detection and Restoration,* Houston,Texas, 13–15 November 1985, National Water Well Association, Dublin, Ohio, pp. 331–357.

PART IV

Remedial Technologies

CHAPTER 13

Biological Treatment of Soils Contaminated by Petroleum Products

Robert N. Block and Thomas P. Clark, Remediation Technologies, Inc., Concord, Massachusetts
Mark Bishop, New England Testing Lab, North Providence, Rhode Island

Biological treatment processes have received increased attention over the past several years as a safe and cost-effective means of treating a wide variety of hazardous and nonhazardous wastes. It has been widely used in the petroleum industry for degradation of oily wastes. This technology has been proved cost-effective for treatment of soils contaminated with petroleum hydrocarbons.

The biodegradation process uses the assimilative capacity of soil to decompose the contaminants through transformation and biological oxidation in the soil treatment matrix. The treatment focuses on the breakdown of the hydrocarbons by aerobic microorganisms through a series of intermediary organic acids ultimately to CO_2 and water. The low cost and apparent simplicity of the technology has made biological treatment very attractive; however, the relatively slow degradation rates and the acreage required for treatment areas has limited application of this technology at some sites.

This technology is being employed at various sites around the country, including the site of a former petroleum products terminal which was used to store gasoline and No. 2 fuel oil. Application of this technology at this site has led to treatment of almost 30,000 cubic yards to date, as well as to improvements in typical biological treatment operations. Operations at this site have also identified a significant complication associated with the TPH analytical protocol (U.S. EPA Method 418.1) and an alternative analytical protocol has been developed.

This chapter briefly summarizes the operations at this site, including the improvements in biological treatment developed. More detailed discussions of full-scale operations at this site are presented in other papers.[1,2] The primary focus of this chapter is associated with the problems encountered with the analytical protocol commonly employed to characterize contamination by petroleum hydrocarbons.

Treatment Operations

Soil treatment operations are conducted in several treatment areas which have been graded to expedite treatment operations. Specific regrading operations include construction of perimeter berms around the treatment areas for storm water management, regrading to improve drainage within the treatment areas, recompaction of the subgrade to support equipment, and placement of a marker layer to allow easy identification of the subgrade during unloading/reloading operations.

Treatment operations commence by loading a particular treatment area with contaminated soils. Soils are end-dumped and leveled. Amendments including lime and fertilizer are subsequently spread over the contaminated soils. A horizontally mounted auger manufactured by the Brown Bear Company mixes the soils and amendments and forms a windrow approximately two feet high and four feet wide. Windrows are placed toe to toe without a travel lane between, allowing for the ratio of soils in treatment to the size of the treatment area to be 2,500 cubic yards per acre.

Soils are aerated approximately twice per week with the Brown Bear. Aeration operations begin at the windrow at the edge of the treatment area which is moved from the right side of the auger to the left. During this operation, the soils are thoroughly mixed and aerated.

Soils are amended with fertilizer during initial construction of the windrows to correct the carbon nitrogen phosphorous ratio to 25:1:0.5. During the first year's operation, manure was used for fertilizer. During 1989 operations, inorganic fertilizer was used. Samples of the contaminated soils are obtained at various times during treatment and analyzed for C:N:P to ensure the target ratio is maintained.

Measurements of pH are regularly performed. Soil conditions are initially acidic, typically in the range of 6.0 to 6.5. As treatment progresses, the soils become more acidic, necessitating the additions of agricultural lime to restore the pH to between 6.0 to 8.0. Irrigation of the soils is periodically required to maintain the moisture content between 70% and 80% of field capacity. A tensiometer is used to monitor moisture content and direct irrigation operations.

Once analytical results indicate the cleanup criteria (0.25 ppm BTEX, 50 ppm TPH, <1.0 ppm Naphthalene) have been achieved, the soils are removed and the treatment area reloaded with contaminated soils. Calculations of kinetics for

Table 1. Summary of Half Lives

Compound	Half-Life (days)
Toluene[a]	5.8– 6.5
p-Xylene[a]	8.6– 10.2
C_{12}Alkane	11 - 15
C_{13}Alkane	10
C_{14}Alkane	8 - 18
C_{15}Alkane	23 - 25
C_{16}Alkane	7 - 30
C_{17}Alkane	23 - 29
C_{18}Alkane	23 - 27
C_{19}Alkane	23 - 27
C_{20}Alkane	23 - 28
C_{21}Alkane	9 - 32
C_{22}Alkane	10 - 23
C_{23}Alkane	11 - 27
C_{24}Alkane	11 - 27
C_{25}Alkane	9 - 45
C_{26}Alkane	13 - 34
Naphthalene[a]	9.1– 13.9
2-Methyl-naphthalene	14 - 23
1-Methyl-naphthalene	12
2,6-Dimethyl-naphthalene	7
1,3-Dimethyl-naphthalene	7 - 9
2,3-Dimethyl-naphthalene	9 - 11
1,2-Dimethyl-naphthalene	10
Anthracene	<9 - 53
Phenanthrene	<6 - 43
Benzo (a) anthracene	63 -231
Chrysene	41 -116

[a]Corrected for volatilization loss.
Source: References #3 and #4.

degradation of BTEX from the field data indicate a half-life of 6.2 days, comparable with literature data as summarized in Table 1. The calculated half-life for TPH is 23 days. No. 2 fuel oil consists of numerous compounds, the largest percentage of which are straight chain alkanes between C_{12} and C_{26}. The values of half-lives reported in the literature vary significantly for these constituents, ranging from as low as 7 days to as much as 45 days. Given the initial concentrations of BTEX and TPH compounds at this site, degradation of the TPH compounds controls treatment operations.

The amount of contaminated soils treated per acre is controlled by the aeration equipment used. Biological treatment of contaminated soils and sludges is typically performed by a thin spreading operation. Soils are spread in an eight inch thick lift and aerated with a rototiller. This equipment can typically aerate soils to a depth of eight inches, which translates to a treatment rate of approximately 1,000 cubic yards per acre. The use of the horizontally mounted auger allows treatment of approximately 2,500 cubic yards per acre, significantly greater than the rate achieved using a standard agricultural tiller. However, use of this equipment significantly increases the cost of aeration as the total operation and maintenance expense exceeds $100 per hour. Other equipment options are available which allow deep tilling, but these alternatives carry similar O/M expenses.

Analytical Considerations

Selection of an appropriate analytical protocol and critical interpretation of test results are fundamental in the evaluation and optimization of the bioremediation process. The analytical protocol for process monitoring must provide reliable and complete petroleum hydrocarbon quantification.

Several methods have been developed for measuring petroleum hydrocarbons. The simplest method (EPA Method 503) involves gravimetric measurement of solvent extractable constituents in the sample. This method is usually modified to include a "silica cleanup" to remove polar solvent extractable components which are naturally present in most soils. This method is useful in measuring gross contamination of soils by relatively high boiling petroleum fractions. However, low boiling components volatilize in the solvent elimination step.

To eliminate this problem, the solvent removal and gravimetric measurement steps have been replaced by infrared (IR) quantification (EPA Method 418.1). This method employs a measurement of infrared radiation absorbance at the characteristic carbon-hydrogen bond stretching frequency. Although infrared spectroscopic technique is very sensitive and commonly employed for characterization of soils contaminated with petroleum hydrocarbons, infrared quantification technique was found to be ineffective for measuring low to medium level petroleum hydrocarbon contamination in soils at this site.

Specifically, initial results from treatment process monitoring showed decreasing petroleum hydrocarbon concentrations; however, a trend of increasing petroleum hydrocarbon contamination was noted as treatment progressed. These inconsistent results lead to duplicate analysis of both contaminated and noncontaminated soil samples by the IR method and by gas chromatography (GC). Table 2 shows a comparison between total petroleum hydrocarbon (TPH) concentrations as measured by IR and by GC for windrow samples and topsoil samples taken from offsite locations.

Many soils contain low level nonpetroleum hydrocarbon interferences. These interferences are presumed to be either naturally occurring humic materials or by-products of the petroleum hydrocarbon biodegradation process. The interferences are not effectively removed by the silica gel cleanup, but they do contain carbon-hydrogen bonds. The presence of this interference falsely indicated that a healthy treatment process had stalled at some low to medium petroleum hydrocarbon level.

Both the gravimetric method and the infrared method are fast, relatively simple methods to perform. Because they are soil matrix extensions of established EPA-approved methods for wastewater (EPA Method 418 and Standard Methods

Table 2. Comparison of TPH Analyses IR and GC Methods

Sample	Location	TPH By IR	TPH By GC
1-AB.1.5	Treatment window	317	69
1-CD.1.5	Treatment window	148	109
1-EF.1.5	Treatment window	136	48
1-GH.1.5	Treatment window	160	65
1-IJ.1.5	Treatment window	167	105
1-K.1.5	Treatment window	172	60
1-L.1.5	Treatment window	110	67
1-M.1.5	Treatment window	153	114
5-A.1.1	Treatment window	250	75
5-B.1.1	Treatment window	144	109
PBGD-4	Park adjacent to site	76	9
PBGD-5	Park adjacent to site	nd	10
PBGD-6	Park adjacent to site	33	10

503), they have widespread regulatory acceptance. However, they do not provide valuable quantitative contaminant information for biological treatment process monitoring. Additionally, the low end-of-treatment petroleum hydrocarbon level required at this site makes the interference problems encountered in these methods a serious obstacle.

Capillary gas chromatography was found to be a successful alternative to the infrared technique. This method is well established in the petroleum industry and the environmental community. The GC method involves separation and detection of the various components comprising the petroleum hydrocarbon contaminant in the sample. The method of component detection can be a simple conventional detector such as a flame ionization detector (FID), or, when considerable quantitative information is needed, a mass spectrometer. This technique enables elimination of interferences and provides quantitative information when treatment problems are encountered.

Obtaining reliable quantitative information of petroleum contamination with either the IR or GC techniques requires the selection of an analytical standard which is similar to the contaminant in the sample. The TPH protocol was originally developed for characterization of contamination by spills of petroleum products. In these incidents, a sample of the product was typically available for use as a standard. In those incidents where no product was available, a surrogate standard consisting of a 3:2:2 mixture of n-hexadecane, isooctane and chlorobenzene is proposed by Method 418.1.

If the petroleum product which initially contaminated the soil can be determined, then that material should be considered for use as a standard. Frequently, however, no single standard can be identified since numerous products may have been stored at the site, the mix of constituents in any specific petroleum product may vary, or the facilities at the site may have been emptied of product. Additionally, weathering of the product once released in the environment changes the mix of constituents. Therefore, a fundamental step in selecting the standard when none is available is to obtain well resolved chromatographic profiles for the petroleum hydrocarbons present in the soil matrix. Gas chromatographic/mass spectroscopic techniques may be used to obtain detailed sample characteristics.

After the samples are well characterized, a standard can be selected which has a similar component profile. Although these standards may be custom blended, they are usually prepared from commercially available petroleum products such as #2, #4, or #6 fuel oil.

In the case of this specific remediation effort, both gasoline and No. 2 fuel oil were stored at the site over a period of over 60 years. Evaluation of chromatographs from contaminated soil samples indicated that the primary contaminant was a middle distillate petroleum product similar to No. 2 fuel oil or diesel fuel. As remediation efforts progressed, a small quantity of free product was encountered which appeared to be weathered No. 2 fuel oil.

Figure 1 presents the component profile for the free product encountered on-site. A chromatogram for the petroleum hydrocarbon contaminant in the soil is shown in Figure 2. Inspection of these chromatograms indicates a good match in the two profiles. Although these component profiles could have been obtained with an FID, a mass spectrometer was used for detection in this case. This enabled further identification of certain components in these samples. Several observations can be made:

1. The alkyl benzene, naphthalene, and alkyl naphthalene content of both samples is low (less than 1%). This analysis could be extended to include: indans/tetralins, indenes, acephthenes, and acephthalenes. Because of the comparatively high toxicity of these compounds, this information can be useful in establishing prudent cleanup criteria.
2. The predominant components of both samples are normal alkanes, and branched alkanes. Some noncondensed polycycloaliphatic hydrocarbons and monocycloaliphatic hydrocarbons were detected.
3. The ratios of certain predominant components are presented in Table 3. This data provides further confidence in using the free product for the standard in this application. It also indicates that certain weathering processes had taken place before the start of this evaluation. The branched C_{19} to normal C_{18} ratio and the branched C^{18} to normal C_{17} ratio are well established indicators of petroleum product weathering. These ratios are significantly higher for the soil contaminant. Finally, the ratio of branched alkanes to normal alkanes is relatively high for both samples when compared to common fuel oil products, further indicating a need for a specialized standard.

After initial characterization, the free product standard was used in quantifying the petroleum hydrocarbon levels throughout the bioremediation process. Chromatographic analytical data provided valuable qualitative insights throughout the treatment process.

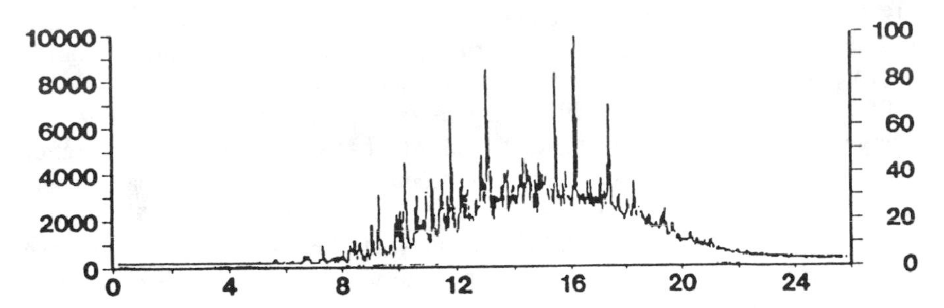

Figure 1. Free product chromatogram.

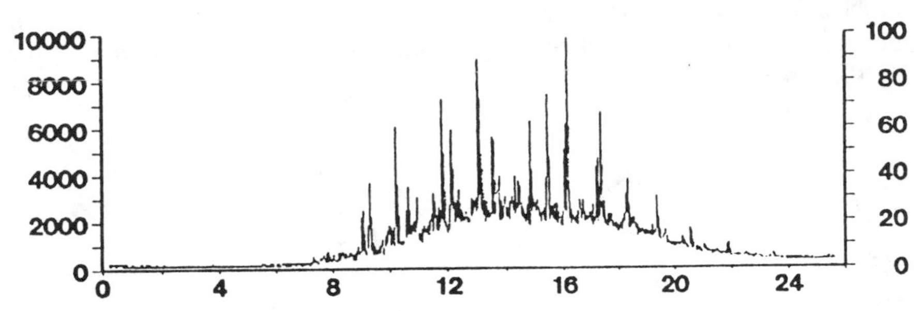

Figure 2. Soil contaminant chromatogram.

Table 3. Component Ratio (Area Component/Sum of Areas Considered)

Component	Soil Contaminant	Free Product
Normal C12	3.7	4.0
Normal C14	8.5	4.8
Normal C16	8.9	2.9
Normal C17	17.7	24.1
Normal C18	9.8	14.6
Branched C14	12.4	12.6
Branched C15	12.8	15.7
Branched C16	13.1	16.7
Branched C18	7.1	2.5
Branched C19	6.1	2.0

Figures 3 and 4 present chromatograms for the soil contaminant after several weeks of treatment. The concentration of petroleum hydrocarbons at this stage was 60 mg/kg; a reduction from the initially measured level of 400 mg/kg (Figure 2). Figure 3 presents this profile at the same scale as Figure 2. This chromatogram illustrates the significant reduction in the levels of petroleum hydrocarbons after treatment.

Figure 4 is a replotted presentation of Figure 3 which reveals the remaining petroleum profile. The same component ratios presented in Table 3 are presented for the initial and treated soil profiles in Table 4. These ratios, together with the general chromatographic patterns, indicate that the component profile after treatment is still approximated by the free product standard. The chromatographic profile indicates that components eluting in the 20 minute plus region of the chromatogram (corresponding to C_{22} plus alkanes) are becoming enhanced relative to the earlier eluting components as treatment progresses. The ratios presented in Table 4 suggest that the components with higher starting concentrations are being removed at a faster rate than those with the lower concentrations.

Analysis of these chromatograms shows a ''healthy'' remediation process is taking place. Further analysis reveals the low aromatic levels present at the start of the evaluation are below the practical quantitative level after treatment. Also, the profiles indicate that no principal components are being selectively not-treated by the process.

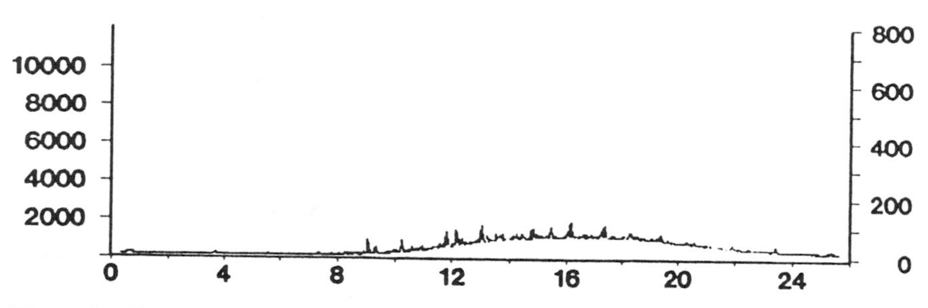

Figure 3. Treated soil chromatogram.

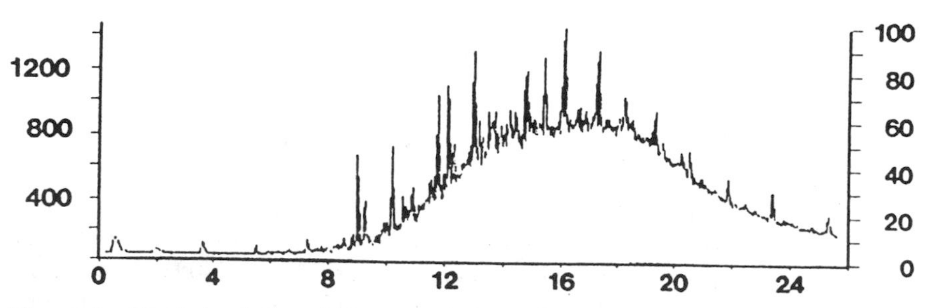

Figure 4. Treated soil chromatogram (enlarged scale).

Table 4. Component Ratio (Area Component/Sum of Areas Considered)

Component	Initial Profile	Treated Profile
Normal C12	3.7	9.7
Normal C14	8.5	11.7
Normal C16	8.9	8.8
Normal C17	17.7	13.8
Normal C18	9.8	10.8
Branched C14	12.4	10.8
Branched C15	12.8	11.8
Branched C16	13.1	10.8
Branched C18	7.1	5.9
Branched C19	6.1	5.9

CONCLUSIONS

Biological degradation is a very effective treatment technology for remediating petroleum contamination in soils. Significant improvements in treatment capacity can be achieved with specialized aeration equipment. Additional research is presently underway to improve degradation rates.

Application of this technology requires careful evaluation of the analytical protocol used to determine cleanup criteria and careful selection of the standard to be employed. The IR method has been shown to be susceptible to interference and is not recommended.

The information gained by process monitoring with a gas chromatographic technique is considerably more useful than simple IR quantification of Total Petroleum Hydrocarbons. In addition to being relatively resistant to interferences, the technique enables the unique characteristics of a given remediation site to be considered in executing remediation efforts.

REFERENCES

1. Block, R. N., and T. P. Clark. "Bioremediation at a Petroleum Products Terminal," presented at the Third annual Conference on Underground Storage Tank Management and Hydrocarbon Contamination Cleanup, Sturbridge, MA, 1989.
2. Block, R. N., T. P. Clark, and M. Bishop. "Biological Remediation of Petroleum Hydrocarbons," presented at the Sixth National RCRA/Superfund Conference on Hazardous Wastes and Hazardous Materials, New Orleans, LA, 1989.
3. American Petroleum Institute. "The Land Treatability of Appendix VIII Constituents Present in Petroleum Refinery Wastes: Laboratory and Modeling Studies," Washington, DC, 1987.
4. Loehr, R. C., J. H. Martin, E. F. Neuhauser, R. A. Norton, and M. R. Malecki. "Land Treatment of an Oily Waste-Degradation Immobilization, and Bioaccumulation," EPA/600/2-85/009, 1985.

CHAPTER 14

In Situ Biological Remediation of Petroleum Hydrocarbons in Unsaturated Soils

Dennis Dineen, Jill P. Slater, Patrick Hicks, and **James Holland,** McLaren Environmental Engineering, Santa Ana, California
L. Denise Clendening, Chevron Oil Field Research Company, La Habra, California

In situ biological remediation of unsaturated soils is a treatment technology that utilizes naturally occurring soil microorganisms to degrade petroleum hydrocarbons to carbon dioxide, water, and humus. Indigenous microorganisms present in the soil are stimulated by providing those elements, usually oxygen and nitrogen, that are limiting the degradation of the petroleum hydrocarbons. Because in situ remediation does not involve excavation, the costs and disruption of excavating soil are eliminated.

Petroleum hydrocarbons with high vapor pressures can be removed from soils very efficiently using in situ vapor extraction. These lighter weight hydrocarbons, which include gasoline, and aromatic additives such as benzene, toluene, and xylene, are extracted as a vacuum is applied to dry wells spaced throughout the contaminated soil. However, petroleum hydrocarbons with lower vapor pressures such as diesel, fuel oil, and crude oil are not easily extractable by this technology.[1]

The in situ biological remediation described in this chapter was developed to remediate soils where excavation would be too expensive or impractical, and where the chemicals are not easily removed by vapor extraction. This in situ bioremediation differs significantly from in situ bioremediation technologies which use "infiltration galleries" to saturate the vadose zone with water containing nutrients and hydrogen peroxide, and then recirculate or discharge the treated water. This in situ bioremediation technology delivers oxygen and nitrogen to the soil in the vapor phase rather than the dissolved phase. Delivery of oxygen and nitrogen in the vapor phase has several distinct advantages over delivery in the dissolved phase:

- Vapor phase delivery maintains unsaturated conditions throughout the affected soil and minimizes the downward migration of chemicals that would occur under saturated conditions.
- Vapor phase delivery provides the soil with atmospheric levels of molecular oxygen (over 20%) compared to the relatively low concentrations of oxygen (less than 800 parts per million) provided by hydrogen peroxide.
- Vapor phase delivery utilizes anhydrous ammonia as a source of reduced nitrogen, thereby reducing the risk of migration of nitrates into the groundwater which could result from excessive application of nitrate fertilizers in water.

Adding oxygen and nitrogen to the soil in situ is accomplished by pulling air through vadose zone wells connected to aboveground pumps and blowers. Oxygen is provided in the air that is pumped from the surface, and reduced nitrogen is provided by injecting low concentrations of anhydrous ammonia into the air stream which passes through the soil.

The design of a successful in situ biological remediation depends on five subsurface parameters:

- Soil microbiology—Petroleum degrading microorganisms must be present throughout the zone where petroleum hydrocarbon concentrations exceed the cleanup standard.
- Soil chemistry—Concentrations of soil nutrients other than oxygen and nitrogen must be adequate to maintain microbiological growth, and no toxic levels of salts or heavy metals can be present.
- Soil physics—Soil air permeability must be adequate to allow movement of oxygen and nitrogen to the affected soil and movement of carbon dioxide away from the affected soil.
- Soil morphology—Soil stratification throughout the affected zone should be well understood to design an effective delivery system.
- Hydrogeology—The depth to groundwater, groundwater flow direction and gradient, the presence or absence of floating product, and petroleum hydrocarbon concentrations in the groundwater should be understood prior to implementation of the in situ bioremediation, to avoid recontamination of cleaned soil from the groundwater.

This chapter describes the results of the bench-scale studies, in field measurements, and the full-scale design of three in situ biological remediation systems in southern California.

SOIL MICROBIOLOGY

The crucial biochemical steps in the breakdown of petroleum hydrocarbons are the oxidation of the straight chained or branched alkanes and the breaking of aromatic rings by oxygenase enzymes.[2] No plants or higher animals are known to have this ability, and relatively few microorganisms possess the enzyme systems necessary to perform these crucial steps.

Surface soils with adequate carbon, oxygen, and nutrients typically contain about ten million to one billion (10^7 to 10^9) microorganisms per gram. Of these, approximately 0.1 to 1.0% are petroleum degraders (10^5 to 10^6). After exposure to petroleum hydrocarbons, the microbial ecology of the soil adjusts so that the number of petroleum degraders increases from 100 to 1,000 times higher (10^6 to 10^8).[3]

Two critical questions for designing an in situ biological remediation in soils in which petroleum hydrocarbons extend to depths up to 50 feet or greater are:

1. Are viable microorganisms present in the soil at the same depths as the petroleum hydrocarbons; and
2. Are these microorganisms capable of degrading petroleum hydrocarbons?

Sites which were candidates for in situ biological remediation were screened to determine whether viable petroleum degrading microorganisms were present throughout the zone of contamination. The following sections describe the methodologies and results of the microbiology screening at a typical site.

Methods and Materials

Samples were collected at 5-foot intervals from the center of the contaminated zone and from an uncontaminated site on the same soil series. Samples were collected in sterile 6-inch brass tubes inside a 2.5-inch diameter split spoon sampler through an 8-inch I.D. hollow stem auger. Upon retrieval of the samples at the surface, soil inside the brass tube was extruded into a sterile bag and stored on ice.

Upon delivery to the laboratory, 10-gram subsamples were extracted and mixed according to standard protocol in a 500-ml flask containing a 0.9% sterile saline solution.[4] One-milliliter aliquots were then taken from the solution and plated on complete agar. Microbial isolates from agar plates used for total cell counts were streaked for purity and further investigated by microscopic analysis for shape and motility. Isolated cultures were then plated on selective media, to allow generic determination of the isolates.

Results and Discussion

At all three sites the total cell counts ranged from 10^2 to 10^5 cells per gram wherever total petroleum hydrocarbon concentrations exceeded 1,000 ppm. Data from uncontaminated soils of the same soil series had cell counts ranging from non-detected to 10^2 cells per gram. There is an apparent correlation between the presence of a carbon source and the presence of aerobic microorganisms at all depths. One notable exception was finding that no microorganisms were present in buried drilling muds, even though the drilling muds had large hydrocarbon concentrations (130,000 ppm). Analyses of these samples for salts and heavy metals did not show any concentration above background levels. The absence of microorganisms in the drilling muds is most likely attributed to the very low oxygen tension in the drilling mud resulting from its low permeability, estimated at 10^{-9} cm/sec. Another possible explanation is the toxicity to microorganisms reported at concentrations above 5% to 10% petroleum hydrocarbons.[5] Subsequent investigation of the isolates indicated that the microorganisms present in the soil were members of the genus *Pseudomonas* and the genus *Arthrobacter*. Members of these genera are the most common petroleum degrading microorganisms.[3]

The results of the initial soil microbiological investigations indicated that indigenous petroleum degrading microorganisms were present throughout the contaminated soil. Additional investigations were then conducted to determine whether these microorganisms could be stimulated to degrade the petroleum hydrocarbons present in the soil.

SOIL PHYSICS

Stimulating indigenous microorganisms to degrade petroleum hydrocarbons depends on the ability to deliver oxygen and reduced nitrogen throughout the contaminated soil. The key parameter controlling vapor movement in the soil is permeability to air (K_{air}). Permeability to air is a function of soil texture (soil particle size distribution), soil moisture, and bulk density. Of these, data on soil texture are most readily available and provide a reasonable indicator of permeability.

Laboratory studies with soil columns have experimentally demonstrated that lighter hydrocarbons are removed rapidly using soil venting techniques. It has also been demonstrated that preferential air paths affect the removal rates.[6] Currently, there is very little predictive capability for evaluating soil venting or air injection systems to determine the field placement of vapor injection or extraction wells. Recently, a few predictive models have been developed to determine the performance of soil venting. However, these models have not been experimentally verified in the field,[7,8] and can only provide guidelines for vapor extraction system design. Therefore, pilot studies were conducted to design the injection and extraction process for the in situ bioremediation systems discussed here.

Predictions of vapor injection and extraction rates were made on the basis of empirical data and on the effective radius of influence of vapor extraction systems.[9] Three 50-foot vadose wells were constructed bisecting the site at a distance of 0, 15, and 45 feet through the contaminated soil. Samples were analyzed for texture, bulk density, moisture content, and permeability. Field tests were then conducted to document the actual permeability in situ.

Methods and Materials

Samples in 6-inch brass tubes taken immediately below the samples analyzed for microbial numbers were subjected to soil physical analyses. In situ permeability tests were conducted by injecting sulfur hexafluoride (SF_6) as a tracer gas into the central well and measuring breakthrough at the 15-foot and 30-foot wells. The tracer gas was injected in the airstream, which was designed to deliver 100 cubic feet of air per minute at a pressure of three to five pounds per square inch. Breakthrough was measured using a specialized gas chromatograph developed for analyzing SF_6 at concentrations as low as 5 parts per trillion (ppt).

Results and Discussion

Vapor phase permeabilities were calculated for two soil strata using data from soil samples which were analyzed for permeability to air in the laboratory using Hazen's equation.[10] Laboratory results predicted an air permeability of 3 cm/sec in the sand stratum and 1 cm/sec in the loamy sand stratum. These permeabilities would result in a transit time of approximately 3 minutes over 15 feet in the sand stratum and 11 minutes in the loamy sand stratum.

Actual breakthrough times of injected air in a second well 15 feet away were measured using SF_6. These data, presented in Figure 1, show breakthrough times of 10 minutes to two hours. The extended breakthrough time of two hours

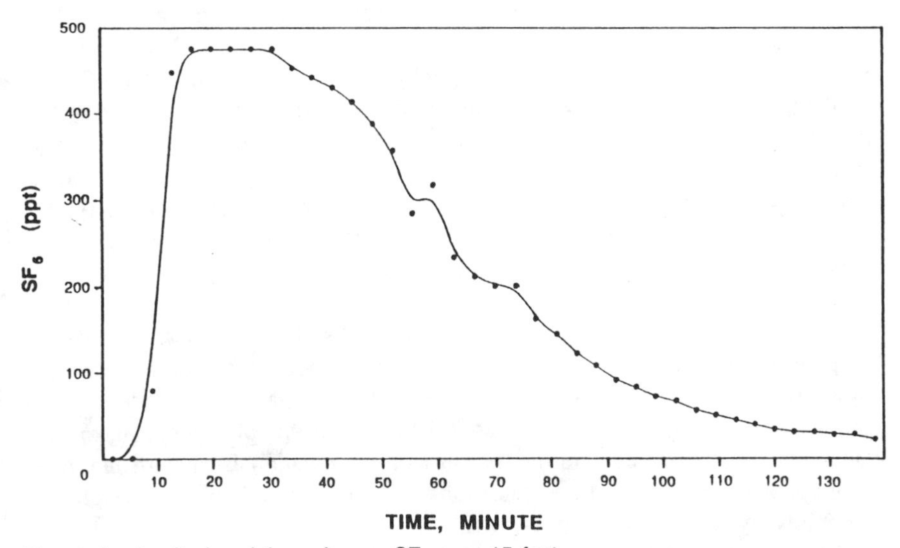

Figure 1. In situ breakthrough over SF_6 over 15 feet.

reflects the multiple pathways taken by air injected into the well. Air moving through the largest pores and/or pores with the least tortuosity travel 15 feet in 10 minutes (0.8 cm/sec), whereas the air moving through the smaller pores and/or the more tortuous pores requires over two hours to travel the same distance (0.06 cm/sec). These data are in general agreement with the empirical data from soil vapor extraction systems.

SOIL CHEMISTRY

In a soil containing petroleum hydrocarbons, biological degradation occurs naturally until the available oxygen and nutrients are consumed. In situ biological remediation stimulates the indigenous petroleum-degrading microorganisms by providing those elements which may be limiting in the soil, notably oxygen, nitrogen, and phosphorus. Once these elements are available in the soil, biological degradation can proceed.

Oxygen is required at a rate of approximately 3.1 pounds of oxygen per pound of hydrocarbon degraded. The maximum amount of oxygen in a well aerated soil is approximately 20%, or 200,000 parts per million (ppm). The maximum amount of oxygen in a saturated soil is approximately 8 ppm. If hydrogen peroxide is used to carry oxygen into the saturated soil, the levels of dissolved oxygen can be increased to 200 to 800 ppm.[11] Since oxygen is usually the most limiting element in contaminated soils, the most efficient system for delivery of oxygen to the soil is in the vapor phase.

Nitrogen is required at a rate of up to one pound of reduced nitrogen per 160 pounds of hydrocarbon degraded.[12] Nitrogen is typically added to the soil as urea or ammonium nitrate, which dissolves in the soil water as ammonium (NH_4^+) and nitrate (NO_3^-). If oxygen is supplied in the vapor phase, reduced nitrogen can also be added in the vapor phase as anhydrous ammonia gas (NH_3). When anhydrous ammonia in the soil air contacts the soil water, the ammonia is dissolved as ammonium ion (NH_4^+). Anhydrous ammonia has been used as a nitrogen source in agriculture for over 40 years. In agricultural applications, ammonia is routinely applied at a rate of 100 to 200 pounds per acre (approximately 100 to 200 ppm) by injecting anhydrous ammonia while disking the soil. Ammonia is toxic to soil microorganisms at concentrations above about 300 ppm.[13]

Phosphorus is generally considered to be the other limiting element in soil bioremediation and is routinely added to above-ground soil bioremediation projects. Phosphorus is very insoluble in most soils, and phosphorus availability decreases below pH 5.5, and above pH 7.0. In southern California, where these demonstrations were conducted, soil pH was 7.5 to 8.5. In this pH range, phosphate availability is decreased even further because of precipitation as calcium and magnesium phosphates. Under these conditions, it is difficult to increase phosphorus availability by adding phosphorus fertilizer.

Other elements are usually not limiting, and there are usually adequate levels in the soil to provide the basic requirements. Bench-scale tests conducted at these sites determined that no elements other than nitrogen and oxygen were limiting bioremediation. Changes in soil microbiology and soil nitrogen levels were measured throughout the treatment.

Methods and Materials

Samples from each of the two major strata were composited separately. Subsamples of approximately 150 grams were transferred from each stratum into 500-mL flasks. One group of 12 replicates from each substratum was a killed control. One additional group of eight replicates from each substratum was treated with 100 ppm anhydrous ammonia and constantly aerated with air, which was bubbled through water to maintain high relative humidity. A third group was similarly aerated, but without the anhydrous ammonia. Four replicates were analyzed on Day 0 for total cell counts. Four replicates from each group were harvested at two weeks and four weeks, and analyzed for total cell counts. In addition, soil was analyzed after four weeks for soil nitrogen to determine the effectiveness of anhydrous ammonia as a nitrogen source.

Results and Discussion

The results of the soil chemical and microbiological studies are shown on Figures 2 and 3. Figure 2 shows the concentrations of all forms of soil nitrogen in the treatment with anhydrous ammonia compared to the treatment with air only. These results show an increase of approximately 50% of all forms of soil nitrogen in the soils treated with anhydrous ammonia. Figure 3 shows the corresponding changes in total cell counts over a four week period. Table 1 summarizes the results of the chemical and microbiological studies.

Data from soil which was treated with 100 ppm anhydrous ammonia in the air stream showed that the use of low concentrations of anhydrous ammonia is an effective mechanism to provide reduced nitrogen to soil microorganisms. Data on changes in microbial populations with treatment show that adding oxygen increased microbial counts by a factor of 10, and that adding both oxygen and reduced nitrogen increased microbial counts by a factor of 100.

Table 1 summarizes the results of the studies to date. Initial concentrations of petroleum hydrocarbons in the soil are 2,000 ppm in the sand, and 6,000 ppm in the loamy sand. Initial cell counts in the sand were 10^5 cells/gram compared to 10^4 cells/gram in the loamy sand. This suggests that the lower petroleum hydrocarbon concentration in the sand is due to the tenfold higher number of petroleum degrading microorganisms. Adding oxygen to the system increased cell counts tenfold in the loamy sand, but not in the sand. This suggests that oxygen was not limiting in the more permeable sand, but was limiting in the deeper, less permeable loamy sand, resulting in a tenfold increase in cell counts.

Adding anhydrous ammonia to the soil increased the cell counts in both the sand and loamy sand. This suggests that nitrogen was limiting at both soil depths,

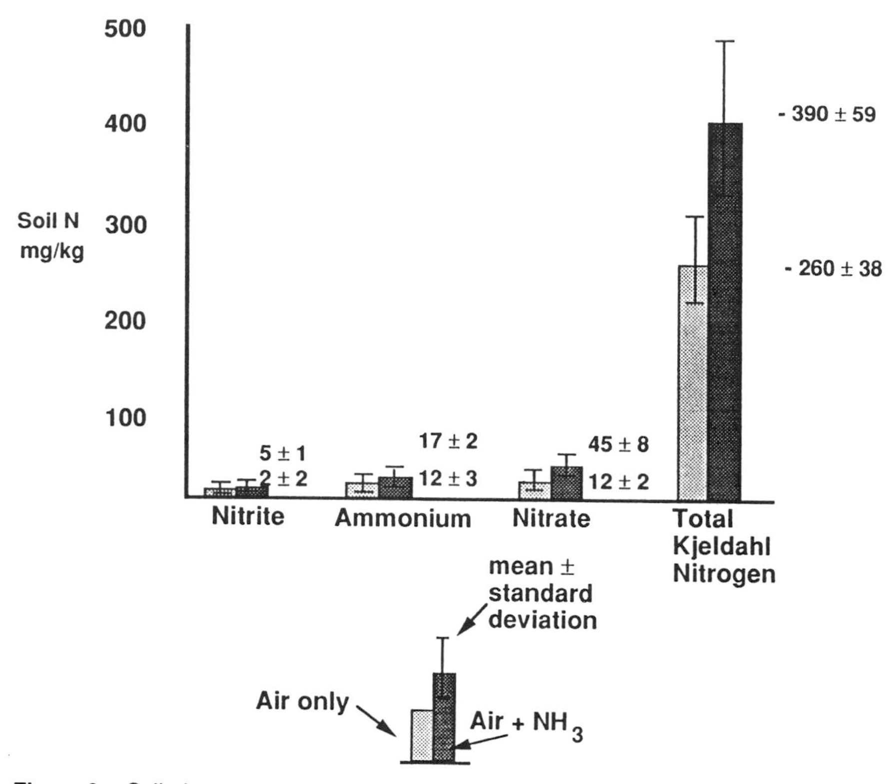

Figure 2. Soil nitrogen concentration with treatment.

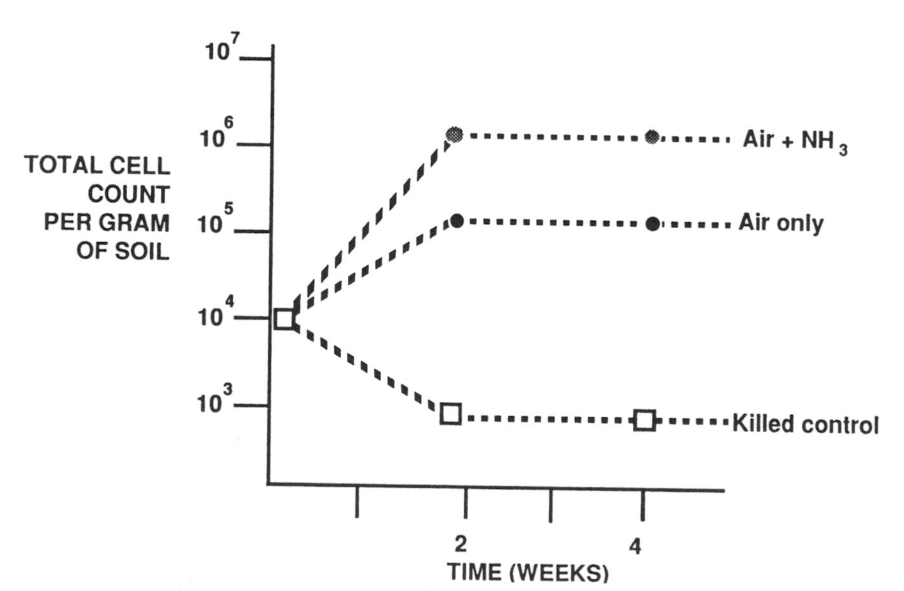

Figure 3. Change in microbial populations with treatment.

Table 1. Summary of Results of In Situ Bioremediation Microbial Studies

Depth	USDA Soil Texture	Soil TPH	Initial Microbial Counts	Microbial Counts After Treatment with Air only	Microbial Counts After Treatment with Air + NH_3
20–35 feet	Sand	2,000 ppm	10^5 cfu/gm	10^5 cfu/gm	10^6 cfu/gm
35–50 feet	Loamy sand	6,000 ppm	10^4 cfu/gm	10^5 cfu/gm	10^6 cfu/gm

cfu/gm = cell forming units/gram soil.

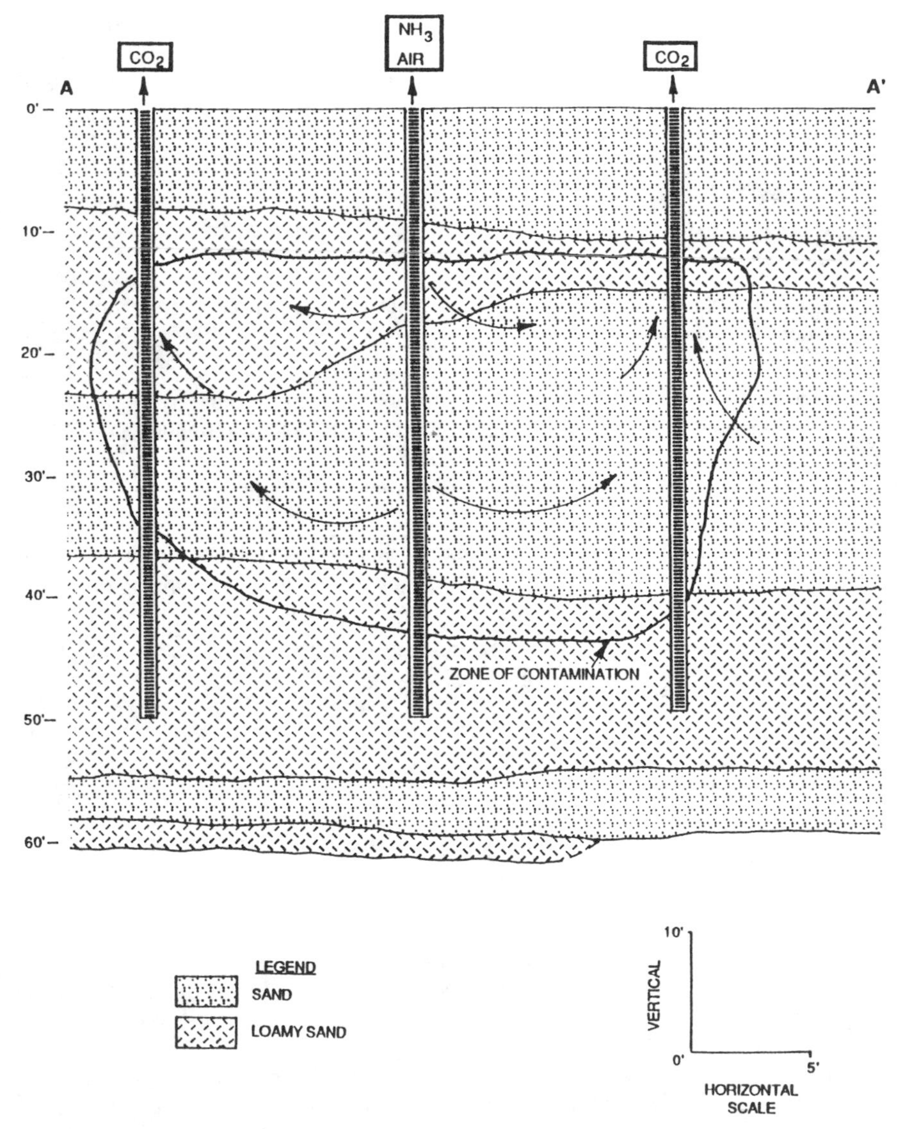

Figure 4. Design of in situ bioremediation using vapor phase application of oxygen and nitrogen.

and that addition of anhydrous ammonia eliminated the nitrogen deficiency, resulting in a tenfold increase in cell counts compared to the well aerated soil.

FULL-SCALE DESIGN AND IMPLEMENTATION

Based on the data from these studies and in situ field measurements, a full-scale bioremediation is being implemented. Air is injected into the soil at a pressure of approximately three pounds per square foot, and anhydrous ammonia is added weekly at 100 parts per million. Additional vapor wells are being installed to cover the entire contaminated area, and to monitor the movement of injected air in the soil.

A cross section of the contaminated soil and schematic of the treatment system are shown in Figure 4.

SUMMARY AND CONCLUSIONS

In situ bioremediation of unsaturated soils involves three very simple, well documented technologies:

- the ability of indigenous microorganisms to degrade petroleum hydrocarbons;
- the use of aboveground pumps and blowers to move vapors through the unsaturated soil; and
- the use of anhydrous ammonia as a source of reduced nitrogen.

Results presented here document that adding anhydrous ammonia to an air stream through the unsaturated soil increases soil oxygen and nitrogen levels, resulting in a hundredfold increase in microbial count. Maintaining viable cell counts at the level of 10^6 to 10^7 is expected to result in a decrease of petroleum hydrocarbons in situ to a cleanup level of 100 ppm.

REFERENCES

1. Hinchee, R. E., D. C. Downey, and E. J. Coleman. "Enhanced Bioreclamation, Soil Venting and Ground-Water Extraction: A Cost-Effectiveness and Feasibility Comparison," *Proceedings of the Conference on Petroleum Hydrocarbons and Organic Chemicals in Ground Water: Prevention, Detection, and Restoration*. National Water Well Association/American Petroleum Institute, 1987.
2. Singer, M. E., and W. R. Finnerty. "Microbial Metabolism of Straight-Chain and Branched Alkanes," in R. M. Atlas, ed., *Petroleum Microbiology* (Macmillan Publishing Co., Inc., 1984).
3. Bossert, I., and R. Bartha. "The Fate of Petroleum in Soil Ecosystems," in R. M. Atlas, ed., *Petroleum Microbiology* (New York: Macmillan Publishing Co., Inc., 1984).

4. Wollum, A. G. "Cultural Methods for Soil Microorganisms," in *Methods of Soil Analysis*. Agronomy Monographs. No. 9, Part 2, 1982.
5. Dibble, J. T., and R. Bartha. "Effect of Environmental Parameters on the Biodegradation of Oil Sludge," *Appl. Environ. Microbiol.* 37:729–739 (1979).
6. Rainwater, K., B. J. Claborn, H. W. Parker, D. Wilkerson, and M. R. Zaman. "Large Scale Laboratory Experiments for Forced Air Volatilization of Hydrocarbon Liquids in Soil," *Proceedings of Petroleum Hydrocarbon and Organic Chemicals in Groundwater: Prevention, Detection, and Restoration*. National Water Well Association/American Petroleum Institute, 1988.
7. Baehr, A. L., and C. E. Hoag. "A Modeling and Experimental Investigation of Induced Venting," *Proceedings of Petroleum Hydrocarbon and Organic Chemicals in Groundwater: Prevention, Detection and Restoration*. National Water Well Association/American Petroleum Institute, 1988.
8. Johnson, P. C., M. W. Kemblowski, and J. D. Colthart. "Practical Screening Models for Soil Venting Applications," *Proceedings of Petroleum Hydrocarbon and Organic Chemicals in Groundwater: Prevention, Detection, and Restoration*. National Water Well Association/American Petroleum Institute, 1988.
9. Krishnayya, A. V., M. J. O'Connor, J. G. Agar, and R. D. King. "Vapour Extraction Systems: Factors Affecting Their Design and Performance," *Proceedings of Petroleum Hydrocarbon and Organic Chemicals in Groundwater: Prevention, Detection and Restoration.* National Water Well Association/American Petroleum Institute, 1988.
10. Burmister, D. M. "The Importance and Practical Use of Relative Density in Soil Mechanics," in *ASTM,* Vol. 48, Philadelphia, PA, 1948.
11. Ward, C. H., J. M. Thomas, S. Fiorenza, H. S. Rifai, P. B. Bedient, J. T. Wilson, and R. L. Raymond. "In Situ Bioremediation of Subsurface Material and Groundwater Contaminated with Aviation Fuel: Traverse City, Michigan," in *Hazardous Waste Treatment; Biosystems for Pollution Control.* Air and Waste Management Association/Environmental Protection Agency Conference, Pittsburgh, PA, 1989.
12. "Manual on Disposal of Refinery Wastes." American Petroleum Institute, Washington, DC, 1980.
13. Tisdale, S. L., and W. L. Nelson. *Soil Fertility and Fertilizers.* (New York: Macmillan Publishing Co., Inc., 1975).

CHAPTER 15

Integrated Zero-Emission Groundwater and Soil Remediation Facility at Lockheed, Burbank

Ron Derammelaere, AWD Technologies, South San Francisco, California
Ron Helgerson, Lockheed Aeronautical Systems Company, Burbank, California

The Lockheed Aeronautical Systems Company (LASC) has over 200 acres of aircraft manufacturing facilities located in Burbank, California. Among the famous aircraft that have been assembled at this facility are the P-38 Lightning, the F-104 Starfighter, the U-2, and the L-1011.

In late 1987, solvent-contaminated soil and groundwater were identified near Building 175. As a result, the Los Angeles Regional Water Quality Control Board (RWQCB) issued a Cleanup and Abatement order requiring soil and groundwater remediation to commence by August 1, 1988, and October 15, 1988, respectively.

LASC selected AWD Technologies, Inc. (AWD) to design, install, and operate a treatment facility to meet the requirements of the RWQCB. AWD is a corporation wholly owned by The Dow Chemical Company. AWD provides a wide range of services for remediation of contaminated soil and groundwater.

THE TECHNOLOGIES

Two technologies were integrated in an innovative way: AquaDetox, a low-pressure steam stripping technology developed by Dow Chemical to extract volatile organic compounds (VOCs) from the groundwater, and Soil Vapor Extraction

(SVE) treatment of the VOCs in the vadose zone. The following paragraphs describe the unique features of these technologies. Their integration will be described in a subsequent section.

AquaDetox

Over the past several years, an effort has been under way to improve the efficiency of air stripping in removing contaminants from groundwater. This work has led to the development of the AquaDetox technology, which surpasses more conventional approaches to air stripping in terms of reduction efficiency. In most cases, AquaDetox can reduce contaminants in groundwater to below Maximum Contaminant Levels (MCLs) without liquid-phase carbon bed treatment. Moderate vacuum and deep vacuum AquaDetox steam stripping go even further, allowing the near total recovery of contaminants for possible recycling.

AquaDetox technology can be used to remove a wide variety of volatile compounds, and many compounds that are normally considered "nonstrippable" (i.e., those with boiling points in excess of 200°C). The application of AquaDetox for the removal of compounds with boiling points greater than 200°C and the use of vacuum are patented by The Dow Chemical Company.

Stripping is commonly defined as a process to remove dissolved, volatile compounds from water. A carrier gas, such as air or steam, is purged through the contaminated water, with the volatile components being transferred from the water into the gas phase. While the physical principles involved are straightforward, the practice of stripping has undergone considerable development since the early 1970s.

Dow's effort has focused on:

1. Development of the proper theoretical relationships that provide a clear understanding of the stripping process.
2. Application of these relationships, along with the correct hardware, to attain higher levels of contaminant removal than previously possible.
3. Development of the proper scale-up parameters to go from pilot units handling <1 gpm to production units handling over 3000 gpm.
4. Development of the conditions under which compounds with very high boiling points (e.g., 200°C) can be stripped from water.
5. Compilation of a vapor-liquid equilibrium database with special emphasis on Environmental Protection Agency (EPA) priority pollutants.

The effort necessary to address these criteria has been carried out by the Separations Section of the Applied Science and Technology Department of Dow. The research and development has been under the direction of Dr. Lanny Robbins.

By the early 1980s, the result of this effort was the AquaDetox process, an innovative technology for the high efficiency stripping of organic contaminants from water.

AquaDetox is capable of effectively stripping over 90 of the 110 volatile compounds listed in CFR 40, July 1, 1986, by the EPA (see Table 1). The ability of

Table 1. Strippable EPA-Designated Priority Pollutants

Volatiles

acrolein
acrylonitrile
benzene
bromoform
carbon tetrachloride
chlorobenzene
chlorodibromomethane
chloroethane
2-chloroethylvinyl ether
chloroform
dichlorobromomethane
1,1-dichloroethane
1,2-dichloroethane
1,1-dichloroethylene
1,2-dichloropropane
1,3-dichloropropylene
ethylbenzene
methyl bromide
methyl chloride
methylene chloride
1,1,2,2-tetrachloroethane
tetrachloroethylene
toluene
1,2-trans-dichloroethylene
1,1,1-trichloroethane
1,1,2-trichloroethane
trichloroethylene
vinyl chloride

Acid Compounds

2-chlorophenol
2,4-dichlorophenol
2,4-dimethylphenol
p-chloro-m-cresol[a]
pentachlorophenol
2,4,6-trichlorophenol

Base/Neutral

acenaphthene
acenaphthylene
anthracene
benzidine
benzo(a)anthracene
benzo(a)pyrene
3,4-benzofluoranthene
benzo(ghi)perylene
benzo(k)fluoranthene
bis (2-chloroethoxy) methane
bis (2-chloroethyl) ether
bis (2-chloroisopropyl) ether
bis (2-ethylhexyl) phthalate
4-bromophenyl phenyl ether[a]
butylbenzyl phthalate
2-chloronaphthalene
4-chlorophenyl phenyl ether
chrysene
1,2-dichlorobenzene
1,3-dichlorobenzene
1,4-dichlorobenzene
3,3′-dichlorobenzidine[a]
di-n-butyl phthalate
2,4-dinitrotoluene
2,6-dinitrotoluene
di-n-octyl phthalate
1,2-diphenylhydrazine (as azobenzene)[a]
fluoranthene
fluorene
hexachlorobenzene
hexachlorobutadiene
hexachlorocyclopentadiene
hexachloroethane[a]
indeno(1,2,3-cd)pyrene[a]
isophorone
naphthalene
nitrobenzene
N-nitrosodimethylamine[a]
N-nitrosodi-n-propylamine[a]
N-nitrosodiphenylamine[a]
phenanthrene
pyrene
1,2,4-trichlorobenzene

Pesticides

aldrin
alpha-BHC[a]
beta-BHC[a]
delta-BHC[a]
chlordane
4,4′-DDT
4,4′-DDE
4,4′-DDD
dieldrin
alpha-endosulfan[a]
beta-endosulfan[a]
endosulfan sulfate[a]
endrin aldehyde[a]
heptachlor
heptachlor epoxide
PCB-1242[a]
PCB-1254[a]
PCB-1221[a]
PCB-1232[a]
PCB-1248[a]
PCB-1260[a]
PCB-1016[a]
toxaphene

[a]Needs further pilot study to determine treatability.

AquaDetox to efficiently attain low levels of contamination in the effluent represents a major breakthrough. Conventional strippers will normally achieve only 95% to 98% removal of the contamination, whereas AquaDetox can achieve up to 99.99%.

Another major concern raised regarding conventional stripping systems is that they simply transfer contaminants from the water to the air. The contaminated air is usually treated over carbon beds, but still releases sometimes significant amounts of contaminants into the atmosphere. The AquaDetox steam stripper (moderate or deep vacuum) condenses the contaminated steam to form a multiphase liquid from which the liquid phase contaminant can be decanted for possible recycling. Only a small stream of noncondensable gases is emitted following carbon treatment.

There are three versions of the basic AquaDetox technology:

1. Air Stripping AquaDetox
2. Moderate Vacuum AquaDetox (requires source of steam)
3. Deep Vacuum AquaDetox (does not require source of steam).

Typical schematic flow diagrams for each of the types of AquaDetox technology are included in a paper by Street, Robbins, and Clark.[1]

Soil Vapor Extraction

Soil vapor extraction (SVE) is a technology commonly applied for the in situ removal of VOCs from soil. A vacuum is applied to vadose zone extraction wells to induce air flows within the soil toward the wells. The air acts as a stripping medium which volatilizes the VOCs in the soil. Soil-gas from the extraction wells is typically treated in carbon beds before release to the atmosphere. Alternatively, the treated soil-gas is reinjected in the soil to control the direction of air flow in the soil.

THE PROJECT

On February 1, 1988, LASC awarded AWD a contract for pilot testing, design, and installation of an integrated 1200 gpm groundwater treatment plant and 300 SCFM SVE system. Fast-track project techniques were used, and seven and a half months later all systems of the $4 million project were operational.

Under AWD management, process engineering and design were performed by Dow Chemical engineers, the SVE conceptual design and permit acquisition by Woodward-Clyde Consultants, and construction by a division of Guy F. Atkinson.

Integrated System

The integrated system consists of two basic processes: an AquaDetox vacuum stripping tower using low pressure steam, and a soil-gas vapor extraction/reinjection process. The system removes VOCs from the groundwater and soil with no gaseous emissions to the atmosphere. Figure 1 shows a schematic flow diagram of the integrated system.

Integrating the two technologies creates a unique system. While the AquaDetox system extracts and treats contaminated groundwater, an array of SVE wells removes contaminated soil-gas from the vadose zone. The soil-gas is treated by the carbon beds and reinjected into the ground to sweep through the soil and remove additional contamination.

The AquaDetox and SVE systems share a 3-bed granulated activated carbon (GAC) unit. When one of the GAC beds is regenerated, the steam and organic vapors are condensed in the secondary condenser of the AquaDetox system. Condensed organics are pumped to a storage tank for recycle, water condensate is pumped to the recycle tank for further treatment by the AquaDetox process, and noncondensables are transferred to the active GAC bed.

Groundwater Treatment Facility

The groundwater treatment technology at the Lockheed site is the Moderate Vacuum Steam Stripper (MVSS) AquaDetox system. Process flow diagrams are shown in Figures 2 and 3.

Contaminated groundwater is fed from extraction wells to a cross exchanger, where it is heated by the treated water. The heated water then enters the top of the stripping column (9′ diameter × 60′ tall) and flows down the column, contacting the rising vapor flow generated by the introduction of steam to the bottom of the column. Under a pressure of 100 mmHg abs., the contaminants are stripped from the liquid into the vapor stream, which exits from the top of the column. The treated water leaves the bottom of the column. The treated water passes through the heat exchanger, where it is cooled and the contaminated feedwater is heated. The water exiting the treatment facility is thereby controlled to 9 to 10°F higher than the incoming groundwater.

The overhead vapors flow to a water-cooled condenser, where the water vapor is condensed and recycled back to the contaminated feedwater. The water for cooling the condenser is provided by diverting a portion of the cool feed stream through the condenser and back to the main feed stream.

Total condensation of the overhead vapors is not possible, due to noncondensable gases from "vacuum leaks" and dissolved gas contained in the contaminated groundwater. These noncondensable vapors, carrying some water, inert

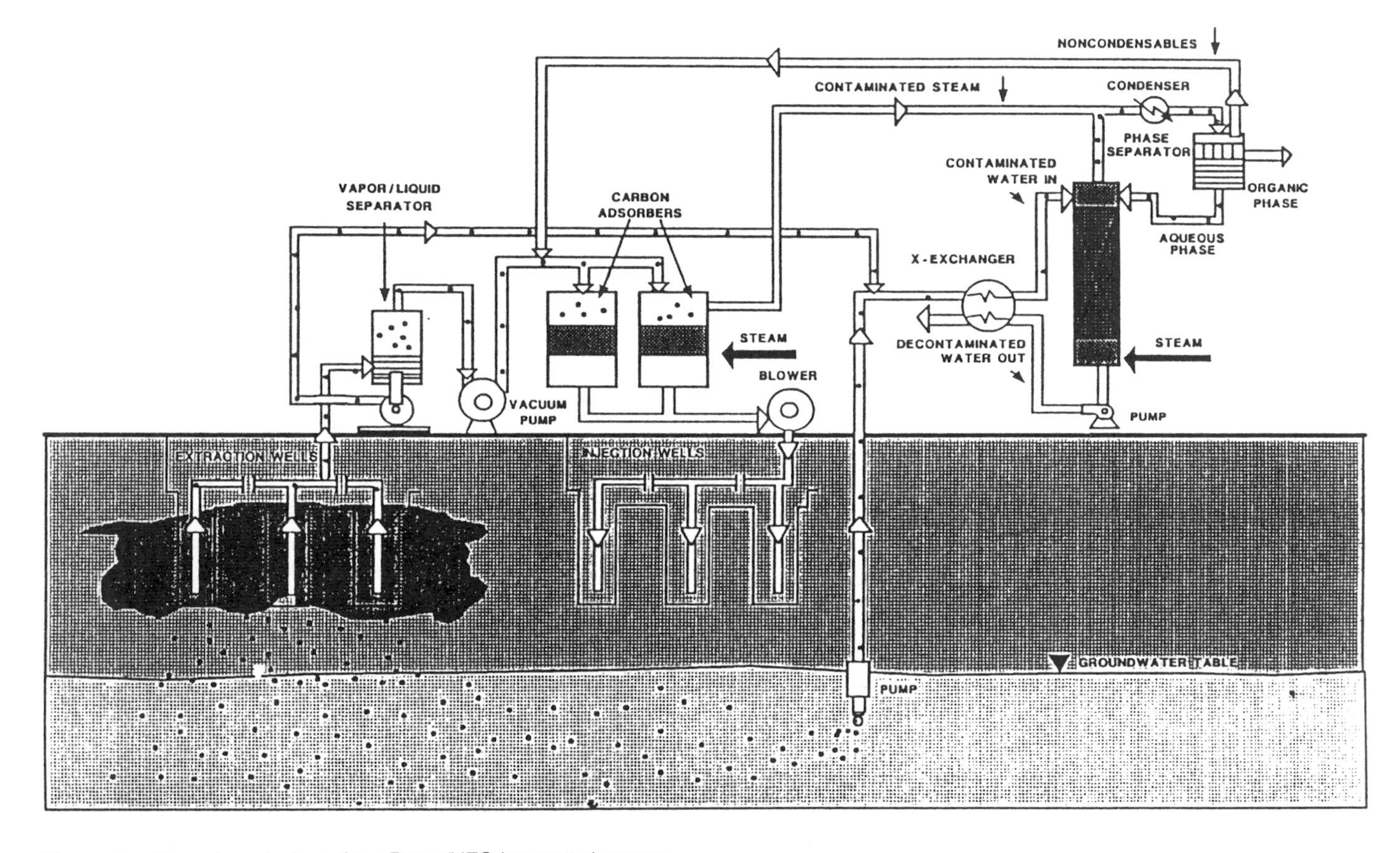

Figure 1. Zero air emissions AquaDetox/VES integrated system.

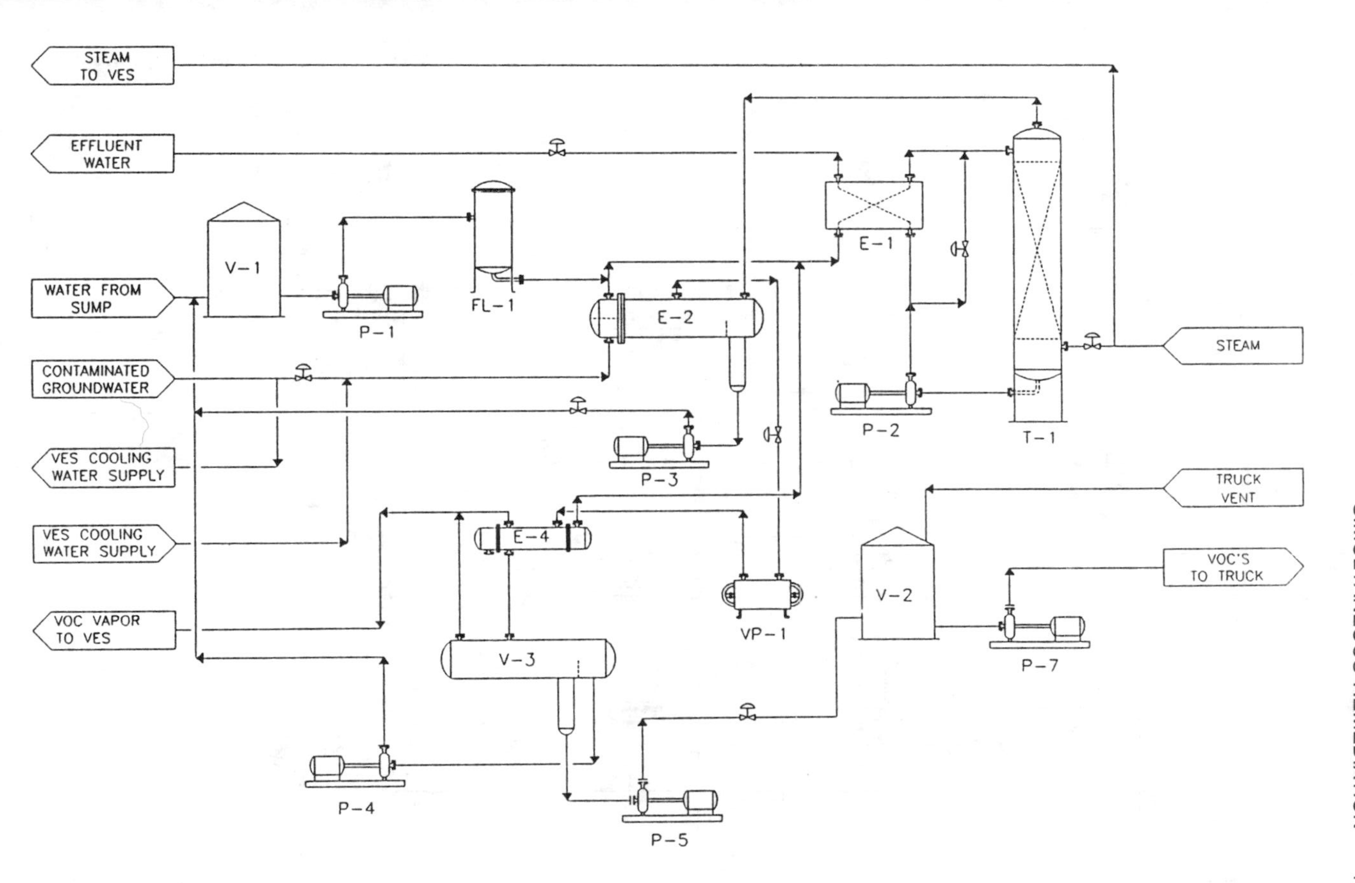

Figure 2. Lockheed AquaDetox flow diagram.

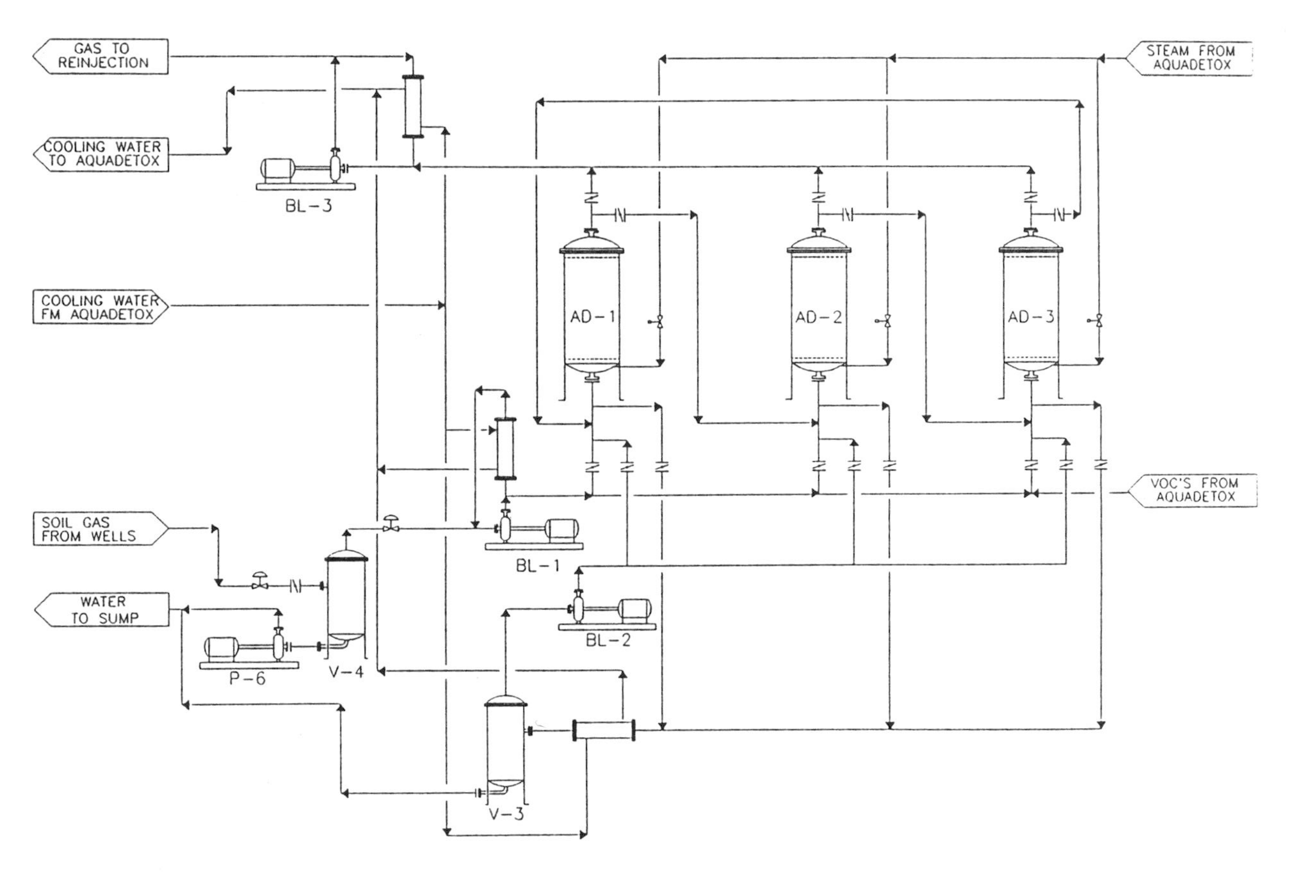

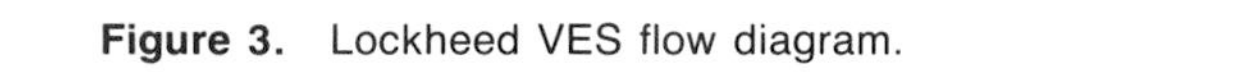
Figure 3. Lockheed VES flow diagram.

gases, and VOCs, enter a vacuum pump where they are compressed to atmospheric pressure. Cooling of this compressed vapor stream results in condensation of water and VOCs.

The water phase is recycled to the contaminated feedwater and the organic solvent phase is withdrawn for reclamation by a contract recycler. The coolant for this secondary condenser is supplied from the feedwater, as is done for the first condensing unit.

The vent stream from the secondary condenser contains the noncondensables and an equilibrium quantity of VOCs. This stream is passed through vapor-phase GAC prior to discharge into the reinjection wells of the SVE system.

Soil Vapor Extraction (SVE) System

Soil vapor extraction (SVE) is being used at the Lockheed site for remediation of contaminated soil because of the relatively volatile character of the reported contaminants, depth to groundwater in the range of approximately 140 to 150 feet, and the predominantly coarse-grained nature of subsurface soils.

Figure 4 shows the locations of the extraction and reinjection wells. The design of the SVE system focused on the distribution of the wells to produce an effective and nondisruptive pneumatic flow regime. "Effectiveness" of SVE was judged to depend on establishing radially inward flow (toward an extraction well) throughout the areas of probable soil contamination; "nondisruptive pneumatic flow regime" refers to injection well placement such that (1) fugitive atmospheric emissions are not created, and (2) soil-gas within the areas of probable soil contamination is not displaced from the zone of extraction well influence.

Extraction wells connected to a common header feed up to 300 CFM of contaminated soil-gas to the system for processing and decontamination via carbon adsorption. Liquids collected in the SVE scrubber sump are pumped to the water recycle tank for processing through the AquaDetox tower. Vapors are exhausted to the GAC beds for hydrocarbon removal prior to reinjection.

Three GAC beds remove chlorinated hydrocarbons from SVE system extraction well soil-gas, along with vent gases from the AquaDetox system. The GAC beds are operated alternately, with two beds on-line in series while the remaining unit is being regenerated. Once each 8 hours, the regenerated off-line bed is placed in service and the spent carbon bed is removed from service and regenerated. Steam is used to strip chlorinated hydrocarbons from the GAC units. The vapors from this regeneration process are condensed and processed in the AquaDetox separator.

Treated soil-gas is reinjected into the ground at depths ranging from 50 to 150 feet through the vadose zone. The soil-gas then sweeps horizontally through the contaminated soil, picking up additional hydrocarbons, and is once again collected in the soil-gas extraction well system, where hydrocarbons are again removed.

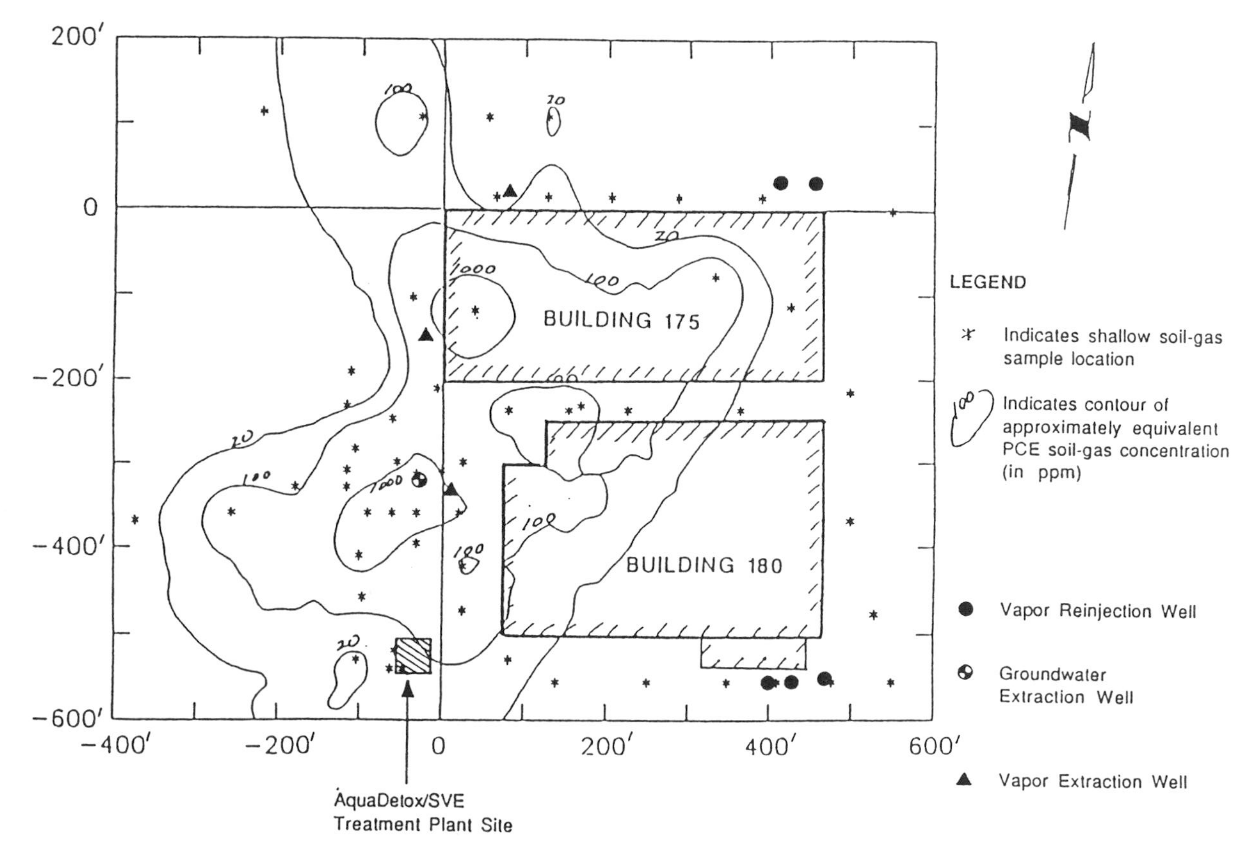

Figure 4. PCE and soil-gas concentrations.

SYSTEM OPERATION

The groundwater treatment plant operates at an average flow rate of 1000 gpm, and the SVE at 170 SCFM. The contaminants treated are listed in Table 2. Initially, total VOC concentrations were 12,000 ppb in the groundwater and 6,000 ppm in the soil-gas. After the integrated system had been operating several months, these concentrations dropped to 5,000 ppb and 450 ppm, respectively. At these levels, the AquaDetox/SVE facility removes more than 70 pounds per day of PCE/TCE from the groundwater and 40 pounds per day from the soil-gas.

Table 2. Integrated System at Lockheed-Burbank Design Criteria and Performance Results

AQUADETOX

Design Contaminants	Design Feed Water Concentration (ppb)	Actual (11/88) Influent Concentration (ppb)	Design Effluent Concentration (ppb)	Actual Effluent Concentration (ppb)
Trichloroethylene	3300.0	2200	4.5	<1
Toluene	180.0	<100	9.5	<1
Tetrachloroethylene	7650.0	11000	3.5	<1
Trans-1,2,dichloroethylene	19.5	<100	15.0	<1
Chloroform	30.0	<100	N/A	<1
1,1-dichloroethane	18.0	<100	5.5	<1
1,2-dichloroethane	4.5	<100	0.8	<1
Carbon tetrachloride	7.5	<100	N/A	<1
Benzene	30.0	<100	0.65	<1
1,1,2-trichloroethane	34.5	<100	N/A	<1
Ethylbenzene	255.0	<100	N/A	<1

SVE

Contaminants	Actual (9/88) Extraction Gas (ppb)	Actual (9/88) Reinjection Gas (ppb)
Total Hydrocarbons	450,000	2,000
Tetrachloroethylene	420,000	365
Trichloroethylene	8,000	60

Table 2 lists the major contaminants in the groundwater feed to the treatment plant. Effluent analyses show that all contaminants have been reduced to below the analytical detection level (1 ppb for most contaminants). This equates to a removal efficiency in excess of 99.99%. The soil-gas treatment by two of three 3,500-pound carbon beds removes VOCs to below 2 ppm before the air is reinjected in the ground. This equates to a removal efficiency of better than 99%.

While the treatment plan has operated consistently at average design flow rates (95% availability factor) and has produced water effluents at nondetectable VOC concentrations, it has not been devoid of typical start-up problems and one operational problem. The start-up problems were typically failures of instrumentation and control software bugs, which have since been resolved. A more persistent problem, however, has been caused by the high alkalinity of the groundwater and resulting calcium carbonate scaling in parts of the treatment plant.

Solubility of the calcium carbonate in the groundwater is reduced in two ways as the water is processed through the AquaDetox system. First, the water is heated and, second, carbon dioxide is removed during the stripping process in the column, thereby increasing the pH. The principal disadvantage of scaling is the reduction in heat transfer efficiency of the cross exchanger, resulting in greater steam consumption. Currently, an antiscalant is injected in the feed water, but cannot totally halt the scaling due to the subsequent removal of carbon dioxide and concomitant pH increase. Periodically the heat exchanger is acidized to maintain its heat transfer properties.

Design is under way to resolve the scaling problem. A sulfuric acid injection system will be installed to control pH and eliminate scaling. The costs associated with the addition of sulfuric acid will be more than offset by: (a) the savings in eliminating antiscalant injection; (b) savings by eliminating phosphoric acid used to clean the heat exchanger periodically; and (c) lowering average steam consumption by improving heat exchanger efficiency. Less than 20% of the steam consumption in the AquaDetox facility is needed to strip contaminants; the other 80% is needed in raising the incoming water to its boiling point of 120°F at 100 mmHg. The cross exchanger helps reduce this steam requirement by using heat from the effluent water. This is a highly energy-efficient and cost-effective approach, and future systems will have even larger cross exchangers.

OPERATING COSTS

Annual operating costs for the AquaDetox/SVE plant are (as shown in Table 3):

Table 3. Aquadetox/SVE Operating Costs, Lockheed-Burbank

		Annual $	Cost Per 1000 Gallons[a]
Labor:	$8,000/month	$ 96,000	0.18
Steam:	3,840 lb/h x 8,760 h x 0.92 x $5.70/1000 lb	176,400	0.34
Chemicals:	$1,800/month	21,600	0.04
Power:	88 kW x 8,760 h x 0.92 x 0.07 $/kWh	50,400	0.10
Supplies:	$5,000/month	60,000	0.11
		$404,400	0.77

[a]Total gallons per year = 1,000 gpm x 60 min/h x 8,760 h/yr = 525 x 10^6 gal/yr.

Labor: One individual was initially assigned full-time for the maintenance and operation of the facility, but after the first six months of operation his time was reduced to three days per week. It is expected that after another year of operation, no more than one day per week will be needed. Current labor costs are about $8,000 per month.

Steam: Steam, which is provided by an existing Lockheed boiler, costs $5.70 per 1,000 pounds. At a 1000 gpm flow rate, the steam consumption is 3,500 lb/hr before calcium carbonate scaling shows its effect on the cross exchanger efficiency. An additional 340 lb/hr of steam (equivalent continuous average) is used to regenerate the carbon beds. This results in a total monthly steam consumption of 3,840 lb/hr at a cost of $14,700.

Chemicals: Significant amounts of antiscale and scale-removing chemicals are currently being consumed, with limited success. A more appropriate solution of sulfuric acid injection is currently being implemented, and will result in a monthly cost of $1,800.

Power: The power requirement to operate the treatment plant is 88 kW. At a cost of $0.07 per kWh, this represents a monthly cost of $4,200. This does not include the power consumption for the groundwater extraction well.

Supplies: Miscellaneous supplies such as oil, replacement gauges, pump seals, spare parts, etc. cost about $5,000 per month.

Based on the above breakdown, monthly operating costs average $33,700, or $.77 per 1000 gallons. Further reductions (particularly labor) are anticipated with time, and savings in steam costs can further be accomplished by installing larger cross exchangers.

REFERENCES

1. Street, G., L. Robbins, and J. Clark. ''AquaDetox Stripping System for Groundwater Remediation,'' presented at HazMat Central 1989, Chicago, IL.

CHAPTER 16

Study of Possible Reuse of Stabilized Petroleum Contaminated Soils as Construction Material

Sibel Pamukcu, Hazem M. Hijazi, and **H. Y. Fang,** Department of Civil Engineering, Lehigh University, Bethlehem, Pennsylvania

In 1984 it was estimated that several hundred thousand underground storage tanks, used for storage of petroleum products, are leaking.[1] Amendments to CERCLA (Comprehensive Environmental Response, Compensation and Liability Act) in 1986 show increased recognition of the problem of leaking underground storage tanks (UST). Out of over 3,000,000 USTs in use throughout the country, as many as 500,000 may be leaking petroleum liquids in the ground.

Most petroleum hydrocarbons are considered immiscible with water; therefore, they are primarily transported in the unsaturated or vadose zone in the soil. However, gasoline-range hydrocarbons contain significant quantities of certain compounds which are partially soluble in water. Some of these compounds are carcinogenic and/or EPA-listed hazardous waste components (e.g., benzene, toluene, and xylenes). The presence of such compounds in the subsurface environment poses a significant health hazard to the public and environment.

Once a spill or a leakage occurs, the hydrocarbon liquid, under gravity, will move downward to the groundwater table, partially saturating the soil in its path. The degree of saturation will depend upon the contact angle between the liquid and the soil particles, as well as the physical and chemical properties of the soil itself. Upon reaching the groundwater table, the liquid will spread horizontally by migration within the capillary zone. At this stage there are three major tasks that need to be performed for remediation and reclamation of the contaminated area. The first is the control of horizontal migration of the contaminant away from the spill or leak source; the second is the cleanup of the groundwater; and the third is the cleanup of the contaminated zone of soil. In general, cleanup of the groundwater contaminated with gasoline-range hydrocarbons consists of pumping water from a well and removal of the floating material. Cleanup of hydrocarbon contaminated soils is usually more complicated. Partially hydrocarbon saturated soil can be a persistent source of contamination of groundwater for decades, as water percolation from the surface or groundwater table fluctuation promotes solution of the more soluble compounds. Therefore, cleanup or removal of the contaminated soil is a significant part of the overall remediation and reclamation action.

A petroleum hydrocarbon either degrades or remains unaltered in soil. Degradation comes about by microbial metabolism in which the hydrocarbon may be oxidized to carbon dioxide and water. Hydrocarbons are also often attenuated on clay surfaces by adsorption. For example, when an advancing front is retained by clay lenses, often only a minor product layer forms over the groundwater table. If the soil is contaminated with gasoline-range hydrocarbon liquids, made up of a mixture of volatile hydrocarbons, the liquid state of the hydrocarbon remains in equilibrium with its vapor state. If the soil can be ventilated, more of the liquid state would pass into vapor state and theoretically the soil can be eventually decontaminated. However, the permeability of the soil and the presence of water are two major limiting factors in accomplishing this type of remediation.

An alternate and economical method of cleanup is physical removal of the contaminated soil from the vadose zone. This is most viable if the contamination is close to the ground surface and thus, would not require very large quantities of soil to be excavated. The technique may also be advantageous when remediation is not feasible because of tenacious retention of the contaminant on the soil, and when other physical and chemical parameters of the soil and the contaminant limit effective removal. Once the contaminated soil is removed, it has to be disposed of safely as a waste material. This often results in creation of new landfills which ultimately may not serve the purpose of land reclamation and rehabilitation.

The study presented here has been intended to deal with this aspect of the remediation; namely, stabilizing the excavated contaminated soil to render it a useful material and thus provide an economical and beneficial solution to the problem of cleanup of petroleum contaminated soils.

The benefits of the approach are expected to be twofold. First, there is a critical need to reclaim and rehabilitate land in parts of the country where population density and value of land is high. Creating more landfills and waste containment sites (a) uses up available land and (b) threatens the fresh water supply in such areas. The latter will also reduce the utility of land, even those far from the contamination site. Therefore re-use of waste in an environmentally safe and technically sound way is the most attractive solution to waste management problems. The second benefit is that there is great interest in industry to find reliable and economical substitutes for conventional virgin materials to counteract the increasing cost of obtaining or producing these. Therefore, effective utilization of a waste material, which is usually abundant and cheaper, can present a viable solution provided that: (1) there are no adverse environmental effects, and (2) the material performs as well as the one it replaces.

In this study, a form of a stabilization/solidification method was applied to petroleum contaminated soil to bind the hydrocarbons in a structure formed by the cementing and conditioning action of pozzolanic and earthen materials to produce chemically and physically stable and mechanically handleable new products. These products, if stable and durable, can be studied for determination of possible use in construction of large-scale earthen structures such as sub-bases, embankments, barrier systems, or as filler material in other types of construction units. The parameters that are involved in such a determination are strength, permeability, compressibility, plasticity, and durability of the product. In this study, only the strength and plasticity parameters of typical stabilized specimens were evaluated.

BACKGROUND

Solidification/Stabilization

Stabilization of a waste is generally defined as chemical modification of the material to detoxify its waste constituents, which may or may not result in improved physical properties of the material. However, factors such as durability, strength, and resistance to leaching play important roles in predicting long-term performance of the new material. Therefore, improvement of physical properties is essential for long-term integrity of the material, especially if it is being considered for re-use. As defined in several EPA publications[2-4] "stabilization," "solidification," and "fixation" refer to waste treatment which produces the combined effects of: (1) improvement of physical properties; (2) encapsulation of pollutants; and (3) reduction of solubility and mobility of the toxic substances. Although each one of the above terms may emphasize one or more of these effects, for all practical purposes the terms have been used interchangeably with little or no error.

There are a number of solidification/stabilization techniques used in the industry for different types of waste materials. Each solidification/stabilization process is formulated to be compatible with the specific waste constituents. They can be divided into the following groups:

1. cement-based techniques
2. silicate-based techniques
3. thermoplastic techniques
4. sorbent techniques
5. organic polymer techniques
6. encapsulation techniques
7. vitrification

Among these, cement and silicate based techniques involve well-known pozzolanic or cementation reactions which result in compounds that act as natural cement. These techniques may offer economy over others, since they often utilize other waste products such as fly-ash, blast furnace slag, or cement kiln dust as pozzolanic additive.

Silicate-based processes cover a wide range of methods in which the siliceous or pozzolanic material is mixed with other alkaline earths, such as lime or gypsum. Although these processes are generally used to stabilize inorganic industrial wastes, they have also been shown to stabilize organic and oily wastes effectively.[5-8] The solidification processes using silicate based materials generally involve pozzolanic reactions between SiO_2, Al_2O_3, Fe_2O_3, and available calcium in lime. These reactions produce very stable calcium silicates and aluminates which act as natural cement similar to portland cement. In stabilization of organic wastes, both cement and silicate based techniques produce a microencapsulating matrix in which the organic component is bound by a combination of chemical reactions and physical isolation. The hydrocarbon component would essentially be fixed or immobilized in such a matrix, which would restrict internal fluid movement also. If the stabilized material is soil-like, a degree of mechanical stabilization, such as compaction, is necessary to ensure low density and formation of a continuous matrix.

The sorbent techniques involve use of certain clay minerals with high specific surfaces to fix and retain hydrocarbon molecules. This retention can take place on outer or inner surfaces of clay minerals. Some of the clay minerals used to retain hydrocarbons are sodium-montmorillonite, vermiculite, and needle-like structured attapulgite.

Geotechnical Properties of Stabilized Product

The main objectives of waste solidification are: (1) to remove free liquid and thus minimize leachability, (2) to render waste physically stable for handling and placement, (3) to reduce permeability and thus minimize leachability. There are a number of geotechnical parameters used by regulatory agencies to assess performance of stabilized/solidified wastes. Among these parameters are unconfined compressive strength, permeability, compressibility, durability, creep potential, and dry density.

Unconfined compressive strength (UCS) is often a good indicator of the integrity of the solidified material, its trafficability, and degree of ease of handling and placement. In addition, UCS measured over time, during the period of curing, may provide insight to the ongoing cementation and pozzolanic reactions. Furthermore, UCS measurements before and after a mechanical improvement, such as compaction or preloading, may also provide information on the effectiveness of such methods in some solidification processes. Compressibility and degree of saturation are also good indicator parameters to assess long-term behavior and effectiveness of improvement methods. For purposes of comparison, unconfined compressive strengths of lime stabilized natural soils range from 80 psi to 1,100 psi, depending on the amount of pozzolanic material present, and curing period.[9] Soaked unconfined compressive strength of soil-cement range from 200 psi to 1,200 psi, depending on soil type and curing period.[9] Freshly prepared solidified material from liquid containing concentrated brine, has been shown to attain 24-hr strength ranging from 7 psi to 21 psi with increased lime content.[10] The 24-hr strength of compacted solidified FGD sludge using SFT Terra-Crete process is on the order of 20 psi, whereas its 28 day strength is about 300 psi.[11] These values indicate the variability of the products of various processes involving different waste materials but similar additives for solidification.

Permeability is a parameter used to minimize potential groundwater contamination. The value of $k=1\times10^{-7}$ cm/sec pertains to a median value for low permeability clay soil systems. Studies done on mean conductivities of pure compacted clays of water permeability of 10^{-8} cm/sec, showed marked increases to hydrocarbons, such as xylene, gasoline, kerosene, and diesel fuel. Most solidification processes result in material permeability of 10^{-5} to 10^{-6} cm/sec. These numbers can be reduced further if the material is "soil like" or granular, and would undergo volume compression under pressure. For materials that set up fast and form a continuous, more like a "crystalline," matrix, application of pressure may have little or no effect. It may also result in crushing the matrix under high compactive or compressive efforts. In such cases, compression or compaction may be done when the material is in semisolid state. Addition of additives such as absorbent clays or other cementatious products may result in reduction of the excess liquid in the mixture. The additives will then take up the pore space, thus increasing the density and reducing the permeability of the material.

Re-Use of Stabilized Material

The stabilized material can be recommended for re-use in the following capacities, depending on the physical and mechanical properties of the material:

- base courses and subbase courses in highway construction
- embankments
- structural backfill
- barrier systems

The critical properties that need to be determined for each one of these applications are permeability, unconfined compressive strength, durability, moisture density relations, and index properties such as plasticity and water content. Each construction may require additional information to design the structure; however, these can either be determined experimentally on a case by case basis, or estimated from existing empirical relationships between various parameters.

INVESTIGATION

Experiments and Results

Local natural soil was obtained near Bethlehem, Pennsylvania. The soil was dried and only the portion passing No. 200 sieve (clay fraction) was utilized in all the tests. This was because clays retain hydrocarbons and their physical and chemical properties appear to be affected more than the other constituents of soil. The intent was to find out how much improvement can be accomplished in the clay by treatment with additives. The soil sample was divided into three sets in which the first set was contaminated with oil, the second set contaminated with oil and treated with lime, and the third set contaminated with oil and treated with portland cement. In each set, the percentage of oil was varied as 5%, 10%, and 20%, and for each one of these the additive amount was also varied as 5%, 10%, and 20%. Each experiment was conducted on at least three duplicate samples, and the average of those results are reported in this study. The percentages, both in the case of oil and additives, were based on the dry weight of the soil.

As for contaminating hydrocarbon constitute, fuel oil grade No. 2 (ASTM D396) was selected. The reason for this selection was that No 2 fuel oil, or heating oil, has slight volatile characteristics and, therefore, did not require special precautions to preserve its stability during and after mixing with soil. It is a light yellow to amber colored liquid with a kerosene odor. The specific gravity is 0.85, and solubility in water is negligible.

All of the soil specimens were mixed at optimum water content before adding the oil and the additives. The optimum water content and the maximum dry density of the raw sample was found to be 14.5% and 18.4 kN/m^3, respectively. Soil mixed at optimum water content was then mixed thoroughly with the specific amount of fuel oil and left for detention in a closed container for 24 hours. After the 24-hr period, portions of the contaminated soil were treated with various percentages of additives and compacted in a Harvard miniature compaction apparatus to prepare specimens for strength testing. The unconfined compressive strengths are given in Table 1.

Table 1. Unconfined Compressive Strength Data for all the Samples

	Fuel Oil			
Soil and Additive	**0%**	**5%**	**10%**	**20%**
Soil	77.2	—	—	—
Soil and Oil	—	38.7	16.8	6.3
Lime—5% Additive	—	90.5	56.7	23.8
Cement—5% Additive	—	104.3	33.4	9.5
Lime—10% Additive	—	122.0	66.2	28.8
Cement—10% Additive	—	105.8	59.0	12.6
Lime—20% Additive	—	106.9	111.7	36.2
Cement—20% Additive	—	101.9	39.5	21.7

A few of the samples were also subjected to tensile strength tests, the results of which are given in Table 2. The same group of specimens was also tested to find out the variation of plasticity values. Those results are reported in Table 3.

Table 2. Tensile Strength Data of 10% Fuel Oil Mixed Stabilized Soil

Soil Alone ⟶ **4.8 KPa**
Soil and 10% Fuel Oil Only ⟶ **0.8 KPa**

	Lime	Cement
5%	2.3	1.3
10%	3.4	1.6
20%	6.4	1.6

Table 3. Plasticity of 10% Fuel Oil Mixed Stabilized Soil

	LL	PL	PI
Soil	29.2	19.58	9.62
Soil and 10% Fuel Oil	33.5	5.0	28.5
Soil and 10% Fuel Oil and 5% Cement	37.0	26.32	10.7
Soil and 10% Fuel Oil and 20% Cement	36.0	12.4	23.6

Typical stress strain curves from the unconfined compression tests of untreated and contaminated soil are shown in Figure 1. Similar curves for lime and cement treated soil are given in Figures 2 and 3, respectively. The variation in cohesion and internal friction angle was also evaluated for treated soils. These results are illustrated in Figures 4 and 5, respectively.

Discussion of Results

As observed from Table 1, the unconfined compressive strength of oil mixed soil reduces significantly with respect to the uncontaminated soil. The same trend can also be observed from Figure 1, where the stress-strain curves indicate a significant loss of stiffness and softening of the soil with increasing percentage of oil. The interesting observation is that with addition of only 5% of oil, the stiffness and the strength reduces to approximately half the original values, and the reduction continues with the same rate as percentage of oil is doubled subsequently. In Table 2, the tensile strength of 10% fuel oil added soil is reduced significantly with respect to that of uncontaminated soil. These results clearly illustrate the adverse effect of fuel oil on the strength parameters of clay soils. The plasticity of 10% fuel oil added soil is also increased significantly, as observed from Table 3.

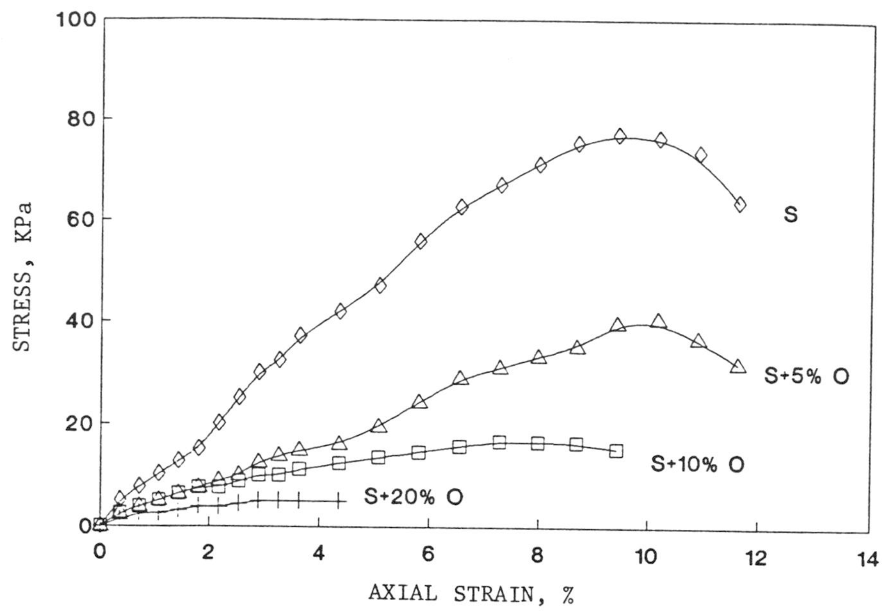

Figure 1. Variation of unconfined compressive strength and stress-strain relation with oil content. S = Soil; O = Fuel Oil.

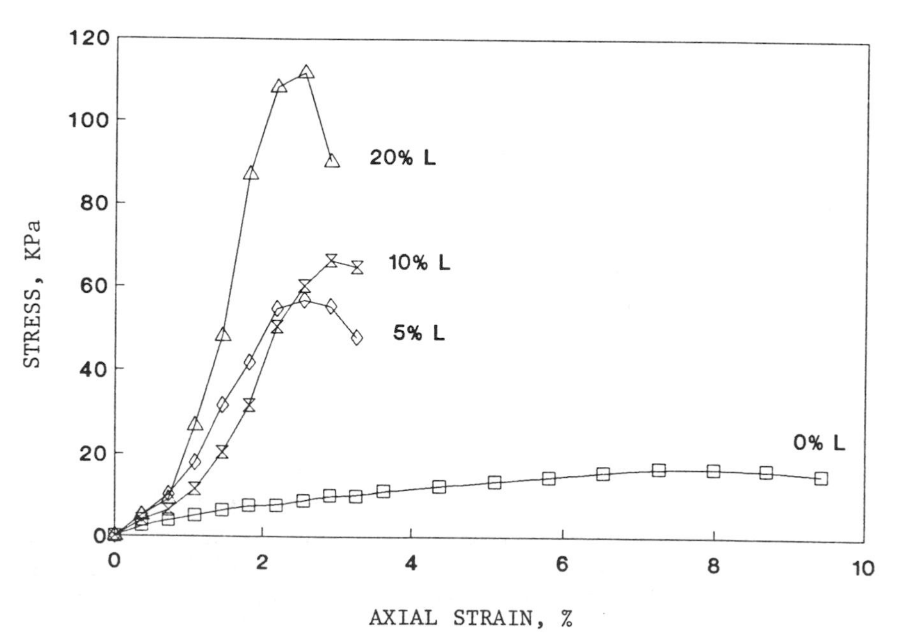

Figure 2. Variation of unconfined compressive strength and stress-strain relation with increased percentage of lime as additive. Fuel oil content = 10%. L = Lime.

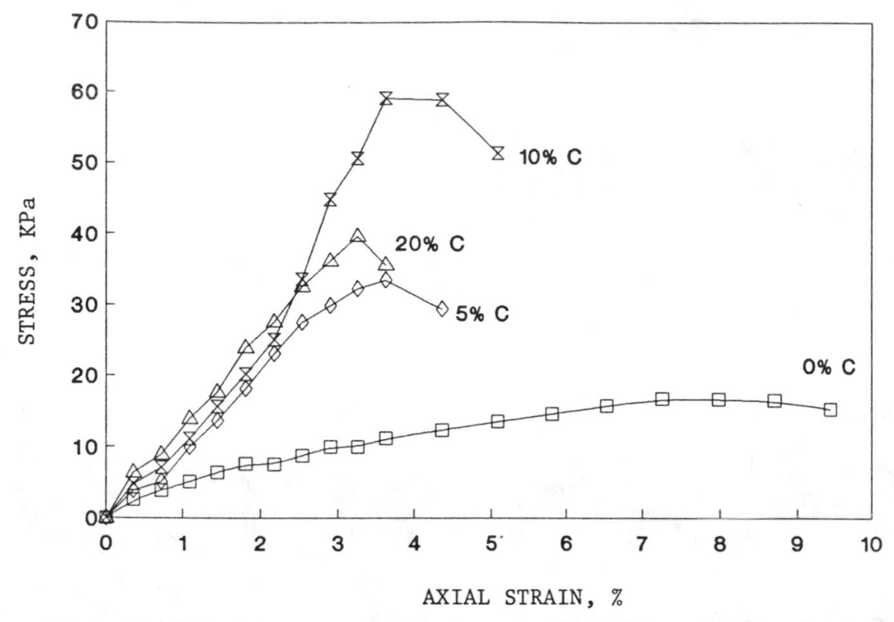

Figure 3. Variation of unconfined compressive strength and stress-strain relation with increased percentage of cement as additive. Fuel oil content = 10%. C = Cement.

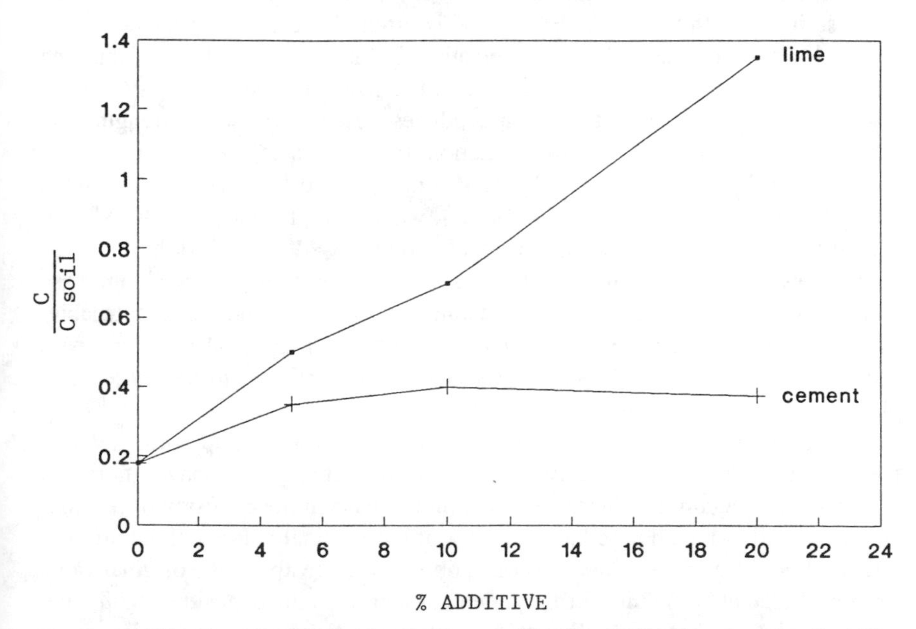

Figure 4. Variation of normalized cohesion, c, of 10% fuel oil contaminated soil with increased percentage of additive.

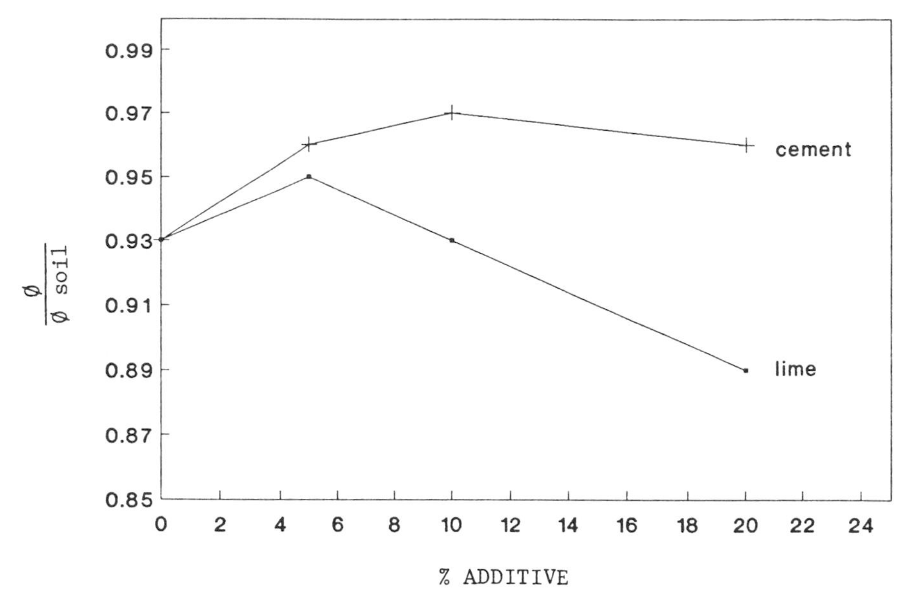

Figure 5. Variation of normalized internal friction angle, Ø, of 10% fuel oil contaminated soil with increased percentage of additive.

When the contaminated soil is treated with lime or cement, there is a marked increase in strength values. This increase is often above the original values for soil for lower percentages of oil contamination. Table 1 indicates that the increased addition of lime creates marked increase in the strength values in all specimens tested. Although the cement addition produces larger increase in strength than lime for low percentage of oil contamination, the same is not true for higher percentages of oil. Moreover, the addition of cement above 10% reduces the gained strength. This is also illustrated in Table 3, where the plasticity is reduced with 5% cement and increases with additional cement.

The stress-strain curves for lime treated contaminated soil show again increase in stiffness of the material. The soil becomes more brittle and less deformable, as can be observed from Figure 2. The same phenomenon is also evident with cement treated soils; however, at 20% cement content, due to the increase in plasticity, the strength drops (Figure 3).

The variation in cohesion of the soil with treatment is shown in Figure 4, where the cohesion values are normalized by the cohesion of uncontaminated-untreated soil. The lime, again, produces a significant increase in the cohesion of the contaminated soil, which had reduced to 20% of the original value with addition of 10% fuel oil. The increase in cohesion is observed to be above the original value for lime added at 20%. Interestingly, cement increases the cohesion slightly and stabilizes shortly thereafter. Figure 5 illustrates the variation of friction angle with the additives. The drop in friction angle is not as substantial as cohesion with addition of fuel oil. Cement appears to increase the friction angle more than lime. Both additional cement and lime decreases the friction angle, as would be expected due to the increase in percentage of fine grained material in the mixture. However, the change in internal friction angle in all cases seems to be slight, as opposed to the change in cohesion of the material.

CONCLUSIONS

The following conclusions have been drawn from this study:

1. Fuel oil reduces the strength and stiffness, and increases the plasticity of clay soils significantly, even at low percentages of addition.
2. Lime appears to produce marked increase in the strength parameters of fuel-oil contaminated soils, possibly due to the adsorbing and binding of oil by lime, which may form an agglomerated and continuous matrix.
3. Cement may be beneficial when added at low percentages. Higher percentages of cement may produce isolated pockets of cement coated with oil, which may act as weak areas and therefore reduce the strength. Increase in plasticity with additional cement may also be a factor in reduced strength or insufficient stabilization.
4. Stabilization and re-use of fuel oil contaminated soils appear to be viable methods of resource recovery, and may produce savings in the construction industry. More research is needed on the permeability and durability of stabilized soils, such as those tested here, to conclude in favor of the proposed approach. However, the results presented here show that there is much potential in stabilizing oil-contaminated soils for use in large constructions, as far as strength of the material is considered.

REFERENCES

1. Dowd, R. M. "Leaking Underground Storage Tanks," *Environ. Sci. Technol.* 18(10) (1984).
2. Bartos, M. J., and M. R. Palermo. "Physical and Engineering Properties of Hazardous Industrial Wastes and Sludges," U.S. EPA Report-600/2-77-139 (1977).
3. "Guide to the Disposal of Chemically Stabilized and Solidified Waste," SW-872, U.S. Army Waterways Experiment Station, U.S. EPA (1982).
4. Cullilane, M. J., Jr., L. W. Jones, and P. G. Malone. "Handbook for Stabilization/Solidification of Hazardous Waste," U.S. EPA Report-540/2-86-001. (1986)
5. Spencer, R. W., R. H. Reifsnyder, and J. C. Falcone. "Applications of Soluble Silicates and Derivative Materials," *Proceedings, Management of Uncontrolled Hazardous Waste Sites,* Hazardous Materials Control Research Institute, Silver Spring, MD (1982) pp. 237–243.
6. Pancoski, S. E., J. C. Evans, M. D. LaGrega, and A. Raymond. "Stabilization of Petrochemical Sludges," *Proceedings of the 20th Mid-Atlantic Industrial Waste Conference,* HMCRI, Silver Spring, MD (1988), pp. 299–316.

7. Pamukcu, S., J. B. Lynn, and I. J. Kugelman. "Solidification and ReUse of Steel Industry Sludge Waste," *Proceedings of the 21st Mid-Atlantic Industrial Waste Conference,* Technomic Press, Lancaster, PA (1989), pp. 3–15.
8. Van Keuren, E., J. Martin, J. Martino, and A. De Falco. "Pilot Field Study of Hydrocarbon Waste Stabilization," *Proceedings of the 19th Mid-Atlantic Industrial Waste Conference,* Bucknell University, Technomic Publishing, Co., (1987), pp. 330–341.
9. Pamukcu, S., and H. F. Winterkorn. "Soil Stabilization and Grouting," Ch. 9 in *Foundation Engineering Handbook* 2nd ed., H. Y. Fang, ed., Van Nostrand Reinhold Co, New York, N.Y. (1989).
10. Myers, T. E. "A Simple Procedure for Acceptance Testing of Freshly Prepared Solidified Waste," Hazardous and Industrial Solid Waste Testing: Fourth Symposium, *ASTM STP* 886 (1986), pp. 263–272.
11. Valiga, R. "The SFT Terra-Crete Process," Ch. 10 in *Toxic and Hazardous Waste,* Vol. 1, R. B. Pojasek, ed., Ann Arbor Science, Ann Arbor, Michigan (1982).

CHAPTER 17

Soil Vapor Extraction Research Developments

George E. Hoag, The University of Connecticut, Storrs
Michael C. Marley, Bruce L. Cliff, and **Peter E. Nangeroni,** VAPEX Environmental Technologies, Canton, Massachusetts

Recently, in situ subsurface remediation processes have been the focus of significant attention by the scientific community involved with the cleanup of volatile and semivolatile environmental contaminants. Of the in situ processes researched to date, vapor extraction holds perhaps the most widespread application to the remediation of these types of organic chemicals frequently found in the subsurface. The vapor extraction process has been successfully employed at many types of sites as a stand-alone technology, and may also be considered a synergistic technology to other types of in situ subsurface remediation technologies such as bioremediation and groundwater pump, skim, and treat.

In the past five years, in situ vapor extraction has been applied at many sites by means of significantly different approaches. These range from "black box" DESIGN WHILE YOU DIG TECHNIQUES to those utilizing sophisticated numerical models interfaced with laboratory, pilot and full-scale parameter determination for design purposes. The extent of success in field application of vapor extraction is varied, in many cases related to monitoring and interpretive limitations employed before, during, and after the remediation. Because application of the technology is quite recent, many remediations are still in progress; thus, final results interpretation and publishing in refereed scientific journals is limited.

Research Milestones in Vapor Extraction

Thorton and Wootan[1] introduced the concept of vertical vapor extraction and injection wells for the removal of gasoline product, as well as vapor probe monitoring for the quantitative and qualitative analysis of diffused hydrocarbon vapors. A further enhancement of this research was published by Wootan and Voynick,[2] in which various venting geometries and subsequent air flow paths were hypothesized and tested in a pilot sized soil tank. In their first study, 50% gasoline removal was achieved, while in their second study up to 84% removal of gasoline was observed.

Local Equilibrium Concept. In controlled laboratory soil column vapor extraction experiments by Marley and Hoag[3] and Marley,[4] 100% removal of gasoline at residual saturation was achieved for various soil types (0.225 mm to 2.189 mm average diameters), bulk densities (1.44 g/cm to 2.00 g/cm), moisture contents (0% to 10% v/v), and air flow rates (16.1 $cm^3/(cm^2\text{-min})$ to 112.5 $cm^3/(cm^2\text{-min.})$ They also successfully developed an equilibrium solvent-vapor model using Raoult's law to predict concentrations of 52 components of gasoline in the vapor extracted exhaust of the soil columns.

Baehr and Hoag[5] adapted a one-dimensional three-phase (immiscible solvent, aqueous, and vapor phases) local equilibrium transport model developed by Baehr[6] to include air flow as described by Darcy's law for compressible flow. This first deterministic one-dimensional model effectively predicted the laboratory vapor extraction results of Marley and Hoag,[3] and provided the basis for higher-dimensional coupled air flow contaminant models for unsaturated zone vapor extraction.

Porous Media Air Flow Modeling. Because local equilibria prevailed in the above studies, a higher-dimensional model, developed by Baehr, Hoag, and Marley[7] was used to model air flow fields under vapor extraction conditions. The three-dimensional radially symmetric compressible air flow model is used to design vapor extraction systems using limited lab and/or field air flow pump tests. A steady state in situ pump test determination of air phase permeability is preferred over laboratory tests because an accounting is possible of the presence of an immiscible liquid, anisotropy, soil surface, variations in soil water conditions, and heterogeneity in air phase permeabilities. The numerical solution developed can simulate flow to a partially screen well, and allow determinations of vertical and horizontal air phase permeabilities. Heterogeneous unsaturated zones can also be evaluated using the numerical simulation. Analytical solutions to radial flow equations, such as one developed early by Muskat and Botset[8] are generally restricted to determination of average horizontal air perambulated determination for impervious soil surfaces.

Removal of Capillary Zone Immiscible Contaminants. Hoag and Cliff[9] reported that an in situ vapor extraction system was effective in removing 1330 L of gasoline at residual saturation and in the capillary zone at a service station, and achieved cleanup levels to below 3 ppm (v/v) in soil vapor, and below detection limits in soils. The entire remediation took less than 100 days. A groundwater elevation and product thickness log for the time period of before, during, and after the vapor extraction remediation is graphically shown in Figure 1. On day 250, only a skim of gasoline was present in this well, and on day 290 no skim was detected. One year after the vapor extraction remediation took place, groundwater samples were nondetected for gasoline range hydrocarbons, reflecting that at least advective dispersive transport and possible natural microbiological activity in the groundwater were mechanisms responsible for this effect.

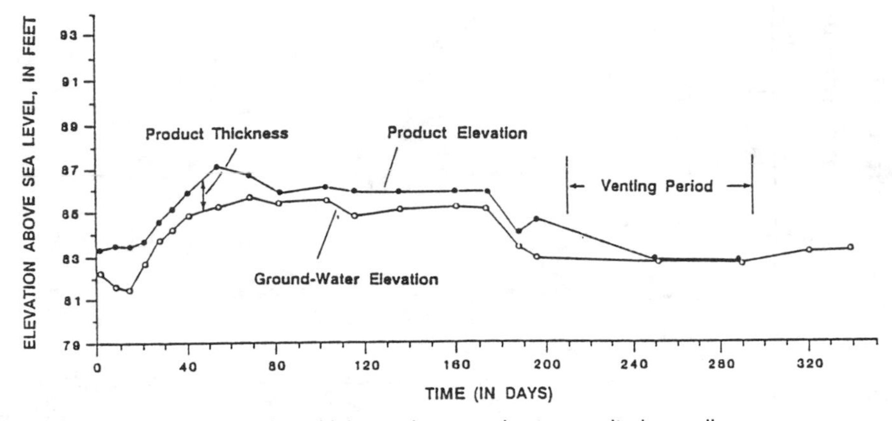

Figure 1. Apparent product thickness in groundwater monitoring well.

Field Application of Porous Media Air Flow Models

Baehr, Hoag, and Marley[7] utilized the above site for a field air pump test to determine the horizontal air phase permeability and to simulate the sensitivity of the model to changes in air phase permeabilities utilizing site geometries and boundary conditions. Based upon a full-scale air flow pump test, the air phase permeability for the site was predicted to be $k=7.0+10$ cm^2 for a mass air withdrawal rate of 11.1 g/sec, and a normalized pressure of $P_s/P_{atm}=0.9$. For reference, 11.1 g/sec, assuming an air phase density of 1.2×10^{-3} g/cm^3, equals about 555 L/min.

To illustrate the sensitivity of the model to a range of air phase permeabilities, the above service station vapor extraction well geometry, depth to water table, and appropriate boundary conditions were used as input, and air phase permeability and mass air withdrawal rates were varied. In Figure 2, the normalized air phase pressure in the well for various mass air withdrawal rates, and air phase permeabilities are shown. Significant increase in the vacuum developed in the wells can be observed for order of magnitude decreases in air phase permeabilities and small increases in the mass air withdrawal rates. Review of Figure 2

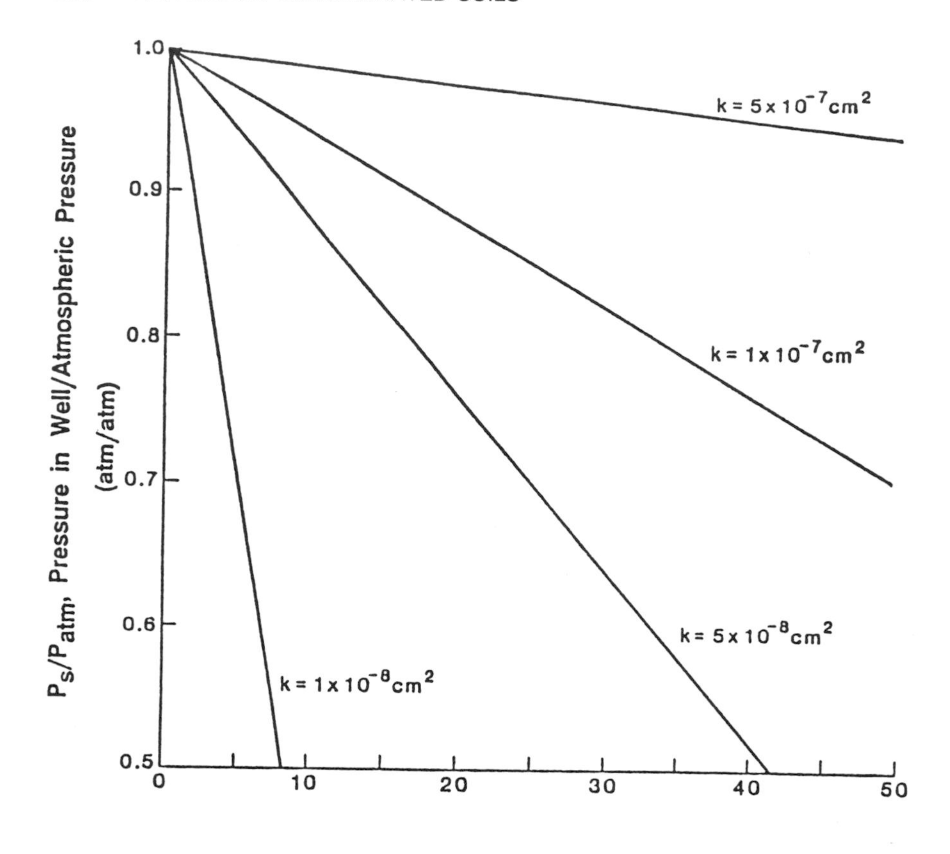

Figure 2. Normalized air-phase pressure at a single well. [NOTE: Various mass air withdrawal rates and air phase permeabilities are simulated (after Baehr, Hoag, and Marley[7])].

indicates that if a mass air withdrawal rate of 40 g/sec was used at the service station site ($k=7.0+10^{-8}$), the normalized air phase pressure at the well would be approximately $P_s/P_{atm}=0.6$, within an acceptable range of operating conditions.

A limitation of the model developed by Baehr, Hoag, and Marley[7] is that it is not coupled to contaminant transport. However, for the design of vapor extraction systems for volatile contaminants, this generally is not a fundamental need and can be accomplished by either laboratory venting tests, similar to those developed by Marley and Hoag[3] or by utilization of the one dimensional coupled model developed by Baehr and Hoag.[5]

RESEARCH NEEDS

A fundamental need of vapor extraction modeling occurs in the area of capillary zone/unsaturated zone interaction when immiscible phases are present on or in the capillary fringe. While air phase modeling alone is probably adequate for most vapor extraction system design purposes, particularly if a full three-dimensional model is used with optimization modeling, a rigorous modeling effort to couple air phase flow and immiscible contaminant transport, particularly in the capillary zone, will provide strategic insight to vapor extraction operation and planning.

To assess research needs in this area, two basic vapor extraction systems applications should be considered: (1) immiscible contaminant with density less that 1.0 (petroleum range hydrocarbons); and (2) immiscible contaminants with density greater than 1.0 (halogenated compounds).

Generalized subsurface phase distributions for immiscible liquids with densities less than that of pure water are illustrated in Figure 3. A typical vapor extraction system installation in this type of subsurface and contaminant condition is found in Figure 4. In the case study presented by Hoag and Cliff,[9] as detailed above, pump and skim was employed at the site for the first 210 days of the remediation with only 300 L gasoline removed (i.e., mostly through manual bailing). Thus, an important question should be: was vapor extraction alone necessary in this case or were both pump and skim and vapor extraction required for optimal, or even effective, remediation of immiscible contaminants? To answer this question requires an understanding of air-immiscible liquid-water three-phase conduction and distribution in the porous media, particularly in the capillary fringe areas at a site. Additionally, the site history of groundwater fluctuation and immiscible contaminant behavior in the capillary fringe is essential information necessary to answer the above question. Parker, Lenhard, and Kuppusamy[10] and Lenhard and Parker[11] provide a parametric model for three-phase conduction and measurements of saturation-pressure relationships for immiscible contaminants in the unsaturated and capillary zones. However, to date, this author is not aware of the coupling of these types of models to air phase and contaminant transport models.

A more in-depth hypothetical examination of the possible relationships near the capillary fringe will illustrate the importance of this zone in determining the need for pump and treat, and the importance of solute mass transfer from the capillary zone into the saturated flow regime. In the case of a recent spill of an immiscible contaminant with density less than water, when relatively steady groundwater flow prevails, a zone may exist on the capillary fringe of floating product, as shown in Figure 5. Infiltrating water, under draining conditions, will reach an equilibrium with the immiscible liquid, resulting in a saturated solute condition. For hypothetical purposes only, if it is assumed that only vertical groundwater flow exists in the capillary zone, the rate of solute input to the saturated zone will be limited by the rate of infiltration and C_s. If it is assumed,

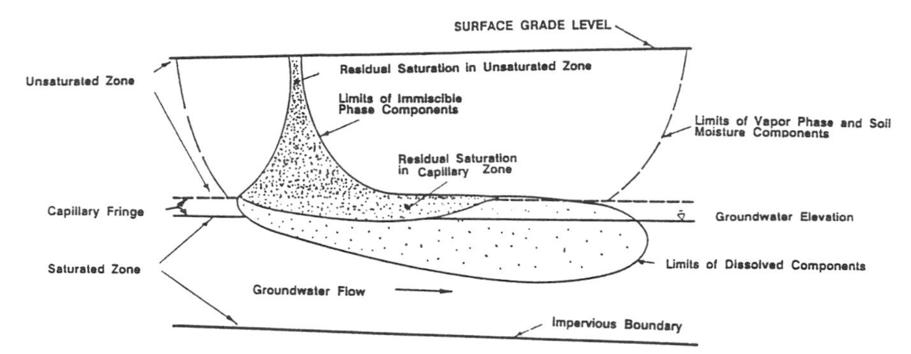

Figure 3. Generalized subsurface phase distribution for immiscible contaminants with densities less than pure water.

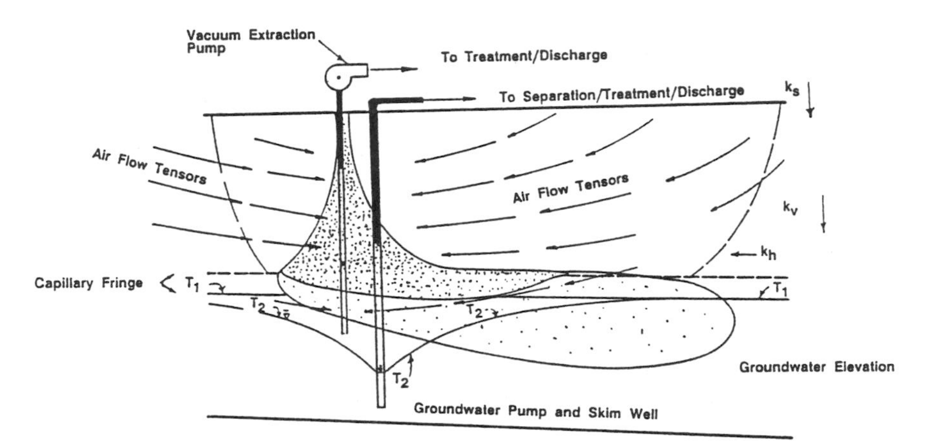

Figure 4. Typical vapor extraction/pump and treat in situ subsurface remediation system for immiscible contaminants with densities less than pure water. [NOTE: T_1 represents groundwater table prior to pump and skin, and T_2 is after pump and skim.]

again for illustrative purposes, that a horizontal flow boundary exists at the groundwater table, then mass transfer of solute from the capillary zone into the saturated zone will have only limited effects on the rate of solute input into the saturated zone. When considering the quantities of water infiltrating through the capillary zone per year in comparison to saturated flow rates, the above assumptions may be valid. The result is relatively inefficient transfer of solutes from the capillary fringe zone to the saturated zone. Thus, in this scenario, pump and treat systems may not be necessary to remove the immiscible contaminants, and advective dispersive dilution may be adequate to protect groundwater resources. Without knowledge a priori of the immiscible liquid distribution and interaction with the capillary zone, and advective-dispersive transport characteristics at a site, this approach may be risky. An alternative, however, may be close monitoring of groundwater in the saturated zone near the spill area, as vapor extraction proceeds. If the scenario in Figure 5 exists at a site, then solute concentrations in groundwater will decrease with time and no pump, treat, and skim system may be necessary to achieve desired levels of remediating in soil and groundwater. If near field transport of solutes from the spill area increases steadily with time, then groundwater pumping may be necessary to employ at that time.

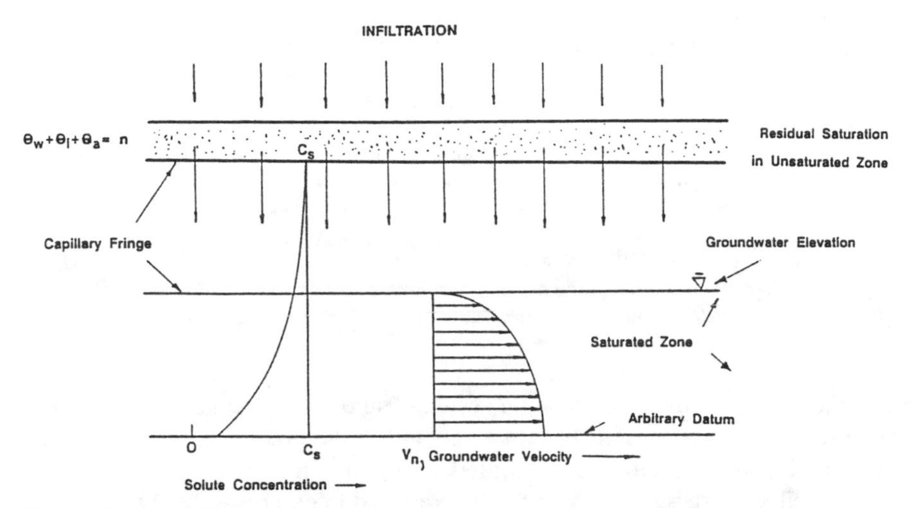

Figure 5. Immiscible contaminant on capillary zone, density less than pure water.

In the case of an immiscible contaminant with a density less than water, with impingement on the saturated zone by penetration of the capillary zone, the potential for solute transfer from the unsaturated zone to the saturated zone is greatly increased. This scenario may be result from the depression of the capillary zone in a spill event where considerable quantities of an immiscible contaminant are spilled, such as that shown in Figure 3. Alternatively, fluctuating groundwater tables may result from a rise in the groundwater table through wetting (imbibition) of the capillary zone, as described by characteristic curves for a given porous media and immiscible contaminant. Remembering that immiscible contaminants become immobilized once at residual saturation, the result of wetting the capillary zone may result in the condition shown in Figure 6. The net result of this scenario is that saturated solute concentration exists at the top of the horizontal flow zone of the saturated zone. This boundary condition enables substantially greater mass transfer of solute into the saturated zone, principally resulting from the upper flow boundary being the immiscible contaminant itself. In Figure 5, the upper boundary was only solute at less than C_s, and solubilization was limited to that achieved through infiltration. Clearly, the difference in these two situations greatly affects the rate of solute input into the saturated zone, and should affect remedial action responses. Unless the solute transport phenomena from an immiscible phase into the saturated zone is understood and physically defined

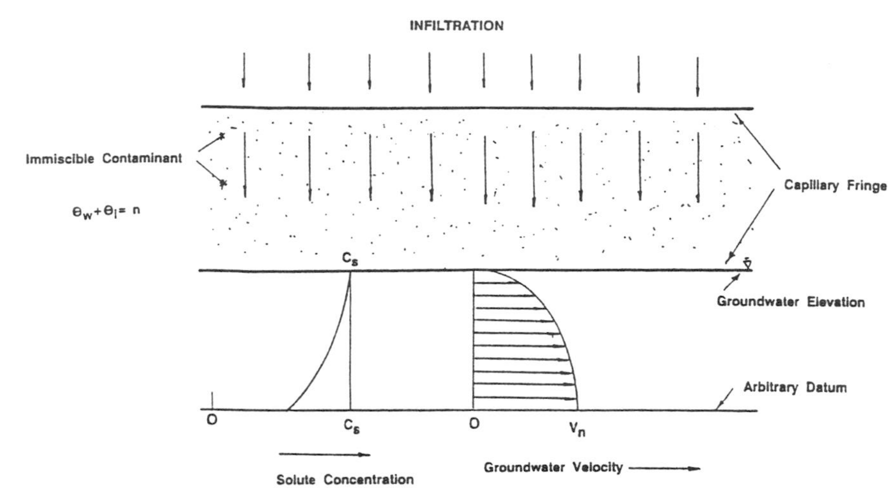

Figure 6. Generalized subsurface condition of immiscible contaminant with density less than water impinging on the water table.

at a site, neither optimal remediation systems can be designed nor can saturated zone solute transport models be effectively utilized to predict the impact of immiscible liquid remediation on saturated zone solute transport.

Immiscible phase boundary conditions presented in Figures 5 and 6 also greatly affect the vapor diffusive flux rates from the capillary zone into the unsaturated zone. Bruell[12] and Bruell and Hoag[13] rigorously investigated the effect of immiscible liquid phase boundary conditions on subsequent hydrocarbon diffusive flux rates of benzene. For a given column geometry (diffusive path length of 47.6 cm), diffusive flux rates for benzene at 20×C for an immiscible phase boundary benzene condition similar to that shown in Figure 5, resulted in benzene diffusive flux rates of 24.9 mg/cm^2-min and 6.1 mg/cm^2-min for dry and wet (i.e., field capacity moisture content) concrete sand, respectively. Thus, moisture content played a significant role in reducing the effective porosity of the concrete sand. When residual saturation immiscible liquid phase boundary conditions were investigated the maximum benzene diffusive flux rate was 26.6 mg/cm^2-min; however, the diffusive path length was only 22.4 cm. The moisture content in the residually saturated zone was 3.2% (v/v). When capillary zone immiscible liquid phase boundary conditions were investigated, the maximum benzene diffusive flux rate was reduced to 4.8 mg/cm^2-min, with a diffusive path length of 22.4 cm. The moisture content in the capillary zone reflected saturated conditions (i.e., O_w=n). This research demonstrated that the immiscible liquid phase boundary condition greatly affects the diffusive flux rates of hydrocarbons that occur in the unsaturated zone. As the moisture content increases, the diffusive flux rates of contaminants will decrease. The net result of these boundary conditions affects the concentrations of hydrocarbon vapors detected using soil gas assessment techniques, and the rates of hydrocarbon recovery utilizing in situ vapor extraction.

With reference to Figure 6, knowing that advective air flow rates also decrease with increasing moisture content, creates a circumstance in the area of the capillary zone where advective air flow may not be in direct contact with the immiscible phase. Thus, diffusion in this case will be the controlling mechanism of contaminant removal during vapor extraction. In the case depicted in Figure 5, it is likely that some advective air flow will contact the immiscible phase, greatly increasing vapor extraction efficiency.

In the case of an immiscible liquid with a density greater than that of water, contaminant distribution is significantly different, given a hypothetical spill to the subsurface. Penetration of the capillary and saturated zones by the immiscible liquid is likely, given sufficient spill volumes as shown in Figure 7. Of great importance is the occurrence of groundwater flow through the immiscible liquid phase in the saturated zone, resulting in substantially greater solubilization rates of the immiscible phase and greater groundwater contamination potential than in the cases presented in Figures 5 and 6.

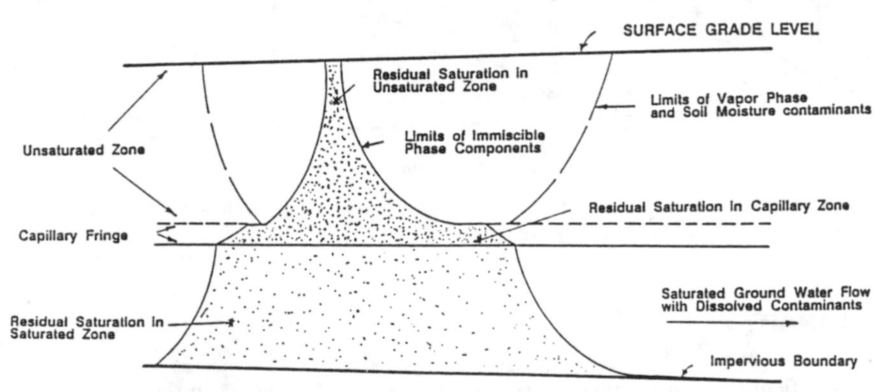

Figure 7. Generalized subsurface phase distribution for immiscible contaminants with densities greater than pure water.

A typical in situ remedial action response to the dense immiscible liquid phase contamination is given in Figure 8. Simultaneous vapor extraction and groundwater pumping are necessary to expose immiscible phase contaminants to advective air flow and to increase diffusive flux rates of contaminants in the vicinity of the groundwater table at time$=T_2$. In this case, dewatering of the saturated zone in the area of immiscible phase contaminants is desirable. Long-term plume management interceptor pumping strategies, such as those developed by Ahlfeld, Mulvey, Pinder, and Wood[14] and Ahlfeld, Mulvey, and Pinder[15] should be implemented to optimally circumvent uncontrolled groundwater contamination and to maximize groundwater contaminant recovery rates. Strategies to maximize saturated zone dewatering in the vicinity of the immiscible phase liquids must be developed to properly implement this approach. Additionally, in situ bioremediation may be considered as an additional technology to further degrade the immiscible liquid, if complete subsurface dewatering is not possible.

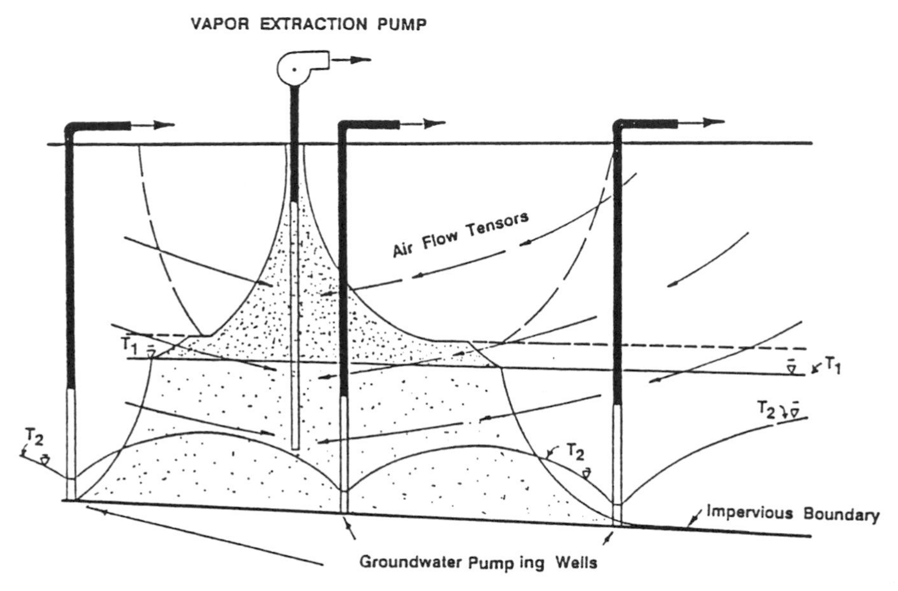

Figure 8. Typical in situ remedial actions to immiscible liquid phase contaminants with densities greater than water.

SUMMARY

Significant advances have been made in the past five years in the understanding of volatile and semivolatile contaminant behavior as related to vapor extraction technologies. Coupled modeling of both contaminant behavior and advective air flow, however, remains limited to one-dimensional systems. Given the significant hydrogeological complexity of porous media and subsequent heterogeneous distributions of immiscible phase contaminants, the design utility of higher-dimensional coupled models is questionable. Higher-dimensional advective air flow models are being used to design vapor extraction systems. These models are generally dependent on site-specific parameters best determined in field air pumping tests, unless uniform hydrogeologic conditions prevail with quantifiable boundary conditions necessary for model design predictions. Three-dimensional models are being adapted to deal with nonradial symmetry, and will be necessary to rigorously model multiple extraction well and extraction well/injection well applications.

Significant modeling and experimental research is needed to further understand immiscible contaminant behavior in the capillary zone and adjacent boundary conditions. The interaction of immiscible phase liquids in the capillary zone with unsaturated zone infiltration and saturated zone transport must be the focus of this research. The approach should include both hydrogeologic characteristics and testing procedures necessary to determine the influencing factors. Chemical fate and transport in the unsaturated zone under natural and advective air flow conditions must also be better understood to more effectively apply optimal in situ remediation processes.

Emphasis should be placed on basic research in the above areas, to be followed at the appropriate time by demonstration level projects. When demonstration level projects precede basic research needs, as has frequently been the case in the past five years, the results generally do not properly reflect necessary parameter control or monitoring, and either inconclusive or misleading results may be generated.

REFERENCES

1. Thorton, J. S., and W. L. Wootan. "Venting for the Removal of Hydrocarbon Vapors from Gasoline Contaminated Soil," *J. Environ. Sci. Health,* A17(1):31–44 (1982).
2. Wootan, W. L., and T. Voynick. "Forced Venting to Remove Gasoline Vapor from a Large-Scale Model Aquifer," American Petroleum Institute, 82101-F:TAV (1984).
3. Marley, M. C., and G. E. Hoag. "Induced Soil Venting for Recovery and Restoration of Gasoline Hydrocarbons in the Vadose Zone," *Proc. of Pet. Hydro. and Org. Chem. in Ground Water: Prevention, Detection and Restoration.* National Water Well Association and the American Petroleum Institute, Houston, Texas, 1984.
4. Marley, M. C. "Quantitative and Qualitative Analysis of Gasoline Fractions Stripped by Air from the Unsaturated Zone," MS thesis, Department of Civil Engineering, The University of Connecticut, 1985.
5. Baehr, A. L., and G. E. Hoag. "A Modeling and Experimental Investigation of Venting Gasoline Contaminated Soils," in *Soils Contaminated by Petroleum: Environmental and Public Health Effects,* E. J. Calabrese and P. T. Kostecki, eds. (New York: John Wiley & Sons, 1985).
6. Baehr, A. L. "Immiscible Contaminant Transport in Soils with an Emphasis on Gasoline Hydrocarbons," PhD dissertation, Department of Civil Engineering, University of Delaware (1984).
7. Baehr, A. L., G. E. Hoag, and M. C. Marley. "Removing Volatile Contaminants from the Unsaturated Zone by Inducing Advective Air-Phase Transport," *J. Contam. Hydrol.,* 4:1–26 (1988).
8. Muskat, M., and H. G. Botset. "Flow of Gas Through Porous Materials," *Physics* 1:27–47 (1931).
9. Hoag, G. E., and B. Cliff. "The Use of the Soil Venting Technique for the Remediation of Petroleum Contaminated Soils," in *Soils Contaminated by Petroleum: Environmental and Public Health Effects,* E. J. Calabrese and P. T. Kostecki, eds. (New York: John Wiley & Sons, 1985).
10. Parker, J. C., R. J. Lenhard, and J. M. Kuppusamy. "A Parametric Model for Constitutive Properties Governing Multiphase Fluid Conduction in Porous Media,"

Virginia Polytechnic Institute and State University, Blacksburg, VA, 1986.

11. Lenhard, R. J., and J. C. Parker. "Measurement and Prediction of Saturation-Pressure Relationships in Air-Organic Liquid-Water-Porous Media Systems. Virginia Polytechnic Institute and State University, Blacksburg, VA, 1986.
12. Bruell, C. J. "The Diffusion of Gasoline-Range Hydrocarbons in Porous Media," PhD dissertation, Environmental Engineering, University of Connecticut, Storrs (1987).
13. Bruell, C. J., and G. E. Hoag. "The Diffusion of Gasoline-Range Hydrocarbon Vapors in Porous Media, Experimental Methodologies," *Proc. of Pet. Hydro. and Org. Chem. in Ground Water: Prevention, Detection and Restoration.* National Water Well Association and the American Petroleum Institute, Houston, Texas, 1987, pp. 420–443.
14. Ahlfeld, D. P., J. M. Mulvey, G. F. Pinder, and E. F. Wood. "Contaminated Groundwater Remediation Design, Using Simulation, Optimization and Sensitivity Theory: 1, Model Development," *Water Resour. Res.* 24(3):431–442 (1988).
15. Ahlfeld, D. P., J. M. Mulvey, and G. F. Pinder. "Contaminated Groundwater Remediation Design Using, Simulation, Optimization and Sensitivity Theory: 2, Analysis of Field Site," *Water Resour. Res.* 24(3):443–452 (1988).

CHAPTER 18

Bioremediation of Heavy Petroleum Oil in Soil at a Railroad Maintenance Yard

Anthony Ying and **James Duffy,** TreaTek Inc., Grand Island, New York
Greg Shepherd, Southern Pacific Transportation Company, San Francisco, CA
David Wright, California Department of Health Service, Sacramento, CA

Numerous locomotive maintenance yards are operated by major railroad companies across the country. In recent years, an increasing number of these railroad yards are being identified as having soils contaminated with various concentrations of petroleum hydrocarbons. The source of the contamination is refueling operations and general locomotive servicing where hydrocarbon-type products like diesel fuel and heavy motor oil are routinely used.

The current available means for cleanup of these oil-contaminated soils is off-site disposal at a secured land disposal facility. In addition to the significant cost of this approach, the inherent liabilities associated with waste transportation and land disposal make this a very unattractive remedial option.

A viable remediation alternative to land disposal is the use of biological treatment. Much has been documented on the success of using microorganisms to degrade oil refinery wastes from the petroleum industry. Through further refinement and development, these microbial techniques are now being applied for specific waste destruction down to low levels in complex soil matrices.

In early 1988, Southern Pacific Transportation Company commissioned TreaTek®, Inc. to develop and design a cost-effective biological treatment solution specific to the soils and petroleum hydrocarbons at these locomotive maintenance locations. This project was partially funded by the Alternative Technology section of California's Department of Health Services, under the California Hazardous Waste Reduction Grant program. TreaTek® is an environmental service subsidiary of Occidental Chemical Corporation and has solid experience in using advanced microbial treatment systems for remediating soils containing organic constituents.

EXPERIMENTAL PROGRAM AND RESULTS

A multistep laboratory treatability program was carried out by TreaTek® using actual oil-contaminated soil samples collected from an operating railyard. The soil contained between 5,000 ppm to 60,000 ppm petroleum hydrocarbons. The program was designed to focus primarily on microbial and chemical additives for enhancing the natural degradation of oil in the soil. The key technical concern centered on the recalcitrancy of the dominant high molecular weight portion of the oil contaminant.

The experimental program contained several preliminary tasks on hydrocarbon characterization, microbial assessment, and inoculum optimization. These initial steps were then followed by two actual bench-scale treatment operations. The first involved 130 model ecosystems (or microcosms) containing 500 gm of contaminated soil each. These microcosms were treated with multiple combinations of different microbial and chemical reagents. Oil degradation in these microcosms was measured and compared over a 16-week period. The second set of treatment simulations involved a much larger quantity of soil—approximately 9 kg each. These scale-up systems enabled actual soil aeration testing, and provided means to measure potential oil removal via volatilization or leachate generation.

Summarized below are some of the key program steps and their corresponding experimental results:

1. Hydrocarbon Characterization

Oil is a chemically complex material consisting of a wide range of hydrocarbons such as branched and straight-chained alkanes, cycloalkane, aromatics, and naptheno aromatics, together with asphaltenes, resins, and waxes. Each of these components exhibit different degradation and mobility characteristics.

The oil extracted from the railyard soil was characterized and found to contain mostly linear and branched alkanes in the C-22 plus range. It is a heavy engine oil with a boiling point over 200°C and an asphaltene content of 0.1% (see Figure 1). This oil is expected to exhibit a fairly slow microbial degradation profile. In addition, its high molecular weight should keep volatilization to a minimum during aerobic treatment.

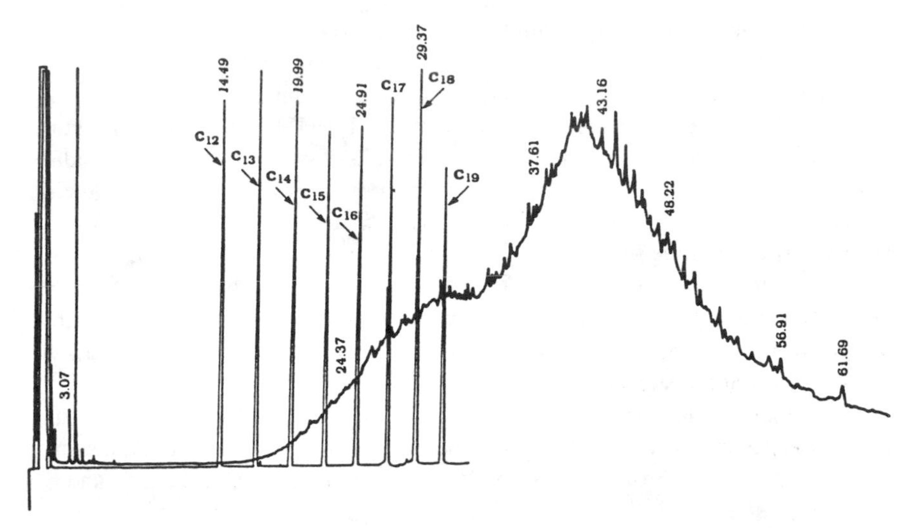

Figure 1. Pretreatment soil.

2. Moisture Optimization

Soil moisture content is an important factor affecting microbial activity. Experiments were carried out measuring carbon mineralization at different moisture level for soil containing 15,000 ppm oil and 35,000 ppm oil. The results are provided in Table 1. The results indicated a microbial activity peak at 15% moisture.

Table 1. Results of Experiments with Carbon Mineralized to CO_2 (mg/100 gm Soil)

Soil with 15,000 ppm Oil (nutrient/glucose supplemented)	
5% moisture	6.9 ± 0.3
10% moisture	52.2 ± 1.2
15% moisture	56.4 ± 0.6
20% moisture	46.5 ± 3.9
Soil with 35,000 ppm Oil (nutrient/glucose supplemented)	
5% moisture	1.8 ± 0.0
10% moisture	19.2 ± 1.2
15% moisture	30.0 ± 6.0
20% moisture	28.8 ± 5.4

3. Surfactant Enhancement

One of the key difficulties often encountered in the persistence of potentially biodegradable contaminants in the environment is their nonavailability to the microflora. In some cases, the microorganisms themselves may produce a surfactant which enhances the solubility of the material, and this alone may suffice for obtaining sufficient quantities in an available form. In other cases, addition of a chemical surfactant to the soil may be required to enhance the bioavailability of the contaminant and so facilitate degradation. TreaTek® has screened a number of surfactants for use in enhancing the bioavailability of oily and tarry residues at both refinery and coal gasification sites, and it was proposed to test a selection of these using the railyard soil. The purpose of such tests was to compare the relative efficacy of the surfactant in leaching the oil contamination from the soil.

Seven surfactants were screened for their oil removal capability. The test consisted of shaking 100 gm of contaminated soil in 100 mL of surfactant solution. The material is then centrifuged to remove the supernatant. The solid residue is then dried and analyzed for total residual hydrocarbon.

The residual hydrocarbon present in the leached soil for each of the seven surfactants, together with the water control and the initial oil concentration, are shown in Figures 2 and 3. From the results, it is clear that surfactant S2, S5, and S6 were most effective in both the medium and high contaminated soil.

In a later test, the surfactants were evaluated for their toxicity effect on the soil microflora. It was found that S5 showed no inhibition, whereas S2 and S3 both showed a slight inhibition of microbial activity. Thus, S5 was chosen as part of the inoculum package for the treatment simulation.

4. Microbial Isolation on Railyard Soil

A slurry of the contaminated material from the railyard was added to a liquid mineral salts medium, supplemented with or without engine oil as sole carbon sources. The cultures were incubated on an orbital shaker at ambient temperature for 10 days, following which microorganisms were isolated from each of the three media onto a solid mineral salts agar medium supplemented with hexadecane, supplied in the vapor phase of sealed petri dishes. After several days of incubation at 15°C, colonies were enumerated (see Table 2).

These results clearly indicated an increase in total microbial biomass relative to the control in media supplemented with oil as the sole carbon source.

Morphologically distinct colonies were picked off with a sterile loop and reinoculated onto fresh plates of hexadecane media. This process was repeated until pure cultures were obtained. A total of eight different isolates were obtained using this method. Growth rates of these organisms were then investigated in liquid mineral salts medium, again supplied with sterile oil as sole sources of carbon.

The cultures were incubated on an orbital shaker at 15°C for 10 days, during which time the optical density was measured as a means of assessing the increase in microbial biomass. The results indicated that two specific isolates, C and H, demonstrated particularly heavy growth in the oil supplemented media (Table 3). These bacteria were retained in the TreaTek® culture collection as ERB085 and ERB086 respectively, for use as prospective oil-degrading inocula.

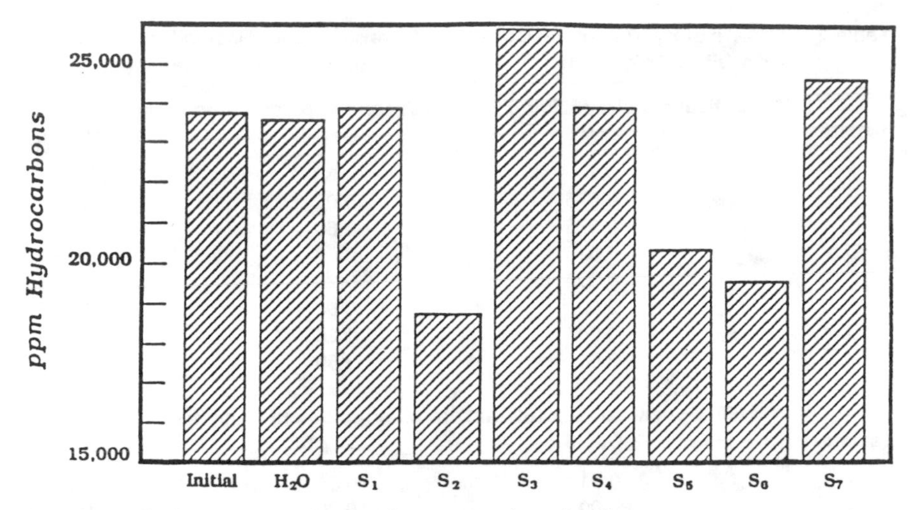

Figure 2. Surfactant screening medium contaminated soil.

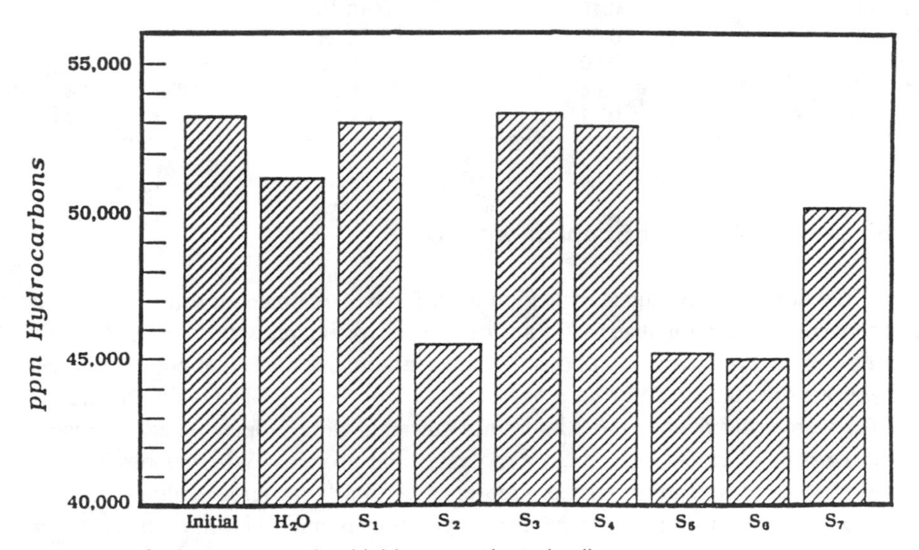

Figure 3. Surfactant screening highly contaminated soil.

Table 2. Numbers of Viable Cells Following 10 Days Enrichment of Inocula from Highly Contaiminated Soil in Liquid Media

Medium	Viable Cells $m1^{-1}$ (mean of ten samples)
Mineral salts only	610 ± 34
Mineral salts plus hexadecane	250,000 ± 25,000
Mineral salts plus oil	500,000 ± 28,000

Table 3. Growth of Hydrocarbon-Degrading Bacteria in Liquid Media Supplemented with Oil as Sole Carbon Source as Measured by Optical Density

	(A) Optical density (620nm) in hexadecane-supplemented media		
Isolate	**Day 3**	**Day 6**	**Day 10**
A	0.33	0.35	0.43
B	0.075	0.08	0.10
C	0.575	1.60	2.22
D	0.023	0.03	0.08
E	0.008	0.22	0.30
F	0.00	0.00	0.00
G	0.007	0.265	0.415
H	2.60	3.50	4.36
	(B) Optical density (620nm) in oil-supplemented media		
Isolate	**Day 3**	**Day 6**	**Day 10**
A	0.41	0.58	0.67
B	0.00	0.00	0.00
C	0.715	1.94	2.25
D	0.15	0.15	0.28
E	0.00	0.21	0.37
F	0.00	0.00	0.00
G	0.06	0.375	0.45
H	0.99	1.72	3.14

5. Organic Nutrient Supplementation

Traditional nutrient supplementation for microbial treatment of soil has involved primarily inorganic nutrients in the form of nitrogen, phosphorous, and potassium. Their effectiveness in boosting and maintaining microbial population has been well documented. From several recent TreaTek® projects, synergistic effects were noticed when these inorganic nutrients were supplemented with various organic nutrients—specifically molasses and proteins. Whereas researchers have used molasses in the past as supplementary carbon source dealing with low level contaminant treatment, the use of proteins as microbial nutrient is rather recent.

Various types of organic proteins were tested for their enhancement value for oil degradation. An example of some recent work is shown in Table 4.

Table 4. Organic Nutrient Study

Organisms/Inorganic Nutrients	Organic Nutrients	Oil (ppm) t=0 wk	t=5 wk
—	—	16,300 ± 599	15,000 ± 745
X	A	16,300 ± 599	13,700 ± 1862
X	B	16,300 ± 539	9,200 ± 1147
X	C	16,300 ± 539	10,500 ± 820

6. Treatment Simulation in Microcosms

From contaminated soil sampled from the railyard, three soils were prepared for use in microcosms. These soils represented conditions of *high* contamination (3% to 4% oil; the H-soil), *medium* contamination (1% to 2% oil; the M-soil) and *low* contamination ($<1.0\%$, the L-soil). Individual soils were mixed thoroughly to enhance sample homogeneity and reduce within treatment variation. Initial concentrations of hydrocarbons (mean ± S.E.M.) were, for each soil, 34193 ± 515 mg kg^{-1} (H-soil), 12803 ± 611 mg kg^{-1} (M-soil) and 6898 ± 611 mg kg^{-1} (L-soil).

The equivalent of 500 gm dry weight of each of the three soils was accurately weighed into 45 Kilner jars. Various combinations of chemical and microbial supplements were added to produce nine different treatments (1–9), together with sufficient water to raise the moisture content to exactly 15% by weight. The selection of chemical and microbial supplements were determined in previous screening experiments. Each treatment was replicated five times, so that 135 individual microcosms were established (3 soils × 9 treatments × 5 replicates).

Microcosms were maintained at a constant temperature of 15 × C to simulate field conditions, and were periodically stirred to improve oxygenation, and remoistened as necessary. The microcosms were sampled after 5, 10, 13, and 16 weeks of incubation by withdrawing approximately 50 gm wet weight of soil from each. The soil was subsequently analyzed for total oil hydrocarbon content by EPA method 418.1. A total of 555 soil analysis were carried out. The two best treatment results (with statistical significance) for each of the three soils are depicted graphically in Figures 4, 5, and 6.

For the L-soil of initial starting concentration 6898 ppm, several treatments resulted in reductions significantly less than the control by week 16. Two treatments produced reductions to below 500 ppm total oil hydrocarbons. The treatment producing the lowest end concentration of oil (Treatment 3; ERB086 + all chemical supplements) has a final reading of 409 ppm, which is equivalent to a reduction of 94%. Treatment 8 (repeated nutrient addition) is also a strong method for hydrocarbon degradation. This may be a higher cost approach than the single inoculum method of Treatment 3. Finally, the apparent increase in hydrocarbons in the control soil between 0 and 10 weeks is probably due to insufficient initial sample mixing. This highlights the difficulties in working with soil matrices and the importance of statistical sampling.

In the M-soil, Treatment numbers 2 and 9 gave the greatest reduction to below 4500 ppm from an initial concentration of 12431 ppm (equivalent to 64%). Treatment 2 involves the single application of ERB085 plus all chemical supplements, while Treatment 9 is nutrients only, without organisms or surfactant. The effectiveness of this nutrient-only treatment is surprising, especially in the first 5 weeks of incubation, where a reduction of almost 45% was recorded. One possible explanation is poor initial mixing of the starting material. It is conceivable that the soil associated with Treatment 9 was inherently lower in hydrocarbon concentration at the start.

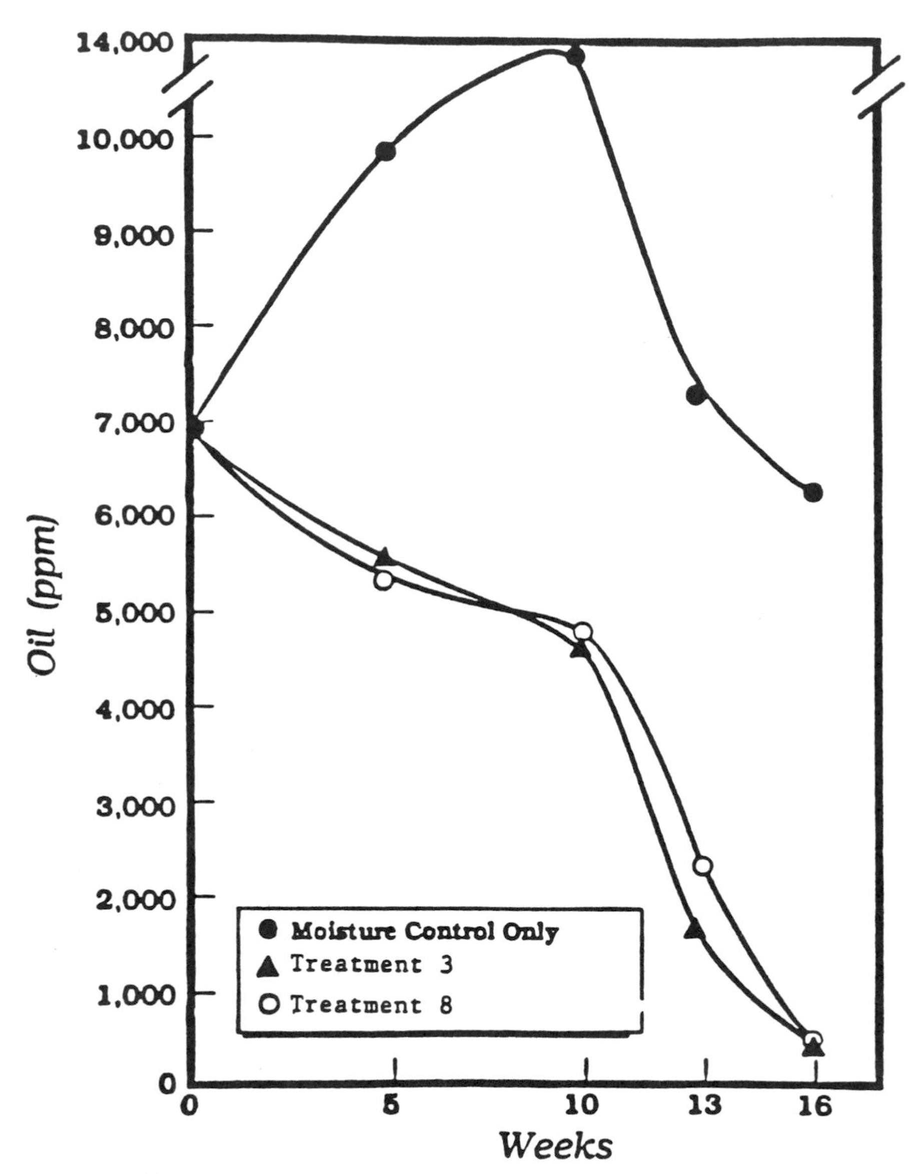

Figure 4. Microcosm results; low contamination.

In the H-soil, the greatest reduction in oil hydrocarbons was 48%. This occurred in Treatment 8 (repeated nutrient addition), where, from an initial concentration of 34,193 ppm, oil had been reduced to 17,629 ppm after 16 weeks. The application of additional organic and inorganic nutrients after 10 weeks appeared to have a favorable effect on the H-soil. The graph in Figure 6 clearly shows that in Treatment 8 (multiple nutrient addition), the hydrocarbons remain fairly constant until 13 weeks, when the treated soil displays a dramatic decline.

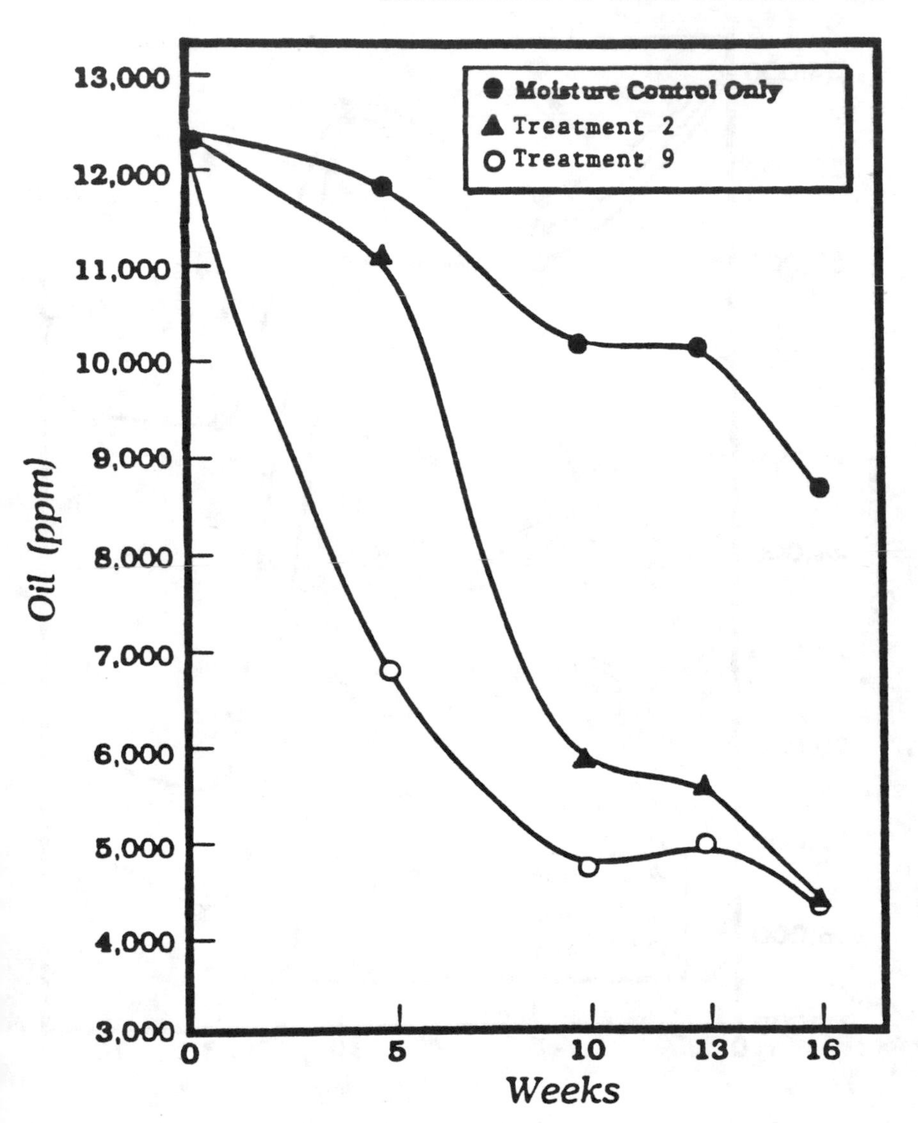

Figure 5. Microcosm results; medium contamination.

Overall, treatments producing the lowest end concentrations of oil hydrocarbons in each of the three soils were not the same. This is likely to be due to the fact that nutrient and microbial additions to the treated soils were on a "per weight of soil" basis, rather than according to the oil concentration in the soil. Therefore, in the three soils containing different concentrations of oil residues, treatments will have resulted in different oil/nutrient/microbial inoculum ratios according to the contamination level of the soil. This will have affected the nutrient/substrate conditions for biodegradation of the oil to different extents.

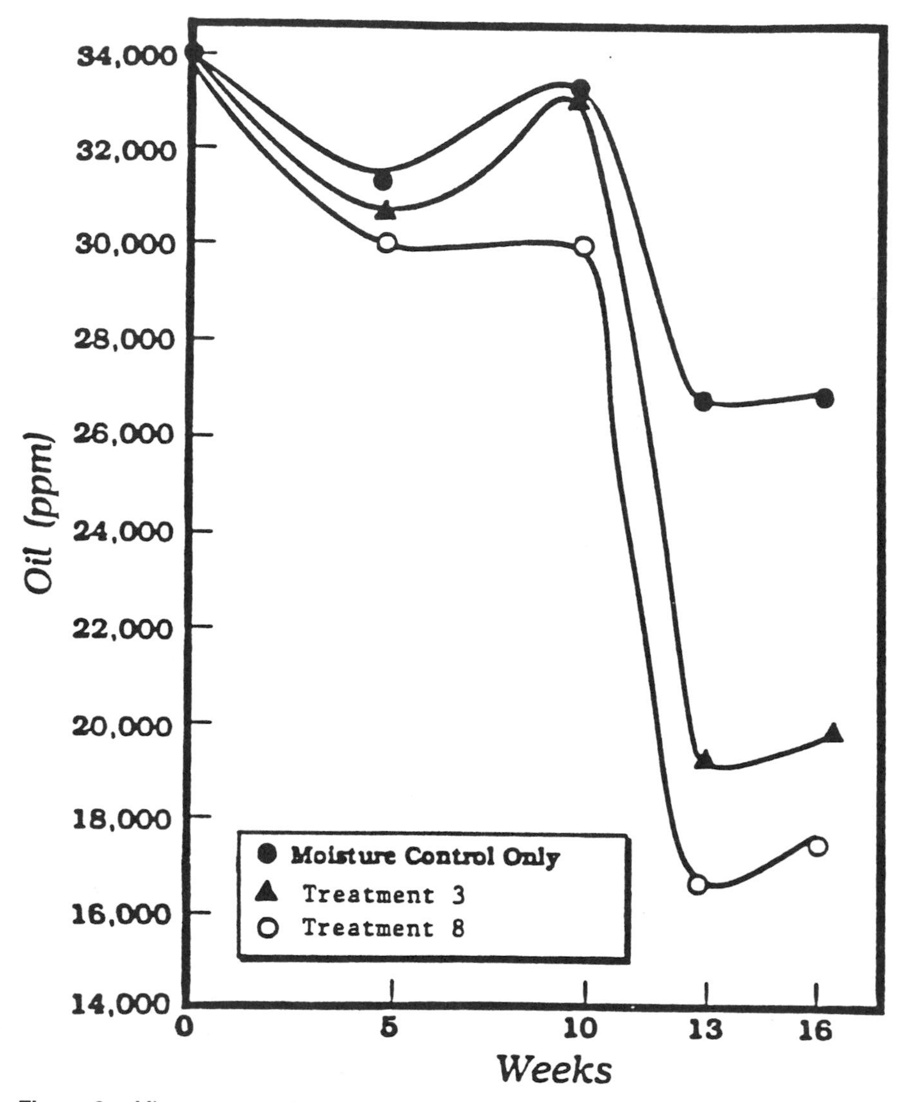

Figure 6. Microcosm results; high contamination.

From these results it appears that oil contamination up to 12,000 ppm is most amenable to microbial degradation at the inoculum concentration of 0.1% inorganic nutrients and 0.5% organic nutrients. Thus, boosting the nutrient addition level is key when dealing with higher oil contaminant levels.

7. Scale-Up Treatment Simulations

At the midpoint of the soil microcosm experiments, several large-scale treatment simulations were initiated in the laboratory. These were treatment units containing up to 9 kg of contaminated soil (versus 500 gm in microcosms). The primary purpose of these simulations is to validate the scale-up capability of the microcosm degradation data. In addition, the larger soil testing volume in these simulation units allows more rigorous aeration, and thus better simulates the real impact to contaminant removal by volatilization and other physical forces.

Unlike the microcosms, the simulation units were not treated with different inoculum combinations, but rather a single "best" formulation was used. Moisture-only control units were carried out for baseline data comparison. Since the controls were also aerated on a regular basis, the amount of contaminant removal in those units is an excellent indication of volatilization.

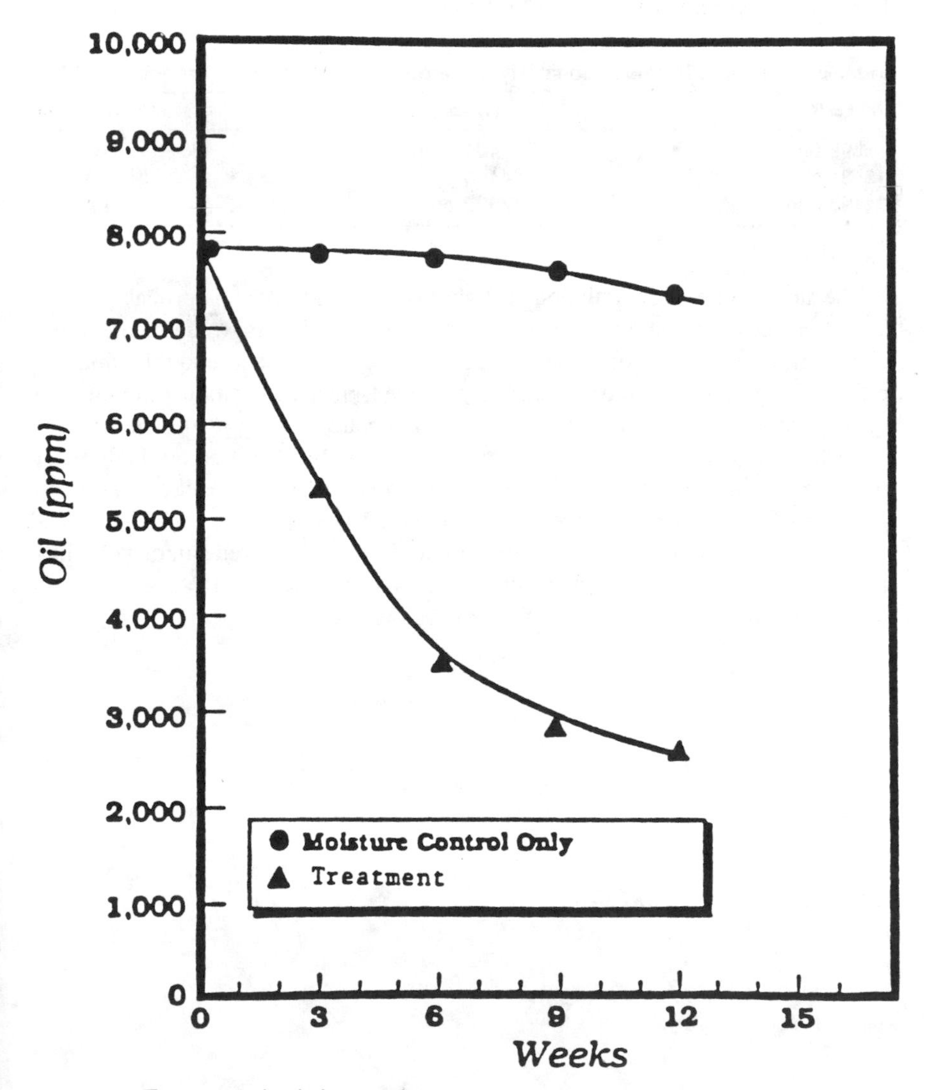

Figure 7. Treatment simulation results.

Shown in Figure 7 is the degradation result from one of the treatment units. Starting at a level of 7,898 ppm oil, a degradation of 62% was recorded over a 12-week period.

CONCLUSIONS

Hydrocarbon containing soils at locomotive maintenance yards can be successfully remediated using a combination of nutrient and microbial supplements. Laboratory data produced at 15°C showed up to 94% degradation of hydrocarbons in soil in only 16 weeks (see Table 5).

Table 5. Results of Remediation of Hydrocarbon Containing Soils (Produced at 15°C)

t=0	t=16 weeks	% Degraded
6898 ppm	409 ppm	94%
12431 ppm	4460 ppm	64%
34193 ppm	17629 ppm	48%

The slower rate of degradation at higher oil concentrations is probably due to an imbalance in the ratio of microbial substrate to nutrient mix, the inherent long chain nature of the hydrocarbon products (heavy oil) present, and the inhibitory effect of high concentration of oil. A pro-rated nutrient addition ratio and frequency based on oil concentration should alleviate this minor deficiency.

Scaled-up treatment simulation units were successfully carried out in duplicating degradation data produced in the microcosm experiments. Heavy oil decomposition of up to 62% was recorded in only 12 weeks.

Rigorous aeration of the soil in the simulation units showed minor volatilization of the contaminant at high concentration material (soil with >3% oil), while no volatilization was observed with soil below 1% oil.

CHAPTER 19

Soil-Induced Decomposition of Hydrogen Peroxide: Preliminary Findings

Bernard C. Lawes, E.I. Du Pont De Nemours & Company (Inc.),Wilmington, Delaware

Hydrogen peroxide is a valuable latent source of oxygen for in situ bioreclamation[1] because of its potential to decompose in contact with soil according to the following equation:

$$2H_2O_2 \longrightarrow 2H_2O + O_2$$

The peroxide-released oxygen can then be used for aerobic microbial metabolism.[2] Because hydrogen peroxide can be pumped as a liquid containing hundreds of parts per million of latent oxygen, it can, as demonstrated in the field by Yaniga and Smith,[3] provide higher levels of oxygen farther into an underground formation than can air sparging.[4] Raymond et al.[5] have suggested that the optimal rate of peroxide decomposition can vary, depending on the oxygen needs of bacteria at the moving front of maximum microbial activity.

Very little is known about how fast hydrogen peroxide can decompose in contact with soil, and just how much variation exists between various soils. One recent study[6] suggested that excessive hydrogen peroxide decomposition can occur in an infiltration gallery from the buildup of too much bacterial catalase. For a rapid evaluation of many soil samples, a simple test is needed that will not be influenced by changes in soil permeability, and will not be particularly sensitive to soil aging or minor variations in test conditions. Other questions for which at least preliminary answers are needed are: what soil characteristics influence peroxide decomposition rate; and, how much hydrogen peroxide is wasted in oxidation reactions that do not yield molecular oxygen.

METHODS AND MATERIALS

Topsoil samples, from the Research and Development Division of the Du Pont Agricultural Chemicals Department, were stored at 4°C. Subsurface samples, shipped nonrefrigerated from various sources, were also stored at 4°C. Chemicals for analytical tests were of standard reagent quality. Dilute 0.1% hydrogen peroxide solutions were made by dilution of TYSUL® WW50 hydrogen peroxide from Du Pont, and these were added directly to soils taken directly from 4°C storage. In some experiments a stainless steel paddle stirrer was used for stirring supernatant only at 20 rpm. In control runs, there was no measurable loss of hydrogen peroxide in the absence of soil.

The standard static batch test developed for this study was run as follows. A 20.0 g portion of soil was placed in a 400 mL beaker. To the soil was rapidly added 150 g of 0.1% H_2O_2. The mixture was not further agitated after the peroxide solution was added, though spot-check experiments showed up to about a 25% increase in rate when the mixture was stirred initially for up to 30 seconds to improve admixture.

Samples of known weight or volume (4–6 g or mL) were pipetted from the supernatant from the same approximate location, just below the surface of the supernatant near the beaker sidewall, typically at 0.5, 1, 2, 3, 4, 5, and 6 hours. The withdrawn sample was added immediately to approximately 100 mL of distilled water, already acidified with approximately 10 mL of 4:1 water:sulfuric acid. To the sample in the acidified water was added 10 mL of 10% potassium iodide, a few drops of ammonium molybdate solution, and a few mL of starch solution. The resulting dark mixture was titrated immediately, and as rapidly as practicable, with 0.01N sodium thiosulfate. The end point (to the absence of a blue starch complex color) generally occurred within two drops of titrant. Spot-check comparisons with permanganate titration gave essentially the same result. The weight percent H_2O_2 was based on the 150 g of dilute hydrogen peroxide solution, and did not include the 13% to 28% moisture in the nine topsoil samples.

$$\text{Wt. } \%H_2O_2 \text{ Left} = \frac{\text{N of titrant} \times \text{mL of titrant} \times 17.01}{\text{g of sample} \times 10}$$

$$\%H_2O_2 \text{ Destroyed} = \frac{\text{Wt. } \%H_2O_2 \text{ Left} \times 100}{\text{Initial } \%H_2O_2}$$

In a few runs the decomposition was followed by oxygen gas release. The weighed soil sample was charged into a 250 or 500 mL erlenmeyer flask with a ground-glass joint. Then, within a 5 second period, was added a fixed weight, typically 100.0 g of the dilute hydrogen peroxide solution, followed by an immediate closure of the flask with a silicone greased ground-glass one-hole stopper connected via thick latex rubber tubing to an L-shaped piece of glass tubing, the other end of which was already inserted into the bottom of an inverted 50 mL burette below the surface of water in a trough, and with the water initially at the zero mL mark of the burette.

For controls, about 30 mg of catalase was used in place of soil, except that the peroxide solution was added first, and the catalase was dropped from inside the stopper after closure—by placing the catalase on the inside of the stopper, inserting the greased stopper into the flask inclined as much as possible, and then placing the closed flask into an upright position to drop the catalase into the peroxide solution. These precautions prevented losing an initial rush of oxygen gas from the rapid catalase-catalyzed decomposition. With modest stirring, decomposition in the control was complete in 5 to 10 minutes, as determined by the absence of color change on a Clinistix indicator strip, capable of detecting H_2O_2 to 2 ppm.

$$\%H_2O_2 \text{ Destroyed} = \frac{\text{cc gas collected in run} \times 100}{\text{cc gas collected in control}}$$

RESULTS AND DISCUSSION

Batch-Test Options to Measure Peroxide Decomposition

A batch, rather than a column, flow-through test was selected, because preliminary tests in our laboratory with glass columns confirmed a report[7] that hydrogen peroxide substantially decreases the permeability of clay soils. For example, the addition of 0.1% H_2O_2 to percolant caused gravity flow to stop completely within minutes for some clayey soils, even with mild suction applied by flex-tube pumping to the column exit.

Most tests used static contact (other than initial mixing from adding liquid to soil), with a high ratio of liquid (0.1% H_2O_2) to soil; namely, 7.5:1. This provided a relatively large supernatant over a relatively small amount of "mud" on the bottom of a beaker. Though most observations were based on this "high-ratio" test, a few tests were run at low ratios of liquid to soil, to simulate better actual in situ conditions in an aquifer. Low-ratios would typically be 1:3 to 1:4.4 for sandy soil, the latter ratio providing a de minimus supernatant. Minimal supernatant was provided by a close to 1:2 ratio for a loamy topsoil (Farmington), and by a close to 1:1.2 ratio for a hard clay that had first been shredded into small curlicues with a cheese grater. Ratios were based on as-received, nondried soil, containing no visibly drainable liquid.

If the purpose of using a batch contact test is to simulate in situ conditions, a suitable test for a sandy soil, for instance, would use 300–400 g of soil and 100 g of 0.1% H_2O_2, under conditions described above. This quantity of peroxide, if completely decomposed according to the equation above, would yield 35 to 40 mL of collectable oxygen gas. For low-ratio tests, only oxygen gas collection could be used to follow the disappearance of hydrogen peroxide. Supernatant titration gave substantially lower values for residual peroxide than oxygen gas collection, apparently because of slow liquid transfer from a deep mud, where presumably faster peroxide decomposition was taking place, into a shallow supernatant, whereas for high-ratio tests, at least for the first few hours, supernatant titration and oxygen gas collection gave very similar indications of residual hydrogen peroxide. More will be said of gas collection later in connection with the mechanism of decomposition.

Figure 1 shows test results using both a low-ratio (1:3) and a high-ratio test on a sandy soil from the Traverse City site where the U.S. Environmental Protection Agency has been evaluating in situ bioremediation on hydrocarbon fuel contamination. Faster rates from the low-ratio test were typical for other soils, though not all soils showed such a fast rate; i.e., peroxide half-lives of less than 15 minutes. Sandy soil samples from three states, for example, gave peroxide half-lives of about 6 hours in a low-ratio test.

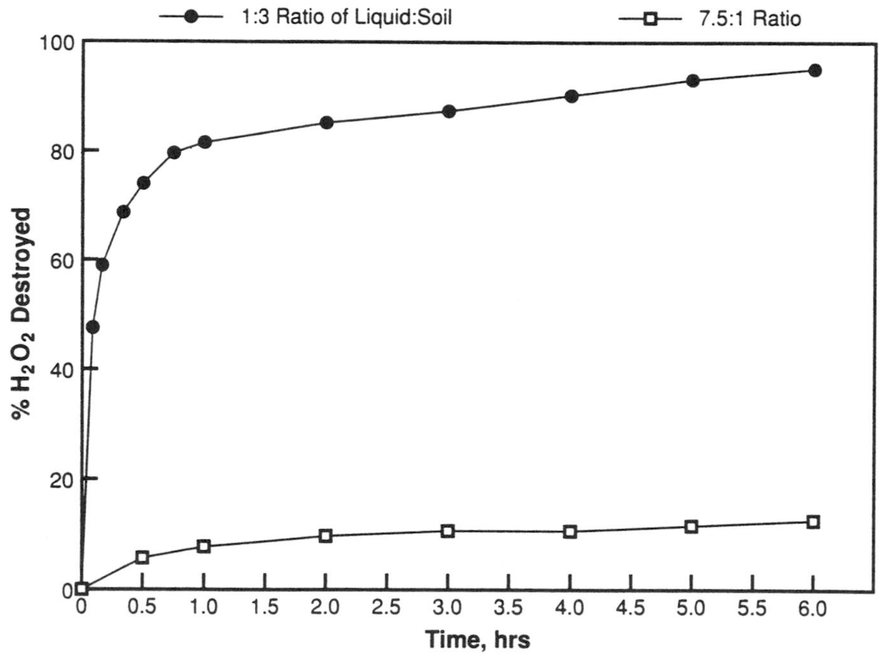

Figure 1. Decomposition of 0.1% H_2O_2 by Traverse City sandy soil.

Variation in Peroxide Decomposition for Soils

Though less simulating of in situ conditions, the high-ratio test was more useful for quick comparative screening. Figure 2 summarizes the decomposition of 0.1% hydrogen peroxide using the high-ratio test, for nine topsoils from five different states. Rates varied widely, from about 42% to about 83% of the peroxide being destroyed after 6 hours. The difference was even more evident in the early reaction stages, with the time to 20% peroxide destruction varying from about 2.5 to about 0.5 hours. When the same test was run on several subsurface soils from 6 different states, the spread in rates was even wider, with the most active destroying virtually 100% of the peroxide within about one-half hour, and the least active destroying no more than 10% of the peroxide in 6 hours. This may be a little surprising in that topsoils are generally thought of as containing relatively high concentrations of organic material (and 5 of the topsoils contained 5% to 6% organic material), and relatively high populations of microorganisms, both of which would be expected to correlate in a positive way with peroxide decomposition.

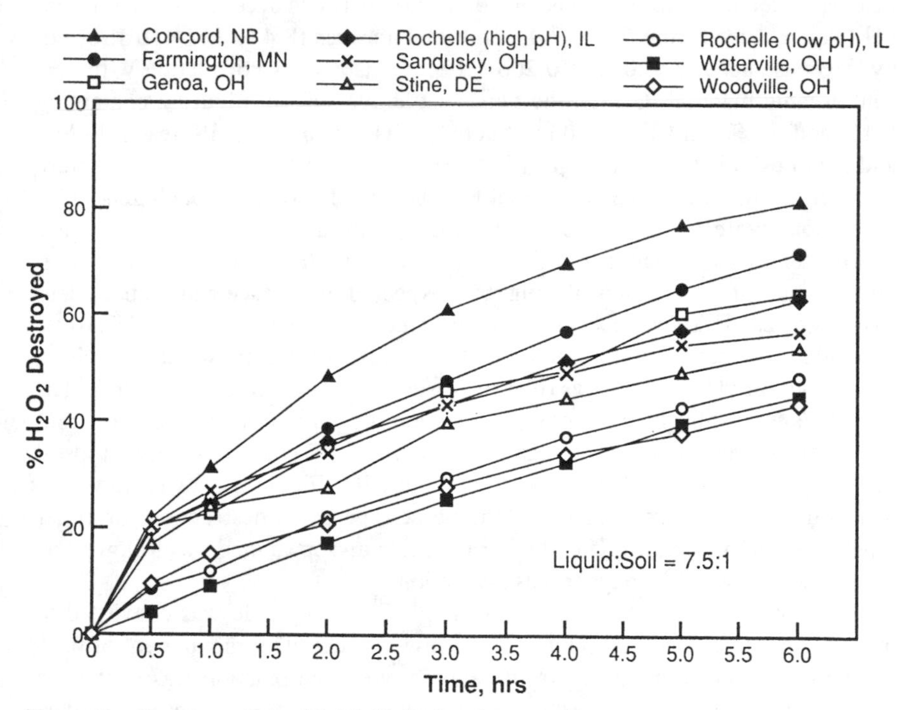

Figure 2. Decomposition of 0.1% H_2O_2 by nine topsoils.

Reproducibility was good for the high-ratio test. For the most and least active and two intermediate topsoils, curves were essentially identical after the soil had aged an additional 7 months at 4°C. Even three months' storage at ambient temperature caused only a small change (increase) in activity for the two topsoils of intermediate activity that were tested. Care was taken to see that factors such as beaker size, and supernatant movement (static versus slow supernatant-only stirring or brief initial stirring after liquid was added) were kept constant during any comparative tests. Changing such factors could change the rate somewhat, though not nearly as much as might be the case by removing diffusion control through complete-mix stirring. Care should also be taken that sufficient samples are taken to properly characterize a site. For a site as a whole, decomposition activity can be fairly uniform or can vary considerably. At two sites, a substantial difference in rate was found for subsamples taken from the opposite ends of approximately 10″ sampling canisters.

Some Factors Affecting Decomposition

For the nine topsoils, soil characteristics such as pH, organic material content, sand/soil/clay composition, or aerobic microbial population did not suggest themselves as reasonable indirect indicators of soil activity for decomposing hydrogen peroxide. There seemed to be no correlation between decomposition rate and pH (which ranged from 6.3 to 7.4), and total aerobes (8.4 to 112×10^7), or dehydrogenase activity (0.015 to 0.266). There was a weak correlation with percent organic material (O.M.), the two most active soils in Figure 2 containing 5.1% and 6.1% O.M., the two least active soils containing 1.6% (each) O.M., and the ones in between containing between 1.2% and 5.5% O.M. Soil texture (sand/clay/silt) was an even less reliable indicator of activity, though generally, sandy soils were less active than those high in silt and clay.

If percent organic matter in soil were an important factor in peroxide decomposition rate, high organic soils might be expected to produce substantially less oxygen (by catalytic decomposition), because more would be reduced to just water by oxidation reactions with organic matter. Thus, difference in decomposition rate as measured by oxygen gas release versus supernatant titration (with the latter not distinguishing between oxygen- and nonoxygen-releasing decomposition) should be greater for an active, high organic soil (6.1%) like Farmington topsoil than for a low organic (1.6%) soil like Woodville. Figure 3 shows, however, that in the first 6 hours of reaction, the percent peroxide destruction for either topsoil was not much different whether peroxide disappearance was followed by supernatant titration or oxygen gas collection.

More complete confirmation that virtually all the peroxide was converted to oxygen gas came from preliminary tests showing that the same amount of oxygen was obtained from soil or from catalase when the reaction was carried to completion under complete-mix stirring conditions. As noted above, the similarity of rate curves from gas collection and supernatant titration did not seem to apply to low-ratio tests, having a relatively deep liquid-saturated soil layer and a relatively shallow supernatant.

Any catalytic decomposition of hydrogen peroxide, can, of course, be from microbial components (e.g., catalase enzyme inside or outside bacterial cells), or from nonbiological materials, such as heavy metal compounds or minerals in the soil. Preliminary work suggests that both can be operative.

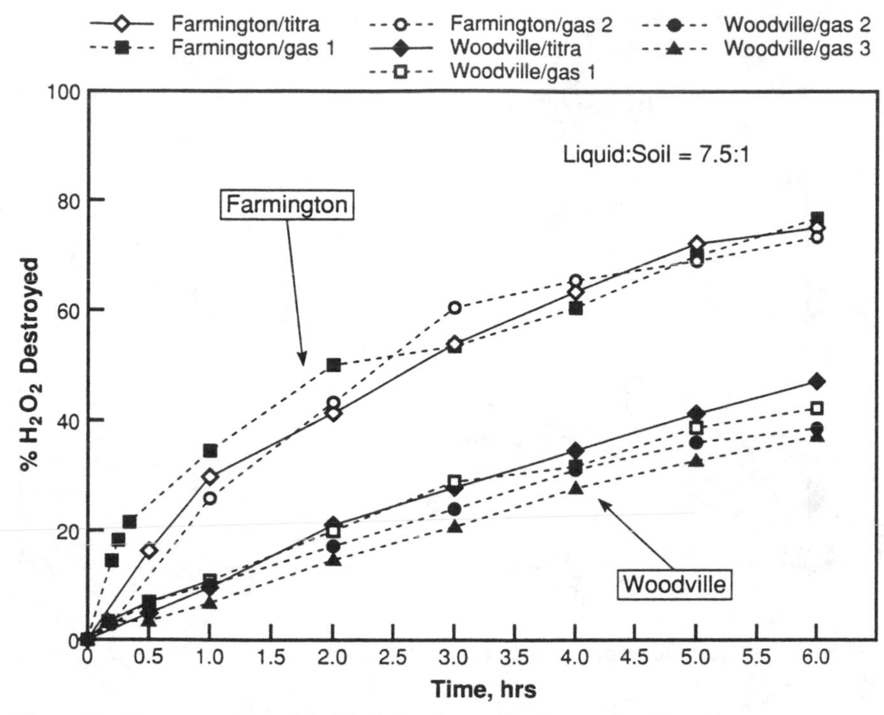

Figure 3. Decomposition of 0.1% H_2O_2: Gas collection vs titration.

Evidence for microbial enzymatic catalysis is in Figure 4, which shows a substantial decrease in decomposition rate when Concord topsoil was autoclaved at 121°C for one hour. This supports the recent Eglin Air Force Base study[6] for enzymatic catalysis, though that case dealt with an infiltration gallery where conditions were especially favorable for the growth of aerobic bacteria. Interestingly, different soils responded differently to autoclaving. For example, for the two least active topsoils (Waterville, Woodville), less than about 10% of the peroxide decomposition activity remained in the autoclaved soil, based on the area under the rate curves for 0 to 6 hrs.

Evidence for nonenzymatic catalysis is in Table 1, which shows a substantial difference in the content of iron and manganese between the two most and the two least active of the nine topsoils, as measured by a soft-digestion procedure, using 1% nitric acid for 15 minutes at 90×C.[8] A California subsurface soil at least as active as the two active topsoils also showed relatively high iron, but unexpectedly low manganese (32 ppm), though only 3% of the total Mn was removed by soft digestion (data not presented). The concentration of other heavy metals whose compounds are known to decompose hydrogen peroxide (Pb, Co, Ni, Cu, Zn, Mo, Wo, and Cr) were also measured by the soft-digestion procedure,

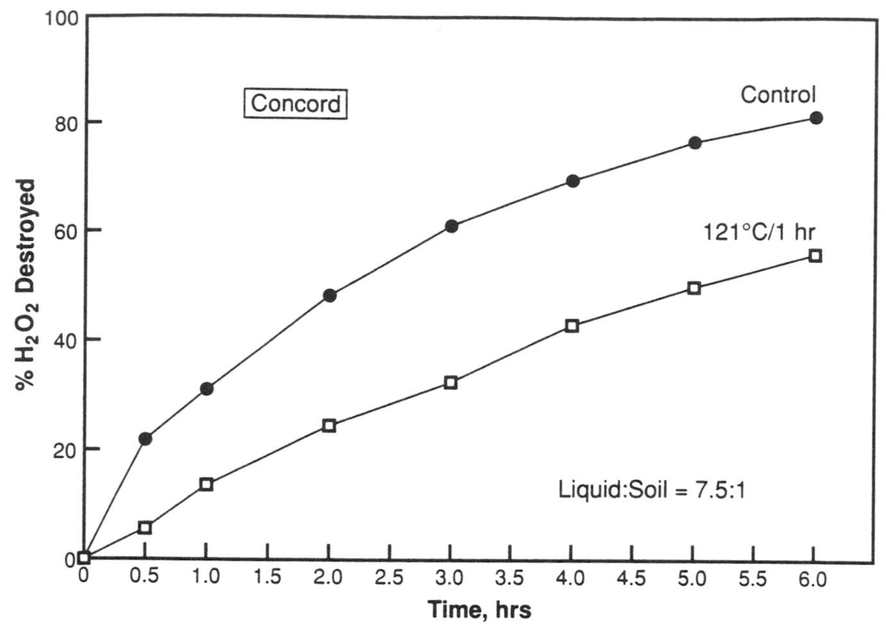

Figure 4. Effect of autoclaving—High activity topsoil.

Table 1. Heavy Metals in Five Soils[a]

	ppm									
	Pb	Mn	Fe	Co	Ni	Cu	Zn	Mo	Wo	Cr
Concord	<5	537	825	<4	3	<4	8	<4	<4	3
Sacramento	95	32	1,060	<4	7	<4	5	<4	<4	4
Farmington	11	315	949	<4	<1	<4	12	<4	<4	8
Woodville	9	43	454	<4	<1	<4	5	<4	<4	<12
Waterville	<5	44	510	<4	<1	<4	<5	<4	<4	<2

[a]Using EPA soft-digestion procedure.

and are shown in Table 1. Excepting a 95-ppm lead level in the subsurface soil (from a gas station site, possibly contaminated with leaded gasoline), no soil showed more than about 10 ppm max of any metal, with most metals being present at less than 4 ppm.

If heavy-metal compounds (or minerals) are implicated in decomposition, it might be expected that phosphates could inactivate catalytically active compounds by converting them to inactive soluble,[9] or in the case of orthophosphate,[10] probably insoluble phosphate compounds, resulting in a decrease in the rate of decomposition. Indeed, this was found to be the case when the active Farmington topsoil was treated with neutral pH orthophosphate. Figure 5 shows an almost 50% reduction in decomposition rate when the 0.1% peroxide contact solution also contained 0.2% sodium phosphate (as PO_4).

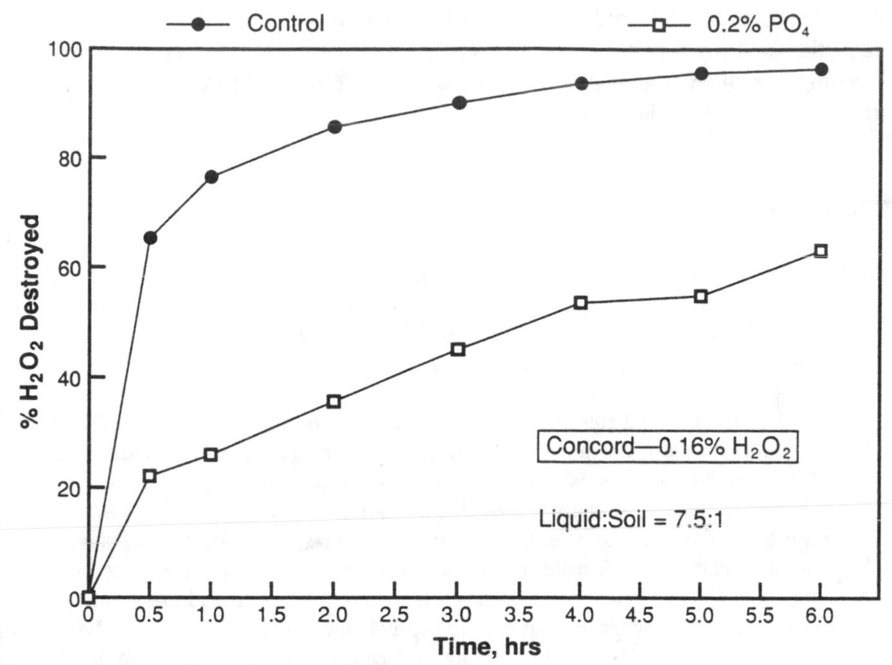

Figure 5. Effect of PO_4—High activity topsoil.

Interestingly, added phosphate had no effect on the rate curves for the two least active topsoils (Woodville, Waterville), the same two soils for which autoclaving (121°C/1 hr), as noted above, provided almost complete removal of decomposition activity. This suggests, as observed earlier,[6] that for some soils, peroxide decomposition can be almost entirely due to microbial enzymes, in which case phosphate treatment would not be expected to help stabilize peroxide against soil-induced decomposition.

It should be cautioned that this study was not intended to show how to obtain basic peroxide decomposition data from the lab that could be used directly for defining operating parameters for in situ application in the field. For one thing, as noted above, the question of optimal decomposition rate is not a simple issue. Decomposition rate can vary not only from site to site, but also with the extent to which bioremediation has progressed, and possibly with other factors such as microbial catalase buildup at injection points. In addition, data are needed to show what happens to soil activity with residence time. If, for example, decomposition catalyst(s) lose activity over time, peroxide would be transported further into a formation before it completely decomposed. Also, the question of peroxide concentration on decomposition could be important, particularly if the initial decomposition rate were to significantly increase with initial peroxide concentration.

Nevertheless, though further work in this whole area is needed, preliminary information on the peroxide decomposition activity of soils, using the simple batch tests described in this study, should provide useful rudimentary information to alert the field practitioner whether a soil is likely to be a fast, medium, or slow decomposer of hydrogen peroxide. This information might help, for example, in determining the placement of wells for injecting hydrogen peroxide.

CONCLUSIONS

1. Individual topsoil and surface soil samples from around the country can differ greatly in how rapidly they decompose dilute hydrogen peroxide solutions, as shown by simple batch contact tests. Individual soil properties were not reliable indicators of peroxide decomposition activity.
2. Though faster decomposition rates from low ratio tests (e.g., 1:3 liquid-to-soil) may be more useful for simulating actual in situ conditions, high-ratio (7.5:1 liquid-to-soil) tests are more useful for quick screening, and provided useful information on the kinetics and mechanism of peroxide decomposition.
3. Oxygen release data strongly suggest that most, if not all, of the hydrogen peroxide was being converted to potentially useful oxygen gas, rather than being wasted in the oxidation of soil materials. Other experiments suggest that the catalysts for this conversion can be both biological and nonbiological, and that iron, and especially manganese compounds, can be implicated in nonbiological catalysis.
4. Even though more lab and field studies are needed before peroxide decomposition data from the lab can be directly related to in situ use in the field, the simple batch tests described in this study may help to determine spacing for peroxide injection by at least distinguishing "fast" and "slow" soils for decomposing hydrogen peroxide.

ACKNOWLEDGMENTS

This contribution would not have been possible without the skillful laboratory efforts of our technician, Herb Doughty, and the valuable consultations with others in Du Pont having an active interest in bioreclamation. These included Drs. Maimu Yllo (analytical chemistry), Carol Litchfield (microbiology), Calvin Chien (hydrology), David Ellis (geology and soil chemistry), Henn Kilkson (kinetics), and Dwight A. Holtzen (mathematics). Many thanks also to Groundwater Technology, Inc., Chadds Ford, Pennsylvania, for providing subsurface samples from three states.

REFERENCES

1. "Field Study of Enhanced Subsurface Biodegradation of Hydrocarbons Using Hydrogen Peroxide as an Oxygen Source," API Publication No. 4448, 1987.
2. Cole, C. A., D. Ochs, and F. C. Funnell. "Hydrogen Peroxide as a Supplemental Oxygen Source," *J. Water Pollut. Control Fed.*, 46: 2579 (1979); J. Houtmeyer,

R. Poffe and H. Verachtest. "Hydrogen Peroxide as a Supplemental Oxygen Source for Activated Sludge: Microbiological Investigations," *European J. Appl. Microbiol.*, 4: 295 (1977).

3. Yaniga, P. M., and W. Smith. "Aquifer Restoration Via Accelerated In Situ Biodegradation of Organic Contaminants," in *Proceedings NWWA/API Conference Petroleum Hydrocarbons and Organic Chemicals in Groundwater.* Houston, TX (1984). National Water Well Association, Worthington, OH. pp. 451–472.
4. Raymond, R. L. "Reclamation of Hydrocarbon Contaminated Ground Waters," U.S. Patent 3,846,290(1954).
5. Raymond, R. L., R. A. Brown, R. D. Norris, E. T. O'Neill. "Stimulation of Biooxidation Processes in Subterranean Formations," U.S. Patent 4,588,506 (1986).
6. Spain, J. C., J. D. Milligan, D. C. Downey, and J. K. Slaughter. "Excessive Bacterial Decomposition of H_2O_2 During Enhanced Biodegradation," *Groundwater* 27:163 (1989).
7. "Chemical Rehabilitation of Soil Wastewater Absorption Systems Using Hydrogen Peroxide: Effects on Soil Permeability," by D. L. Hargett, E. J. Tyler, and J. C. Converse, a 1983 report from the Department of Soil Science, Department of Agricultural Engineering, Wisconsin Geological and Natural History Survey, University of Wisconsin-Madison. Also, 1983 PhD Thesis, D.L. Hargett, The University of Wisconsin-Madison.
8. "Methods for Chemical Analysis of Water and Wastes," U. S. Environmental Protection Agency, Cincinnati, OH.
9. Brown, R. A., and R. D. Norris. "Nutrient for Stimulating Aerobic Bacteria," U.S. Patent 4,727,031 (1988).
10. Britton, L. N. "Feasibility Studies on the Use of Hydrogen Peroxide to Enhance Microbial Degradation of Gasoline," API Publication 4389, May, 1985.

PART V

Risk Assessment

CHAPTER 20

A Critical Evaluation of Indicator Compound Methodologies for No. 2 Fuel Oil

Charles E. Gilbert and **Edward J. Calabrese,** School of Public Health, University of Massachusetts, Amherst

Since it is impractical to analyze for all the constituents in a complex mixture such as No. 2 fuel oil from environmental media or body fluids, the environmental concentration or body burden of mixtures may be estimated from results on a few of the constituents or their metabolites. The use of indicator compounds to demonstrate exposure is based on the assumption that the toxicity of the indicator represents the toxicity of the mixture, or that the retention of the constituent reflects the retention of the mixture.[1]

Indicator compound methodology is intrinsically attractive for a complex mixture such as No. 2 fuel oil since it contains more than 1000 constituents. Conducting a public health evaluation on 1000 or more constituents may be time-consuming and very impractical. Generally the indicator compound should represent those agents that are the greatest potential public health risks at the site. Ideally, the compounds selected to be indicators will be the most toxic, in the highest concentration, the most mobile, and the most persistent in the environment resulting in human exposure.

HISTORICAL CONSIDERATIONS

Using an indicator compound to represent the toxicity of a complex mixture should be done cautiously, as the example here with benzo(a)pyrene indicates.

Benzo(a)pyrene (BaP) has been used frequently as an indicator compound for organic mixtures. Unfortunately, the concentration of BaP by itself does not always reflect the potential carcinogenicity of a mixture. BaP is a potent animal carcinogen, and can be accurately analyzed by standard techniques.[2] Atmospheric levels of BaP have been associated with higher risks of lung cancer for coal tar workers, coke oven workers, and others.[3]

Studies of roofers[4] indicate that their exposure to BaP ranging 14-6000 $\mu g/m^3$, is 10 to 1000 times greater than that of two-pack-a-day smokers whose BaP exposure is 1.4 μg/day. However, the lung cancer rate is much higher for cigarette smokers than for roofers. This suggests that the BaP concentration is not always an accurate indicator of the carcinogenic potential of an organic compound mixture. Other agents acting alone or in concert with the BaP may have a significant impact on the health effect of the mixtures. Therefore, although benzo(a)pyrene is a carcinogen, it does not influence the development of lung cancer as significantly as other constituents in cigarette smoke. This type of interaction should be considered when selecting indicator compounds for complex mixtures.

The bioavailability of compounds should be considered to determine the potential exposure to humans, since it provides a basis for determining the biologically effective dose. The biologically effective dose varies as a percentage of the administered dose and sets the parameters necessary for estimating the chemical's body burden. Bioavailability may be influenced by agents or circumstances that change the release, absorption, distribution, metabolism, or biologic effect. Bioavailability information will provide more insight into the mixture's biological effect and the toxicity of the mixture, especially its interactive effects.

CONSTITUENTS OF NO. 2 FUEL OIL

The analytical determination of the individual chemical constituents of petroleum products such as No. 2 fuel oil is a very difficult process. Isolation, identification, and quantification of the chemical constituents in high temperature boiling fractions of petroleum are particularly complex tasks. The American Petroleum Institute (API) started Research Project No. 6 in 1927 entitled "The Separation, Identification, and Determination of the Chemical Constituents of Commercial Petroleum Fractions."[5] Research for this project was terminated in 1967 after 40 years of work.[6] More sophisticated technology developed since that time has resulted in better identification of the individualized components in petroleum fuels, but there is still not a clear understanding of petroleum constituent concentrations.

Product specifications for petroleum fuels are primarily described in physical characteristics which will guarantee certain handling and combustion properties, whether the products are a virgin distillate, a reformed product, or, as is most often the case, a blend from all of these sources. Different blends with different chemical specifications are used in different geographical areas and under different climatic conditions. Thus, a master list of No. 2 fuel oil chemical constituents including qualitative and quantitative characteristics is not possible. Since physical specifications are sufficient to define the characteristics required for combustion, the more difficult analyses required for compositional identification are generally not conducted.

Fuel oil No. 2 may be produced by straight distillation procedures or through a process known as cracking. Cracking the petroleum component stream involves the conversion of higher molecular weight components above the boiling range of the product into lower molecular weight components by catalytic or high temperature decomposition. The boiling range of components for No. 2 fuel oil is from 177°C to 343°C (360°F to 650°F).

Subclasses of compounds, especially in the aromatic class, are in No. 2 fuel oil (see Table 1). Not all the chemicals in each subclass have been identified, but the principal constituent types are known. Table 2 shows the constituents of No. 2 fuel oil that have been detected by high resolution capillary gas chromatography.

The constituents that make up the higher boiling point fractions of gasoline (>150°C) are also the major constituents of kerosene and No. 2 fuel oil. In addition, the lower boiling point constituents of kerosene and No. 2 fuel oil are in the higher boiling fractions of gasoline.

Wenger et al.[8] evaluated three samples of No. 2 fuel oil and two reports on the chemical constituents of No. 2 fuel oil. The three samples were selected to represent the product marketed as No. 2 fuel oil from refined crude oil. Using U.S. EPA standard procedures (GC/MS), the oil was analyzed for compounds on U.S. EPA's Hazardous Substance List. Only four of the 34 volatile organic

Table 1. Constituents of Crude Oil Derived No. 2 Fuel Oil

Composition	Wt %
SECTION 1	
Saturates	
G1 Paraffins (Alkanes)	30.4
G2 Mononaphthenes (Cycloalkanes)	17.5
G3 Polynaphthenes	14.0
SECTION 2	
Olefins	
G1 Olefins	0.0
SECTION 3	
Aromatics	
G1 Benzene	10.3
G2 Indans/tetralins	7.3
G3 Dinaphthenebenzenes	4.6
G4 Naphthalenes	5.9
G5 Acenaphthenes	3.8
G6 Acenaphthalenes	5.4
G7 Benzothiophenes	0.9

Source: U.S. Department of Energy, Reference #7.

Table 2. Constituents of No. 2 Fuel Oil

Constituent	CAS No.
SECTION 1. PART A. STRAIGHT CHAIN OR n-ALKANES	
1. C_{11}-undecane	1120-21-4
2. C_{12}-dodecane	112-40-3
3. C_{13}-tridecane	629-50-5
4. C_{14}-tetradecane	629-59-4
5. C_{15}-pentadecane	629-62-9
6. C_{16}-hexadecane	544-76-3
7. C_{17}-heptadecane	629-78-7
8. C_{18}-octadecane	593-45-3
9. C_{19}-nonadecane	629-92-5
10. C_{20}-eicosane	112-95-8
11. C_{21}-heneicosane	629-94-7
12. C_{22}-docosane	629-97-0
13. C_{23}-tricosane	638-67-5
14. C_{24}-tetracosane	646-31-1
15. C_{25}-pentacosane	629-99-2
16. C_{26}-hexacosane	630-01-3
17. C_{27}-heptacosane	593-49-7
SECTION 1. PART B. BRANCHED ALKANES	
1. Farnesane (2,6,10-trimethyldodecane)	3891-98-3
2. Pristane (2,6,10,14-tetramethylpentadecane)	1921-70-6
3. Phytane (2,6,10,14-tetramethylhexadecane)	638-36-8
SECTION 2, 3, AND 4. CYCLOALKANE, ALKENE, AND CYCLOALKENE	
Listing not obtainable	
SECTION 5. AROMATICS, mono	
1. Benzene	71-43-2
2. Toluene	108-88-3
3. m-Xylene	108-38-3
4. p-Xylene	106-42-3
5. o-Xylene	95-47-6
SECTION 6. POLYCYCLIC AROMATIC HYDROCARBONS	
1. Naphthalene	91-20-3
Group 2 Methylnaphthalenes	
Group 3 C_2-Naphthalenes	

Source: U.S. Department of Energy, Reference #7.

compounds tested for were found. None of 15 semivolatile acid extractable compounds was found. Three of 53 semivolatile base neutral extractable compounds were found. Benzene and n-hexane were not detected in No. 2 fuel. Aliphatic hydrocarbons C_8-C_{18} were identified to make up the bulk of No. 2 fuel oil. Six other compounds, 2-methylnaphthalene, xylene, naphthalene, phenanthrene, ethylbenzene, and toluene were identified. Wenger et al.[8] evaluated the work of Thomas,[9] whose research was not to conduct a complete analysis of fuel oils, but to determine particular chemical properties of chemicals present in synthetic and natural oils. One sample of No. 2 fuel oil (labeled for use as diesel fuel) was analyzed by Thomas and found to contain 2-methylnaphthalene, naphthalene, phenanthrene, ethylbenzene, and xylene.

Pancirov et al.[10] analyzed only the polycyclic aromatic hydrocarbons (PAHs) present in some fuels. One sample identified as distillate product No. 2 fuel oil was reported to contain benzo(a)pyrene and benzanthracene. Also identified in No. 2 fuel oil were pyrene, fluoranthene, triphenylene, and chrysene. Wenger et al.[8] state that there is no indication that benzene is present in No. 2 fuel oil. They support this through Thomas' (1984) research which did not report benzene as a constituent of No. 2 fuel oil and their own analysis (GC/MS) of three samples which did not detect benzene at 0.02% (detection limit 1250 mg/L). Seven of 102 organic chemicals tested for were found. Wenger et al.[8] noted that xylene was detected at concentrations of 0.15% to 0.42%. Toluene was not reported in the Thomas analysis and was reported to range from 0.025% to 0.11% by Wenger et al.

Octane, an eight carbon chain, was the smallest aliphatic hydrocarbon reported in the Wenger et al. and Thomas studies. Polycyclic aromatic hydrocarbons were detected in the Thomas and the Wenger et al. studies. Chrysene and phenathrene have caused skin tumors on mice but the International Agency for Research on Cancer[2] does not believe that sufficient evidence is available to classify these compounds as carcinogenic. The recognized carcinogens benzanthracene and benzo(a)pyrene were detected by Pancirov et al.[10] at one part per million. However, Wenger et al. and Thomas did not find these chemicals in their samples of No. 2 fuel oil.

Wenger et al. identified quinoline, cresols, and phenol for consideration in the toxicology of No. 2 fuel oil. These chemicals were reported in the No. 2 fuel oil analysis by Thomas in concentrations less than 60 parts per million. Total cresols were found at 54.3 mg/L, quinoline at 9.2 mg/L, and phenol at 6.8 mg/L. The development of skin cancers in workers and animals exposed to creosotes has been reported.[11] Phenol has insufficient data to classify it as a carcinogen. Quinoline may increase the incidence of tumors in rat liver.[12] Quinoline has not been classified as carcinogenic because the data are insufficient to estimate cancer potential.

VARIATIONS IN NO. 2 FUEL OIL CONSTITUENTS

Different blends make up a given fuel, each with its own requirements. Blending is geared toward the fuel use in different geographic areas, and for variations in seasonal and climatic conditions. These requirements contribute to variations in the constituent makeup between different fuel samples.

Fuels produced by chemical transformation, reforming, or cracking have different compositions than fuels distilled from crude stocks. The U.S. Department of Energy[7] reports that chemically transformed stocks contain fewer cancer-linked compounds (e.g., benzo(a)pyrene, benzanthracene) than stocks derived by distillation. This is less important for most potential public exposures since the final marketable fuel is usually a blend of virgin distillates and chemically transformed stocks.

An additional area of concern is the compositional differences between fuels derived from natural petroleum and fuels derived from synthetic petroleums or other synthetic pathways. A heavier fuel like synthetic No. 2 fuel oil may contain a greater concentration of mutagens and carcinogens (e.g., polycyclic aromatic hydrocarbons) than natural petroleum-derived No. 2 fuel oil.[7]

TOXICITY OF INDIVIDUAL CONSTITUENTS IN PETROLEUM MIXTURES

The toxicity of an individual compound in a petroleum mixture may be quite different from the toxicity reported for that same compound acting alone. The carcinogenic effects of one compound affected by the toxicity of another compound can be demonstrated by looking at the initiating carcinogenic, cocarcinogenic, and tumor-promoting properties of several long chain N-alkanes. The middle to upper length N-alkanes of the C_6 to C_{20} series are the principal constituents in No. 2 fuel oil.

An initiating carcinogen is a chemical that causes an alteration in genetic material (mutation) contributing to a change in that cell which can later result in tumor formation. There usually is a latency time period between the modification in the cell and tumor formation. Following the cell's exposure to an initiator, one or more consecutive cellular events must occur to result in tumor development. A cocarcinogen is an agent that is not carcinogenic by itself but enhances the carcinogenic effect when it is given at the same time as the carcinogen. A promoter is an agent that is not carcinogenic by itself, but when given after the carcinogen enhances cancer development.

There are some published examples of initiation carcinogenic, cocarcinogenic, and promotion that may help clarify the role of interactions between the chemical constituents of No. 2 fuel oil. These examples described below include the compounds 7-12-dimethylbenzanthracene, some N-alkanes, and benzo(a)pyrene.

The carcinogen 7-12-dimethylbenzanthracene was applied to mouse skin at less than the carcinogenic dose. Subsequent dermal administration of dodecane and tetradecane resulted in the promotion of skin tumors. Skin tumors did not form on mice with no subsequent dermal treatment or subsequent dermal treatment with hexane or octane. The dermal cancer promotion activity of hexadecane, octadecane, and eicosane were equivocal in these mice pretreated with 7-12-dimethylbenzanthracene. Finally, none of N-alkanes administered to the mice alone with no subsequent treatment produced skin tumors.[13]

While dodecane and n-hexadecane are carcinogenic on mouse skin, at concentrations as low as 20%, octadecane and eicosane induce tumors more rapidly than dodecane and n-hexadecane and are active both as cocarcinogens and promoters.[14] Dodecane and decane have been found to be cocarcinogens with benzo(a)pyrene, and are not carcinogenic when free of polycyclic impurities.[15] Decane, dodecane, and tetradecane were cocarcinogenic when administered with ultraviolet light. Decane and tetradecane produced tumors at wavelengths greater than 350 nm which are considered to be noncarcinogenic wavelengths.[16]

Thus, the constituents of petroleum products can have altered toxicity when they are combined into a mixture such as No. 2 fuel oil. These interactions may enhance a toxic effect, inhibit toxicity, or result in a toxic effect when constituents are combined which never occurred during exposure to individual constituents.

These combined studies provide some guidance in the selection of indicator compounds in No. 2 fuel oils. While none of the N-alkanes listed above have been shown to be initiators of cancer, some are cocarcinogenic and others are promoters of cancer. The indicator compound selection list could include constituents that are potential cancer promoters such as octadecane and eicosane and potential cocarcinogens such as dodecane, n-hexadecane, octadecane, and eicosane.

HEALTH EFFECTS

The health effects of refined petroleum products are due to four major groups of hydrocarbon components: alkanes, alkenes, alicyclic and aromatic compounds. The alkane (paraffin) compounds of refined petroleum products are primarily aliphatic hydrocarbons from C3 to C8. Alkanes have potent narcotic action when inhaled at high doses. Straight chain alkanes are, in general, more toxic than branched chain isomers.[17,18] Acute alkane intoxication involves a transient depression of the central nervous system. Polyneuropathy has developed in animals and humans following chronic intoxication by alkanes.[18] Alicyclic hydrocarbons, which include saturated and unsaturated naphthenes or cycloparaffins, have toxic effects similar to aliphatic hydrocarbons. Alicyclic compounds and alkene (olefin, unsaturated aliphatics) constituents from petroleum products have limited acute toxicity, are anesthetic, and are CNS depressants.

The aromatics, which include benzene, alkyl derivatives of benzene (such as toluene and xylene), and polynucear aromatic hydrocarbons (such as anthracene, phenanthrene), have been considered the most toxic components of refined petroleum.[17,19] Humans are more likely to be exposed to these compounds, since the aromatics are more water soluble than the other hydrocarbon compounds in petroleum products. Benzene has the highest health impact, since it is volatile, water soluble, exerts its toxic effect on blood forming tissues, and is a human carcinogen. Toluene, the xylenes, and the alkylated benzene derivatives are less toxic than benzene but are quite water soluble and still pose significant toxicity to humans.[20]

MINOR FUEL CONSTITUENTS CONTRIBUTION TO TOXICITY

Those constituents making up the minor fraction of No. 2 fuel oil may be responsible for the majority of the toxic effects of these mixtures.[21] Two factors may contribute to this potential; first, the minor constituents may contaminate an environment more rapidly than the major constituents and second, the minor constituents may be more toxic than the major constituents. For example, the toxicity of water contaminated with No. 2 fuel oil depends on what constituents are water soluble and their persistence in the environment. The persistent water soluble constituents resulting from mixing No. 2 fuel oil and sea water are aniline-alkylaniline alkylphenols, and indole-alklindoles.[22] While these compounds are the principal constituents of the water soluble fraction of No. 2 fuel oil, they are not the major constituents of No. 2 fuel oil.

The principal mutagenic and carcinogenic constituents in crude oils are not necessarily the principal quantitative constituents. In natural crude oil a primary class of compounds responsible for mutagenicity and carcinogenicity in experimental animals are the polycyclic aromatic hydrocarbons (PAHs) with four rings or greater.[7,16,23]

Some heterocyclic compounds may be responsible for the carcinogenicity of natural crude oils in animals. The PAHs believed to be responsible for the mutagenicity and carcinogenicity in oil include benzanthracenes, dibenzanthracenes, substituted anthracenes, benzopyrenes, benzofluorenes, pyrene, substituted pyrenes, and chrysenes.[16,23]

Synthetic crude oils have been found to have greater mutagenic potential than natural crude oils. Polycyclic aromatic amines (PAAs) are the constituents responsible for the increased mutagenicity that synthetic crude oils display relative to natural crude oils.[24,25] All the mutagenicity observed in the Ames Assay conducted on natural crude oils are thought to be associated with the PAHs in the neutral fraction of the oil. The basic fraction of natural crude oil, which contains the PAAs, does not show Ames Assay mutagenicity. This suggests that the PAA concentration of natural crude oil is very low. In synthetic crude oil the results were quite different, since the basic fraction (containing PAAs) contributed 14% to 84% of the Ames Assay mutagenicity.[25] These results indicate a higher concentration of PAAs in synthetic crude oil than natural crude oil.

Three different investigations have found that coal-derived synthetic crude oils have greater mutagenic potential than natural crude oil.[23,25,26] The PAAs found in coal-derived synthetic crude oils include amino-anthracenes, amino-naphthalenes, amino-phenanthrenes amino-pyrenes, and amino-chrysenes.[24,25]

PAHs with four or more aromatic rings and heterocyclic compounds are concentrated in the heavier fractions of crude oils, 355° to 540°C for cracked streams, lower for straight distillation streams.[16] Number 2 fuel oil with a boiling range of 343°C would, therefore, contain few if any PAHs. No carcinogen effect occurred in mouse skin treated with distillation fractions from the 343°C boiling range of a commercial fluid catalytic cracking from a petroleum residium. However, 29 to 600 ppm of the PAH benzo(a)pyrene was found in heating oil, diesel fuel, and unused motor oil.[16]

Heavier synthetically derived fuels such as No. 2 fuel oil and kerosene may contain greater concentrations of PAAs,[7] since the concentration of PAAs increases as the boiling temperature of the fuel increases. Some two and three ring

PAAs in synthetic crude oils are mutagens and carcinogens,[25] and since two and three ring PAAs have lower boiling temperatures, they are likely to be found in No. 2 fuel oil and kerosene. The data are limited concerning the PAA content in synthetic crude oil derived No. 2 fuel oil and kerosene, and further research in this area would be beneficial. Chemical reforming reduces PAH and PAA levels, and therefore, fuels derived from chemically reformed fuel stocks should have reduced mutagenic and carcinogenic potential.[24]

CONSTITUENT TRANSFORMATION AND ITS EFFECT ON TOXICITY

Following its release into the environment, the fuel constituents may become transformed, and this may enhance or reduce their toxicity. The burning of the fuel will result in the formation of mutagenic and carcinogenic PAHs. PAHs form during the incomplete combustion of organic compounds. These PAHs can be expected to be produced during the burning of No. 2 fuel oil to heat residences.

Photooxidation products of oils are of significant concern since these products have been shown to have increased toxicity over the parent compound. Photochemical transformation may be the most significant modifying factor of No. 2 fuel oil but the one with the least available information.[27] Photooxidation of No. 2 fuel oil produces water soluble compounds that are more toxic to yeast, algae, fish, and shrimp than water soluble compounds from No. 2 fuel oil that has not been exposed to sunlight.[27-29]

These photooxidation products of No. 2 fuel oil include hydroperoxides, such as, cumene hydroperoxide, tert-butyl hydroperoxide, and tetralin hydroperoxide. These hydroperoxides are thought to be responsible for the enhanced toxicity of photooxidized No. 2 fuel oil. Phenolic compounds and carboxylic acids are also produced through the photooxidation of No. 2 fuel oil.[28]

INDICATOR COMPOUND METHODOLOGY EXAMPLES

Overview

The indicator compound methodology has been applied to a number of situations such as a pesticide plant, mine waste area, and hazardous waste sites. Indicator compound methodologies have been developed by the Ebasco-Army, the Environmental Protection Agency (EPA) for Superfund sites, and the EPA for underground storage tank risk assessment procedures. These applications are discussed below.

Indicator Compounds at a Pesticide Production Plant

Marshall et al.[30] developed an indicator compound selection methodology for a former pesticide production plant using the methodology described in U.S. EPA

Superfund Public Health Evaluation Manual.[31] The selection criteria were applied to 70 chemicals with some migration potential, to reduce the list to a smaller, more manageable number of indicator chemicals which would represent the hazards on this former pesticide production site.

Some of the chemicals on the site were isolated in storage drums which are believed to have prevented the chemicals from entering the air, ground, and water. Seventy chemicals that were not contained in drums were identified as the potentially mobile chemicals.

The 70 potentially mobile chemicals were divided into 15 categories, including 13 organic categories and 2 inorganic categories. Organic chemicals were grouped by toxicological characteristics and environmental behavior. The site distribution patterns and reported concentrations were not used as the primary criteria in this categorization.

If a chemical category contained only one chemical, then this chemical was selected as the indicator compound.

Chemical categories containing more than one chemical were evaluated for carcinogenicity based upon the Annual Report on Carcinogens,[32] from the National Toxicology Program and on the carcinogenic potency reported by the Carcinogen Assessment Group.[33] If only one chemical in a category was carcinogenic, it was selected as the indicator compound.

Chemical groups with more than one known or suspected carcinogen were reviewed to select the compound with the highest reported carcinogenic potency.

Chemical categories with no known carcinogens were ranked according to: (1) acute or chronic toxicity as indicated by the Threshold Limit Values, developed by the American Conference of Governmental Industrial Hygienists,[34] or the U.S. EPA's Ambient Water Quality Criteria or National Interim Primary Drinking Water Standards;[35] (2) water solubility; (3) volatility; and (4) the reported concentration in the soil surface.

Indicator Compounds at a Mine Waste Contaminated Site

LaGoy et al.[36] initiated their selection of indicator compounds by evaluating the environmental laboratory data for a site contaminated with mining waste. They determined that the lead and zinc in soil and mine tailings at the site were greater than background soil samples or normal background levels. Cadmium, copper, and mercury were detected at levels approximately one order of magnitude above expected background levels. Barium, manganese, and arsenic were somewhat elevated compared to background soil. Zinc, copper, barium, and manganese were considered to be toxic to humans only at high levels of exposure and were not included as indicators. Mercury was considered to be both low in concentration and less toxic in its inorganic form and, therefore, was not included as an indicator. These decisions resulted in lead, cadmium, and arsenic being selected as indicator compounds for this site contaminated with mining waste.

Indicator Compounds at a Hazardous Waste Site

Monitoring Data and Preliminary Indicators

Brett et al.[37] modified the procedure for selection of indicator chemicals outlined in the EPA Superfund Manual,[31] because in this site they were concerned with the volatilization of chemicals from soil during the excavation process and potential inhalation of the vapors by nearby residents.The selection of indicator compounds occurred through a three phase procedure.

1. Onsite chemical concentrations were compared to background concentrations. All inorganic chemicals were found at concentrations similar to background concentrations and were excluded as indicator chemicals. The authors checked this procedure to ensure that omitting the inorganic chemicals would not significantly affect the risk estimate. These calculations were made using the amount of inorganic containing soil suspended in air that would be necessary to exceed the ACGIH TLV for these compounds, assuming that the chemicals were present in soil at the maximum concentration found on site. The calculations showed that the contaminated dust concentrations necessary to exceed the TLV would have to exceed 100 mg/m^3 for each of the inorganic chemicals. This concentration is 10 times greater than the occupational TLV for nuisance dust, and is not likely to be produced at the closest residence during excavation. It was concluded that the concentrations of these inorganic chemicals would not significantly increase the health risk during excavation.
2. Chemicals in trace amounts, those found in only one sample, or tentatively identified were excluded as indicators. These chemicals had concentrations less than low ppb levels with low toxicity. Chemicals found outside the excavation area were excluded.
3. Final indicator compounds were those identified in more than one sample on the site.

The indicator chemical selection process resulted in 13 chemicals from 6 chemical classes used to estimate the risk during the excavation. Some chemicals were excluded as indicators because more toxic members of the same chemical class were present at higher concentrations and were selected. The chemicals selected and the chemical classes are shown in Table 3.

Toxicity Criteria for Indicator Chemicals

Acute, subchronic, and chronic health effects were evaluated for each of the indicator chemicals Brett et al.[37] identified. The toxicity data were compiled into toxicity profiles. Toxicity profiles also documented the development of acute and subchronic acceptable daily intakes, (ADI), and unit cancer risks (UCRs). Subchronic ADIs were used to assess potential risks from noncarcinogens because

Table 3. Selected Indicator Chemicals and Chemical Classes at a Hazardous Waste Site (Brett et al.,[37])

Chemical	Chemical Class
Chlorobenzene	Chlorobenzenes
2-Chlorophenol	Phenols
Cresol[a]	Phenols
1,2-Dichlorobenzene	Chlorobenzenes
2,4-Dimethylphenol	Phenols
Ethylbenzene	Alkylbenzenes
Nitrobenzene	Nitro/amino-benzenes
Phenol	Phenols
Tetrachloroethylene	Chloroalkenes
Toluene	Alkylbenzenes
1,2,4-Trichlorobenzene	Chlorobenzenes
1,2,3-Trichloropropane	Chloroalkanes
Xylene[b]	Alkylbenzenes

[a]Includes both o- and p-cresol (2-methylphenol and 4-methylphenol).
[b]Includes both o- and m-xylene (1,2-dimethylbenzene and 1,3-dimethylbenzene).

it was assumed that the population would be exposed to the contaminated soil through the excavation for four months.

Acute ADIs for the maximum one-day exposure concentration that are unlikely to result in adverse effects were derived for each indicator chemical. EPA one-day health advisories were used to calculate the acute ADIs for ethylbenzene and xylene. ACGIH TLVs were used for the remaining 11 chemicals. All doses were converted to mg/Kg/day.

Acute ADIs were compared with maximum estimated daily exposure to the indicator chemical concentrations resulting from excavation of the most contaminated portions of the site. Unit cancer risks were calculated for carcinogens.

Tetrachloroethylene was the only indicator chemical classified as an animal carcinogen. One chemical, 1,2,3-trichloropropane, is suspected by EPA of being a carcinogen and the National Toxicology Program is considering testing 1,2,3-trichloropropane for carcinogenicity. Brett et al.[37] developed a hypothetical unit cancer risk value of 0.10 mg/kg day for both inhalation and ingestion exposure to 1,2,3,-trichloropropane. This value was derived using the geometric mean of the oral unit cancer risk values of carbon tetrachloride, chloroform, 1,2-dichloroethane, dichloromethane, and 1,2-dibromo-3-chloropropane.[38]

While several potential routes of exposure were identified in the Brett et al.[37] risk assessment, the inhalation of vapor generated as a result of excavation activities was identified as the major contributor to public health risk.

Of the 13 indicator chemicals identified at the hazardous waste site, the three of greatest health concern are tetrachloroethylene, 1,2,3-trichloropropane, and xylene. Estimated releases of 1,2,3-trichloropropane and tetrachloroethylene from the soil contributed to a calculated upperbound risk of cancer to residents of three new cases per lifetime for every 100,000 residents exposed.

Other chronic health effects were estimated to occur in residents due to exposures to 1,2,3-trichloropropane and xylene. Acute toxic effects in local residents were not expected since maximum daily exposures were estimated to be less than 6% of the acute ADIs.

Ebasco-Army Indicator Chemical Procedure

The Ebasco Corporation has developed an indicator chemical selection process for the U.S. Army that it uses to: "determine which chemicals will receive a detailed documentation and quantitative risk evaluation and which will be considered for only descriptive/qualitative documentation regarding their toxicity, risk, and significance of their presence. . . ."[39]

The Ebasco process uses the following procedure to assess the number of chemicals which will undergo a detailed assessment and documentation in estimating soil criteria in their Preliminary Pollutant Limit Values (PPLV).

1. The frequency (F) of occurrence of each target contaminant measured is calculated as

$$F = \frac{\#\text{ of samples with contaminant}}{\#\text{ of samples analyzed}} \times 100$$

2. The average contaminant concentration and the associated standard deviation (SD) are calculated.

3. A ranking index, similar to U.S. EPA's exposure index is calculated as:

$$(EI)_R = \frac{\text{Average Contaminant Concentration}}{PPLV_R R} + 2\ SD \times F$$

where $PPLV_R R$ is the preliminary pollutant limit value and is calculated for the rural residential land use scenario.

The PPLVs are calculated for each target chemical as a function of the contaminant concentration in various transport media, intake rates, and intermedia partition coefficients for different exposure routes.

4. The chemical contaminants are then ranked using the $(EI)_R$ value. A decision is then made as to which agents will undergo a detailed documentation and quantitative risk evaluation to be used for the qualitative documentation.

EPA Superfund Indicator Chemical Selection Methodology

The Superfund public health evaluation procedure is based upon selected chemicals, referred to as indicator chemicals, that pose the greatest potential health hazard at a site.[38,40]

Indicator chemicals are selected to represent the most toxic, mobile, and persistent chemicals, and those present in the largest amounts. The EPA's indicator chemical selection procedure is based on algorithms and professional judgment.

The toxicity of each chemical found at the site is evaluated based on the concentrations measured in specific environmental media. The chemical and physical properties that affect the potential for a substance to migrate or persist in the environment are included in the process.

The EPA uses the following algorithm to develop a quantitative ranking Indicator Score for each chemical found at the site.

$$IS_i = (C_ij T_ij),$$

where

IS_i = indicator score for chemical i (unitless)
C_ij = concentration of chemical i in medium j at the site based on monitoring data (units in mg/L for water, mg/kg for soil, mg/m^3 in air)
T_ij = a toxicity constant for chemical i in medium j (units are inverse of concentration units).

The selection of the concentration value may be based on the geometric or arithmetic mean of some or all of the samples. As an alternative, a concentration reflecting a trend over time can be selected. The frequency of detection of the chemicals should be incorporated into the concentration values.

Toxicity constants (T constant values) are derived for each environmental medium and for two types of toxic effects, cancer, and noncancerous chronic effects.

Toxicity constants for noncarcinogens (Tn) are formed from the minimum effective dose (MED) for chronic effects, a severity of effect factor (graded from 1 to 10), and standard factors for body weight and exposure. Toxicity constants for potential carcinogens (Tc) are derived directly from the dose at which a 10% incremental carcinogenic response is seen (ED_{10}) and the same standard intake and body weight factors as used above.

Carcinogens and noncarcinogens are scored and selected independently because of probable differences in dose response mechanisms; that is, nonthreshold vs threshold. Toxicity constants are used only for selecting indicator chemicals and are not used to characterize risk associated with exposure to the selected chemicals.

Five chemical properties also influence the selection factors of the indicator compounds. These five properties that can influence exposure include: water solubility, vapor pressure, Henry's law constant, organic carbon partition coefficient, and persistence in various media. The values of these properties may influence the exposure potential of a certain chemical and may result in the inclusion of that chemical on the indicator chemical list even if the chemical has a low indicator score.

Additional questions that have been recommended to be applied to the indicator chemical selection process include the following:

1. Have outlying data points unrealistically influenced the indicator scores?
2. Have temporal and spatial variabilities of the contaminants been considered?
3. Has the presence of a contaminant in more than one medium been adequately addressed?
4. Have elevated background concentrations and/or offsite sources of contamination been considered?

No. 2 Fuel Constituents Considered by EPA

Fuel oils are classified as middle distillate because of their volatility during their distillation from crude oil.[41] These middle distillate fuels are composed of a variety of hydrocarbons including aromatic hydrocarbons, olefins, paraffins, and cycloparaffins with carbon numbers from C9 to C20.[41] The percentages of hydrocarbon types in a middle distillates fuel sample indicate the composition and/or source of the crude oil and the refining techniques used to create it. Aliphatic compounds include straight and branched alkanes, and cyclic alkanes. Aromatics include alkyl substituted benzenes, naphthalenes, and three to four ring polycyclic aromatic hydrocarbons (PAHs). The range of the hydrocarbon composition, by weight, of middle distillates are 35% to 90% paraffins and cycloparaffins, and 10% to 58% aromatics. Generally, No. 2 fuel oil contains a higher volume percentage of benzenes and naphthalenes compared to kerosene and diesel fuels. Kerosene has the least amount of aromatic hydrocarbons of the middle distillate fuels; benzenes, indanes, and naphthalenes are the major aromatic components. Diesel fuel contains high amounts of naphthalenes, acenaphthenes, phenanthrenes, and anthracenes.

Selection of No. 2 Fuel Indicator Chemicals

Benzene, toluene, and xylene are selected by the EPA as indicator compounds for No. 2 fuel oil in the Petroleum, Underground Storage Tank, Risk Assessment Procedures Manual.[42] Benzene, toluene, and xylene were selected because they have relatively higher toxicity and water solubility than the other petroleum constituents. Benzo(a)pyrene is selected as an indicator compound because of its carcinogenicity in animals, its mutagenicity, and teratogenicity. However, benzo(a)pyrene has not been found to cause toxic effects in humans when it has been the only contaminant. When combined with other PAHs, benzo(a)pyrene, has been linked with human cancer.[17]

Selection of Additional Indicator Chemicals

At some sites it may be necessary to select other chemicals as indicator chemicals if additives or other contaminants are present in the No. 2 fuel oil or in the tank. Factors for selecting additional indicator chemicals include their measured concentration in the petroleum mixture, their toxicity, and physical parameters, or chemical parameters, that may influence the compound's mobility or persistence. The algorithm and media specific constants developed for the Superfund Public Health Evaluation Manual are used to select the additional indicator compounds. Representative and maximum concentrations are multiplied with media-specific toxicity constants and then summed for all media to derive an indicator score.

The chemical, physical, and persistence data characterizing the chemical transport and fate are subjectively considered with the indicator score to determine the final indicator chemical. The EPA has developed a worklist to organize other potential indicator chemicals so they can be evaluated systematically (see Figure 1).

Facility Name: ____________________ Location: ____________________

Type of Petroleum Stored: ____________________ Date: ____________________

Analyst: ____________________ Quality Control: ____________________

INSTRUCTIONS:

1. List all chemicals not listed on Exhibits 2-1 or 2-2 that are of concern for the site.
2. List the maximum and representative concentrations at which these chemicals are found in the ground water and other relevant exposure media.
3. List critical toxicity values for each chemical.
4. List physical and chemical data for each chemical.
5. Qualitatively evaluating the data for each chemical, considering noncarcinogen and carcinogens separately, select additional indicator chemicals and check in final column.

Chemical	Concentration in Ground Water		Concentration in Other Media	
	max. mg/l	repres. mg/l	max.	repres.
1. ________	____	____	____	____
2. ________	____	____	____	____
3. ________	____	____	____	____
4. ________	____	____	____	____
5. ________	____	____	____	____

	Critical Toxicity Value	Water Solubility mg/l	Vapor Pressure (mmHg)	Henry's Law Constant (atm-m^3/mol)	Koc	Indicator Chemical?
1.	____	____	____	____	____	____
2.	____	____	____	____	____	____
3.	____	____	____	____	____	____
4.	____	____	____	____	____	____
5.	____	____	____	____	____	____

Figure 1. Worklist for organizing other potential indicator chemicals for evaluation.

Chemical concentrations above background, not typically found in petroleum products should be considered for inclusion. The concentration values should represent the monitoring data from soil, water, air, or other media. The representative concentrations of constituents are often determined through the use of geometric means. The representative concentration should be based on all site monitoring data depicting potential long-term human exposure including evaluation of locations close to human populations. The detection frequency should also be considered for the representative concentration, since those constituents found less often may have less significant impact.The toxic effect evaluation should include noncarcinogenic and carcinogenic endpoints. Cancer potency factors, threshold doses, and other toxicity values may be obtained from EPA sources such as the Superfund Public Health Evaluation manual.[31]

The physical and chemical attributes of chemicals also influence their inclusion as indicator compounds. The five most significant physical and chemical properties include: water solubility, vapor pressure, Henry's law constant, organic carbon partition coefficient (Koc), and persistence in exposure media. The values of these properties for a chemical on the site may cause a chemical to be included as an indicator compound in spite of a lower toxicity score.

Other chemical properties could affect exposures, but the five listed above are the most significant with respect to environmental transport and fate. In addition, including only these five properties limits the amount of data required to be collected and considered. Some of the properties influence exposure pathways in different modes, and these influences should be considered when evaluating the physical and chemical factors in the selection process. As a result, consideration of the potentially important exposure pathways at a site is important when applying physical/chemical factors in the selection process. The relevance of each characteristic on chemical release, transport, and fate is described below. Further discussion of these properties is available in other references, including Kenaga and Goring,[43] Lyman et al.,[44] Nelson et al.,[45] Maki et al.,[46] and Mackay et al.[47]

Water solubility is the maximum concentration of a chemical that dissolves in pure water at a specific temperature and pH. Water solubility of inorganic compounds can vary broadly, depending on temperature, pH, redox potential (Eh), and the types and concentrations of complexing species present. Solubilities range from less than 1 ppb to greater than 100,000 ppm. The most common organics fall between 1 ppm and 100,000 ppm.[44] Water solubility is a crucial characteristic that affects the environmental fate of chemicals.[48] Highly soluble chemicals can be rapidly leached from contaminated soil and are mobile in groundwater. Solubility is one of the dominant factors affecting leachate strength and chemical migration from contamination sites (adsorption potential, soil type, and water infiltration are also important). Solubility affects the leaching of these contaminants into groundwater and surface water, and highly soluble compounds are usually less strongly adsorbed to soil and therefore more mobile in both ground and surface water. Soluble chemicals also are more readily biodegraded than those with low solubility.[44] Solubility may also affect volatilization of compounds from water. Usually, high solubility is associated with lower volatilization rates.[48]

In some instances chemicals may be measured in concentrations greater than their water solubilities. This situation can result when nonaqueous phase liquids (i.e., liquids that are not dissolved in water) form a second liquid layer, and float on top of an aqueous phase or are held at the top of an aquifer. Almost pure petroleum may be found in a floating, nonaqueous phase. Other contaminants may be dissolved in the nonaqueous phase at concentrations higher than their water solubilities. Special steps in selection of indicator chemicals may be necessary when chemicals are detected at concentrations higher than their water solubilities.

Vapor pressure and Henry's law constant are two influences of chemical volatility and are important in evaluating air exposure pathways. Air exposure pathways are significant if contaminated groundwater seeps into homes and organic chemicals volatilize into the living space. Air exposure pathways may be important if volatile chemicals can be released to the atmosphere in some fashion that would result in human exposure. Vapor pressure is a relative measure of the volatility of a chemical in its pure state.[49] Vapor pressures of liquids range from 0.001 to 760 torr (mm Hg); vapor pressures of solids range down to 10^{-7} torr.[50] Vapor pressure is a significant determinant of the rate of vaporization from contamination sites. Vapor pressure is directly pertinent to exposure pathways involving chemical releases to the air from spills or contaminated surface soils. Henry's law constant combines vapor pressure, solubility, and molecular weight, and is used to estimate releases to air from contaminated water (e.g., ponds, lagoons). Vapor pressure and Henry's law constant are important for those chemicals that volatilize from the plume dissolved in groundwater.

The organic carbon partition coefficient (Koc) indicates the relative adsorption potential for organics. The Koc reflects the tendency of an organic chemical to be adsorbed, and is independent of soil properties.[44] Koc has significant influences on environmental fate for all exposure pathways. Koc is expressed as the ratio of the amount of chemical adsorbed per unit weight of organic carbon to the chemical concentration in solution at equilibrium. The equation is as follows:

$$\text{Koc} = \frac{\text{mg adsorbed/kg organic carbon}}{\text{mg dissolved/liter solution}}$$

Koc values range from 1 to 10^7, with higher values indicating greater adsorption potential.[44] Koc was selected as a chemical property to consider over the other partition coefficients that exist (e.g., Kom, Kd, Kow), because it is chemical-specific; that is, independent of soil conditions, and for organic chemicals it is related to soil and sediment adsorption, which may influence chemical fate processes at many petroleum contaminated sites. The distribution coefficient for a specific soil type (Kd) or the maximum exchangeable mass may be a more appropriate indicator of adsorption potential for inorganic chemicals.

The influence of Koc varies with different exposure pathways. In groundwater, low Koc values indicate faster leaching from the tank or contaminated soil into an aquifer and rapid movement through the aquifer. Koc is directly proportional to the retardation factor used in many groundwater transport models. Therefore, among chemicals with similar concentrations or toxicities, low Koc (high mobility) chemicals would be of more concern.

In surface water applications, Koc is significant for several reasons. A high Koc (low mobility) indicates tight adhesion of a chemical to soil organic matter. This results in less of the chemical leaving in site runoff. Since the chemical has a tighter bond to the organic constituents, this also implies that runoff of contaminated soil particles may occur over a longer time period. In some areas, direct recharge of surface water by groundwater is important and in these locations, because of the groundwater Koc relationship previously discussed, chemicals with high Koc are of relatively less concern. When a chemical gets into surface water, a high Koc may be of great concern because it indicates a potential for bioaccumulation. If aquatic food chain pathways are possible, the association of Koc with bioaccumulation should be considered. The Koc value also indicates the relative amount of sediment adsorption in surface waters.

A chemical may be selected even though it has a low indicator score because a low Koc (high mobility) with a high soil concentration indicates that significant release of the chemical to groundwater is possible in the future.

Persistence in different media is the last chemical property to be considered in the EPA's indicator selection process. This property is an indicator of how long a chemical will remain in a given medium. Processes which significantly remove a chemical from media include phase transfer (e.g., water to air, soil to water), chemical transformation (e.g., hydrolysis, photolysis), and biological transformation. Half-lives of chemicals vary from seconds to thousands of years. In general, chemicals with short half-lives are of less concern, although degradation products that have higher toxicity or enhanced environmental mobility than the parent chemical will raise the probability of including the compound on the indicator chemical list.

An additional factor in the indicator chemical selection process, for potential carcinogens only, is the qualitative weight-of-evidence rating. The weight-of-evidence rating is an expression of the quality and quantity of data underlying a chemical's designation as a potential human carcinogen. The categories of evidence for human carcinogenicity include sufficient, limited, and inadequate. Chemicals on the preliminary indicator list with sufficient evidence of human carcinogenicity (EPA Group A) and chemicals with limited human evidence and sufficient animal evidence (EPA Group B1) should be selected as final indicators unless there are convincing reasons to do otherwise. For chemicals with similar concentration and toxicity values, ones with stronger weight-of-evidence for carcinogenicity should usually be selected.

SUMMARY

Overview of No. 2 Fuel Oil

The constituents of No. 2 fuel oil have not been fully characterized and their identification would increase our understanding of its toxicity. The characterization of No. 2 fuel oil should include the variability of the composition between fuel brands and any seasonal or geographic variations.

The interactions of the individual constituents in No. 2 fuel oil, about which very little is known, are potentially of toxicological significance. The limited data suggest some interactions enhance toxicity and some interactions decrease toxicity. Additional research into No. 2 fuel oil constituent interactions may provide us with a clearer understanding of the risks to human health and the environment.

The toxicity of No. 2 fuel oil or a basic fraction of the fuel or a neutral fraction of No. 2 fuel oil may be greater than the individual toxicities of the constituents making up the major volume fractions. Water soluble fractions may have more toxicological significance than other fractions.

Once No. 2 fuel oil is released into the environment it may be changed through the effects of sun, biological organisms, wind, soil chemistry, and other factors. These environmental influences may alter the toxicity of No. 2 fuel oil through changes in the chemical constituents. The effects of environmental factors such as sun, water, and soil on No. 2 fuel oil should be evaluated further.

Ebasco-Army and EPA Comparison

There are two alternate directions for the selection of indicator compounds for No. 2 fuel oil. The Ebasco-Army and the EPA's indicator compound approaches may be used when chemical contamination is caused by an unknown mixture. The EPA's Petroleum Underground Storage Tank, Risk Assessment Procedures Manual has preselected indicator compounds. This is more applicable when the source of the contamination is known to be No. 2 fuel oil.

An evaluation of the intent of the Ebasco-Army and the U.S. EPA generic indicator compound selection processes indicates that they are quite similar, and possibly identical. Both procedures are designed to provide a reliable, defensible, efficient, and effective method to differentiate between the chemical contaminants with respect to their public health concerns. To accomplish this task, both methodologies must differentiate among agents on the basis of their concentrations, environmental fate potential, and toxicological evaluations. A qualitative evaluation given in Table 4 reveals that both EPA's and Ebasco-Army's methodologies are designed to achieve these goals.

While the two approaches are similar in achieving their goals, the operational methodology proceeds somewhat differently, and this may impact the resulting ranking of chemicals.

Table 4. Factors Considered by Ebasco-Army and EPA Indicator Compound Selection

Do Procedures Consider?	Ebasco-Army Yes	Ebasco-Army No	EPA Yes	EPA No
1. high average concentrations	X		X	
2. range of concentrations	X		X	
3. high maximum contaminant values	X		X	
4. environmental mobility	X		X	
5. public health/toxicity	X		X	

EPA and Ebasco-Army Differences

The Ebasco-Army method is more prescribed and less flexible since it is a more formula driven methodology. The EPA approach allows for a considerably greater amount of site-specific judgment. This is stated in the EPA superfund manual where ". . . the indicator chemical selection process . . . is not supposed to contravene professional judgment." This difference in the two approaches is apparent on issues such as the determination of a representative concentration.

During the determination of a representative concentration the Ebasco-Army refers to an average concentration. The EPA Superfund manual provides considerable discussion on what a representative concentration is and how it should be derived. The Ebasco-Army clearly describes the process by which concentration variation and percentage of positive findings are included in the formula process. The EPA, however, requires that these factors be considered through a professional judgment process.

The Ebasco-Army incorporates environmental partition coefficients within the PPLV derivation process. The EPA methodology provides for environmental partition coefficient to be considered through professional judgment.

The Ebasco-Army and EPA selected different toxicological approaches in their respective Indicator Score process. The EPA employs a Toxicity Constant (Tc) approach for each media. This is a toxicity based assessment. However, it does not use existing health standards such as Maximum Contaminant Levels (MCL), Threshold Limit Values (TLVs), and others. Thus the Tc values of the EPA do not include adjustments based on risk management decisions.

In comparison, the Ebasco-Army PPLV uses recognized national standards such as MCLs which are the result of risk management decisions. Although the Ebasco-Army and EPA differ in their approach here, it is not clear if there would be any significant difference in the ranking process.

The EPA process does not permit a scientific differentiation between potential carcinogens and noncarcinogens, since according to their definition they are using different evaluation scales.

In contrast, the PPLV methodology used by the Ebasco-Army approach is fully capable of making differential, health-based decisions. Again this clear articulation of a difference may not translate into a practical difference in the ranking results. A firm answer to the questions would require further documentation.

The EPA has incorporated an often discussed but not widely adopted *severity factor* concept with respect to noncarcinogens. This concept has important toxicologic utility, and is a useful adjunct to the minimum effective dose (MED) used to develop the Tc. The Ebasco-Army approach, on the other hand, does not have a need for the severity factor concept since it relies heavily on established environmental public health standards which incorporate safety factors for noncarcinogens and may therefore either directly or indirectly address the severity factor issue. However, how precisely it achieves this is unknown and is likely to vary according to the specific compound and standard.

The EPA directs the summation of C.T across the various affected media. While this is generally comparable to the incorporation of single specific PPLVs into a cumulative PPLV, the mathematical technique by which this is done differs markedly between the two approaches. In the case of the EPA it represents a simple and direct summation. However, the PPLV approach of the Ebasco-Army involving separated multiple exposure pathways incorporates their values via a technique identical to adding resistance in parallel.

DISCUSSION

Despite these differences in the chemical indicator methodologies of the Ebasco-Army and EPA, both are designed to flag (i.e., give high ranking scores) contaminants with high site concentrations and high toxicity/carcinogenic potency. The procedures are also designed to be successful in identifying those at the opposite end of the spectrum concern; that is, those of low site concentration and low inherent toxicological concern. However, there are notable differences in the actual conduct of an analysis that may lead to some significant differences in final rankings.

The most significant component in the EPA methodology is the need for experienced, multidisciplinary, and integrated professional judgment. The heavy reliance on professional judgment by EPA demands that the EPA have exceptionally skilled teams in the conduct of site-specific evaluations. On the other hand, the Ebasco-Army methodology is direct and simplified, and could be utilized with relatively little variation by different site assessment terms. It incorporates environmental partition coefficients, site contamination data, and toxicological values into fairly fixed formulae.

A potential concern with the Ebasco-Army approach is that it uses a $PPLV_R R$ value for the denominator of the equation to determine the ranking value. The problem is that it is using the PPLV value to derive a chemical ranking so as to justify deriving a more comprehensively based PPLV. This seems to be problematic, since it allows a specific methodology (i.e., PPLV) a disproportionately dominant role in the overall evaluation. Another concern is how the very preliminary PPLV method derived value is derived, given the use of very little site-specific data at the time of this initial evaluation.

CONCLUSIONS

Both approaches appear capable of assisting site evaluation teams in their ranking of chemical contaminants for No. 2 fuel oil. The EPA procedure places a considerable burden on professional judgment, while the Ebasco-Army has a formula-driven decisionmaking process. It is not possible to conclude which is the best methodology, since the EPA procedure is dependent on professional judgment. Presumably, the application of excellent professional judgment in the EPA methodology offers the opportunity for a significant improvement over the Ebasco-Army's approach due to site-specific fine tuning adjustments. On the other hand, if considerable professional experience is not uniformly available, as is often the case, the procedure of the Ebasco-Army may be more appropriate.

Since the purpose of the EPA and Ebasco-Army approaches are to achieve similar goals using a number of different yet reasonable approaches to the problem, it would be of considerable value for these respective federal organizations to provide a comparative ranking by both methodologies to establish how consistent they are to each other.

GLOSSARY

Additive—A substance added to; e.g., lubricating oils to impart new or to improve existing characteristics.

Aliphatic hydrocarbon—Hydrocarbon in which the carbon-hydrogen groupings are arranged in open chains that may be branched. The term includes paraffins and olefins, and provides a distinction from aromatics and naphthenes which have at least some of their carbon atoms arranged in closed chains or rings.

Alkane—*See* Paraffin.

Alkene—Alkane with one or more double bonds.

Alkylation—The chemical reaction of a low molecular weight olefin and an isoparaffin to form multiple branched paraffins of high octane rating.

Alkyne—Hydrocarbon with one or more triple bonds.

Aromatic—Compound containing one or more benzene rings that also may contain sulfur, nitrogen, and oxygen.

Blending—Intimate mixing of the various components in the preparation of a product to meet a given specification.

Bunker fuel—Heavy residual oil, also called bunker C, bunker C fuel oil, bunker oil.

Cetane—n-Hexadecane, used as a reference fuel for rating diesel fuel ignition quality.

Coking—Thermal or other process yielding; e.g., distillate, gasoline, gas, and nonvolatile residue (coke). Coking also occurs in catalytic cracking. X-ray analyses have shown coke to consist of condensed aromatic structures arranged in a disordered graphitic pattern.

Cracking—A process whereby the relative proportion of lighter or more volatile components of an oil is increased by changing the chemical structure of the constituent hydrocarbons.

Cracking, catalytic—A cracking process in which a catalyst isused to promote reaction.

Cracking, hydro—A cracking process carried out at high temperature and pressure in the presence of hydrogen, and in which a catalyst is used to promote reaction. The process combines cracking and hydrogenation.

Cracking, steam—Thermal cracking of; e.g., naphtha, at high temperatures with superheated steam injection.

Cracking, thermal—A cracking process in which no catalyst is used to promote reaction.

Crude oil—Naturally occurring mixture consisting essentially of many types of hydrocarbons, but also containing sulfur, nitrogen, or oxygen derivatives. Crude oil may be of paraffinic, asphaltic, or mixed base, depending on the presence of paraffin wax, bitumen, or both paraffin wax and bitumen in the residue after atmospheric distillation. Crude oil composition varies according to the geological strata of its origin.

Cycloalkane—*See* naphthene.

Cycloparaffin—*See* naphthene.

Diesel fuel—That portion of crude oil which distills out within the temperature range 200–370°C. A general term covering oils used as fuel in diesel and other compression ignition engines.

Diesel oil—*See* diesel fuel.

Distillate—A product obtained by condensing the vapors evolved when a liquid is boiled, and collecting the condensate in a receiver which is separate from the boiling vessel.

Distillation range—A single pure substance has one definite boiling-point at a given pressure. A mixture of substances, however, exhibits a range of temperatures over which boiling or distillation commences, proceeds, and finishes. This range of temperatures, determined by means of standard apparatus, is termed the 'distillation' or 'boiling' range.

Extract—During solvent refining processes, other than dewaxing or deasphalting, part of the feedstock passes into solution in the solvent and is subsequently recovered by evaporating off the solvent. This fraction is the extract and is generally aromatic in character and thus referred to as an aromatic extract.

Feedstock—Primary material introduced into a plant for processing.

Fractionation—A distillation process in which the distillate is collected as a number of separate fractions, each with a different boiling range.

Fuel oil—A general term applied to an oil used for the production of power or heat. In a more restricted sense, it is applied to any petroleum product that is burned under boilers or in industrial furnaces. These oils are normally residues, but blends of distillates and residues are also used as fuel oil. The wider term, 'liquid fuel,' is sometimes used, but the term 'fuel oil' is preferred.

Gasoline (petrol)—Refined petroleum distillate, normally boiling within the limits of 30° to 220°C, which, combined with certain additives, is used as fuel for spark-ignition engines. By extension, the term is also applied to other products that boil within this range.

Initial boiling point—The temperature (corrected if required) at which the first drop of distillate falls from the condenser during a laboratory distillation carried out under standardized conditions.

Jet fuel—Kerosene or gasoline/kerosene mixture for fueling aircraft gas turbine engines.

Kerosene—A refined petroleum distillate intermediate in volatility between gasoline and gas oil. Its distillation range generally falls within the limits of 150° and 300°C. Its main uses are as a jet engine fuel, an illuminant, for heating purposes, and as a fuel for certain types of internal combustion engines.

Kerosine—European term for kerosene.

Light distillate—A term lacking precise meaning, but commonly applied to distillates, the final boiling-point of which does not exceed 300°C.

Liquefied petroleum gas—Light hydrocarbon material, gaseous at atmospheric temperature and pressure, held in the liquid state by pressure to facilitate storage, transport, and handling. Commercial liquefied gas consists essentially of either propane or butane, or mixtures thereof.

Lubricating oil—Oil, usually refined, primarily intended to reduce friction between moving surfaces.

Lubricating oil distillate—A vacuum distillation cut with a distillation range and viscosity such that, after refining, it yields lubricating oil.

Middle distillate—One of the distillates obtained between kerosene and lubricating oil fractions in the refining processes. These include light fuel oils and diesel fuels.

Naphtha—Straight-run gasoline fractions boiling below kerosene and frequently used as a feedstock for reforming processes. Also known as heavy benzine or heavy gasoline.

Naphthene—Petroleum industry term of a cycloparaffin (cycloalkane).

Octane number—*See* octane rating.

Octane rating (of gasoline)—The percentage by volume of iso-octane in a mixture of iso-octane and n-heptane which is found to have the same knocking tendency as the gasoline under test in a CFR engine operated under standard conditions (also called octane number).

Oil mist—Suspended liquid droplets of oil which are produced when there is condensation of the oil from the gaseous to the liquid state; alternatively, a mist can be produced by dispersion of liquid oil.

Olefin—Synonymous with alkene.

Operator—Within the context of this volume, one of the employees who actually runs (operates) the various units, plant, and equipment that make up a refinery.

Pale—Term of U.S. origin used to describe a lightly refined, low viscosity index naphthenic oil.

Paraffinic oil—A petroleum oil derived from a crude oil with a substantial wax content.

Paraffin (alkane)—One of a series of saturated aliphatic hydrocarbons, the lowest numbers of which are methane, ethane, and propane. The higher homologues are solid waxes.

Reduced crude—The product obtained after removal, by atmospheric distillation, of the light components of crude oil.

Refinery—A plant, together with all its equipment, for the manufacture of finished or semifinished products from crude oil.

Refining—The separation of crude oil into its component parts and the manufacture therefrom of products. Important processes in lubricating oil refining are distillation, hydrotreatment, and solvent extraction.

Reforming—A process for treating light petroleum fractions to yield gasoline with a higher aromatic content and a higher octane number than the feedstock.

Reforming, catalytic—Reforming in which reaction is promoted by a catalyst.

Reforming, thermal—Reforming without the use of a catalyst.

Residual oil—Grade No. 4 to grade No. 6 fuel oils.

Residue (residuum)—The heavy fraction or bottoms remaining undistilled after volatilization of all lower-boiling constituents.

Short residue—The residual fraction from the vacuum distillation of long residue.

Stripping (in catalytic cracking)—Process whereby spentcatalyst from a catalytic cracking unit comes into contact at an elevated temperature with steam, with the aim of desorbing adsorbed hydrocarbons.

Vacuum distillation—Distillation under reduced, as opposed to atmospheric, pressure; e.g., fractional distillation of short residue to produce distillates for lubricating oil manufacture.

REFERENCES

1. "Complex Mixtures Methods for In Vivo Toxicity Testing," National Research Council. (Washington, DC: National Academy Press, 1988).
2. "The Rubber Industry," IARC monographs on the evaluation of the carcinogenic risk of chemicals to humans, Vol. 28. International Agency for Research on Cancer, Lyon, France, 1983.
3. Speizer, F. E. "Assessment of the Epidemiologic Data Relating Lung Cancer to Air Pollution," *Environ. Health. Perspect*. 47:33–42 (1983).
4. Sawicki, E. "Airborne Carcinogens and Allied Compounds," *Arch. Environ. Health,* 14:46–53 (1967).
5. Rossini, F. D., B. J. Mair, and A. J. Streiff. "Hydrocarbons from Petroleum," Reinhold Publishing Corp., New York, 1953.
6. Camin, D. L. "History of Chromatography in Petroleum Analysis," in *Chromatography in Petroleum Analysis,* K. H. Altgelt and T. H. Gouw, eds. (New York, NY: Marcel Dekker, Inc., 1979), pp. 1–12.
7. "Relative Health Effects of Gasoline and Heating Fuels Derived from Petroleum and Synthetic Crudes," U.S. Department of Energy. Envirocontrol Inc., Rockville, MD, 1980.
8. Wenger R., R. Powell, and R. Putzrath. "Analysis of the Potential Hazards Posed by No. 2 Fuel Oil Contained in Underground Storage Tanks for Oil Heat Task Force," Environ Corporation, Washington, DC, 1987.
9. Thomas, B. L. "Determination of Oil/Water and Octanol/Water Distribution Coefficients from Aqueous Solutions from Four Fossil Fuels," Master's thesis/ Richland, WA: Pacific Northwest Laboratory. PNL-5002, UC-90d, 1984.
10. Pancirov, R. J., T. D. Searl, and R. A. Brown. "Methods of Analysis for Polynuclear Aromatic Hydrocarbons in Environmental Samples," in *Petroleum in the Marine Environment,* L. Petrakis and F. T. Weiss, eds. (Washington, DC: American Chemical Society, 1980).
11. IARC Overall Evaluation of Carcinogenicity: An Updating of IARC Monographs, Volumes 1 to 42. Supplement 7. International Agency for Research on Cancer, 1987.
12. Hirao, K. Y., H. Shinohara, S. Fukushima, M. T. Takahashi, and N. Ito. "Carcinogenic Activity of Quinoline on Rat Liver," *Cancer Research,* 36:329–335 (1976).
13. Sice, J. "Tumor-Promoting Activity of N-Alkanes and 1-Alkanols," *Tox. Appl. Pharm.* 9:70 (1966).

14. Horton, A. W., D. N. Eshleman, A. R. Schuff, and W. H. Perman. "Correlation of Cocarcinogenic Activity Among n-Alkanes With Their Physical Effects on Phospholipid Micelles," *J. Nat. Cancer,* 56(2):387–391 (1976).
15. Horton, A. W., D. T. Denman, and R. P. Trosset. "Carcinogenesis of the Skin II." The accelerating properties of aliphatic and related hydrocarbons. *Cancer Research,* 17:758–766 (1957).
16. Bingham, E., R. P. Trosset, and D. Warshawsky. "Carcinogenic Potential of Petroleum Hydrocarbons. A Critical Review of the Literature," *J. Environ. Pathol. Toxicol.,* 3:483–563 (1980).
17. "Drinking Water and Health: Toxicity of Selected Organic Contaminants in Drinking Water," Volume 4, pp. 247–264. National Research Council. (Washington, DC: National Academy Press, 1982).
18. "Criteria for a Recommended Standard . . . Occupational Exposure to Alkanes (C5-C8). U.S. Department of Health, Education, and Welfare, Public Health Service, Center for Disease Control. National Institute for Occupational Safety and Health, Cincinnati, OH (DHEW (NIOSH) Publ. No. 77–151), 1977.
19. "Polynuclear Aromatic Compounds, Part 1, Chemical Environmental and Experimental Data," IARC monographs on the evaluation of the carcinogenic risk of chemicals to humans. Volume 32. International Agency for Research on Cancer, 1983.
20. Beck, L. S., D. I. Hepler, and K. L. Hansen. "The Acute Toxicology of Selected Petroleum Hydrocarbons," in H. N. MacForland, C. E. Holdsworth, J. A. MacGregor, R. W. Call, and M. L. Lane, Eds. *Advances in Modern Environmental Toxicology, Vol. VI, Applied Toxicology of Petroleum Hydrocarbons,* (Princeton, N.J.: Princeton Scientific Publishers, 1984).
21. Goodman D. R., and R. D. Harbison. "Toxicity of the Major Constituents and Additives of Gasoline. Kerosene and No. 2 Fuel Oil," Division of Interdisciplinary Toxicology, University of Arkansas for Medical Sciences, Little Rock, AK, 1988.
22. "Oil in the Sea," National Research Council. (Washington, DC: National Academy Press, 1985).
23. Epler, J. L., J. A. Young, A. A. Hardigre, T. K. Rao, M. R. Guerin, I. B. Rubin, C. H. Ho, and B. R. Clark. "Analytical and Biological Analyses of Test Materials from Synthetic Fuel Technologies. Mutagenicity of Crude Oils Determined by Salmonella-Typhimurium-Microsomal Activation System," *Mutation Research* 57(3):265–276 (1978).
24. Wilson, B. W., R. Pelroy, and J. T. Cresto. "Identification of Primary Aromatic-Amines in Mutagenically Active Subfractions From Coal Liquification Materials," *Mutat. Res.,* 79(3):193–202 (1980).
25. Guerin, M. R., I. B. Rubin, T. K. Rao, B. R. Clark, and J. L. Epler. "Distribution of Mutagenic Activity in Petroleum and Petroleum Substitutes," *Fuel,* 60:282–288 (1981).
26. Rao, T. K., B. E. Allen, D. W. Ramey, J. L. Epler, I. B. Rubin, M. R. Guerin, and B. R. Clark. "Analytical and Biological Analyses of Test Materials from the Synthetic Fuel Technologies. 3. Use of Sephidex LH-20 Gel Chromatography Technique for the Bioassay of Crude Synthetic Fuels," *Mutat. Res.* 89(3):35–43 (1981)
27. Scheier, A., and D. Gominger. "Preliminary Study of Toxic Effect of Irradiated Versus Non-Irradiated Water Soluble Fractions of No. 2 Fuel Oil," *B. Env. Con.,* 16(5):595–603 (1976).
28. Larson, R. A., T. L. Bott, L. L. Hunt, and K. Rogenmus. "Photo-Oxidation Products of a Fuel Oil and Their Antimicrobial Activity," *Env. Sci. Tec.* 13(8):965–969 (1979).

29. Callen, B., and R. A. Larson. "Toxic and Genetic Effects of Fuel Oil Photoproducts and 3 Hydroperoxides in Saccharomyces—Cerevisine," *J. Tox. Env. H.*, 4(5–6):913—917 (1978).
30. Marshall T. C., M. Dubinsky, and S. Boutwell. "A Risk Assessment of a Former Production Facility," in D. J. Paustenbach Ed., *The Risk Assessment of Environmental Hazards.* (New York, NY: Wiley-Interscience, 1989), pp. 461–504.
31. Superfund Public Health Evaluation Manual. EPA 540/1-86/060. U.S. Environmental Protection Agency, Office of Emergency and Remedial Response, Washington, DC, October 1986.
32. Third Annual Report on Carcinogens, NTP 82-330. Public Health Service, USDHHS, Washington, DC, 1983.
33. Health Assessment Document for Polychlorinated Dibenzo-p-Dioxins, EPA 600/8-84-0014F. U.S. Environmental Protection Agency, Office of Health and Environmental Assessment, Washington, DC, 1985.
34. TLVs: Threshold Limit Values for Chemical Substances and Physical Agents in the Workroom Environmental Assessment, American Conference of Governmental Industrial-Hygienists, USEPA, Washington, DC, 1985.
35. National Interim Primary Drinking Water Regulations. U.S. Environmental Protection Agency, Washington, DC, 1982.
36. LaGoy P. K., I. C. T. Nisbet, and C. O. Schulz. "The Endangerment Assessment for the Smuggler Mountain Site Pitkin County Colorado: A Case Study," in D. J. Paustenbach, Ed., *The Risk Assessment of Environmental Hazards.* (New York, NY: Wiley-Interscience, 1989), pp. 505–525.
37. Brett S. M., J. S. Schlesinger, D. Turnbull, and R. J. Machado. "Assessment of the Public Health Risks Associated with the Proposed Excavation of a Hazardous Waste Site," in D. J. Paustenbach, Ed., *The Risk Assessment of Environmental Hazards.* (New York, NY: Wiley Inter-Science, 1989), pp. 427–458.
38. Superfund Public Health Evaluation Manual, OSWER Directive 9285-4-1. U.S. Environmental Protection Agency, Office of Emergency and Remedial Response, Office of Solid Waste and Emergency Response, Washington, DC, 1986.
39. Endangerment Assessment, Draft Final Technical Plan, Ebasco, Denver, CO, 1987.
40. Zamuda, C. "Superfund Risk Assessments: The Process and Past Experience at Uncontrolled Hazardous Waste Sites," in D. J. Paustenbach, Ed., *The Risk Assessment of Environmental Hazards.* (New York, NY: Wiley-Interscience, 1989), pp. 266–295.
41. Liss-Sutter, D. "A Literature Review-Problem Definition Studies on Selected Toxic Chemicals. Environmental Aspects of Diesel Fuels and Fog Oils SGF No. 2 and Smoke Screen Generated from Them," Volume 8. The Franklin Institute Research Laboratories, 1978.
42. ICF, Incorporated. "Petroleum Underground Storage Tank Risk Assessment Procedures Manual, ICF, Inc., Washington, DC, 1987.
43. Kenaga, E. E., and C. A. Goring. "Relationship Between Water Solubility, Soil Sorption, Octanol/Water Partitioning, and Bioconcentration of Chemicals in Biota," in *Aquatic Toxicology,* ASTM STP 707, J. Eaton, P. Parris, and A. Henricks, Eds. American Society for Testing and Materials, Philadelphia, PA, 1978.
44. Lyman, W. J., W. F. Reehl, and D. H. Rosenblatt. *Handbook of Chemical Property Estimation Methods.* (New York: McGraw-Hill Book Company, New York, 1982).
45. Nelson, D. W., D. E. Elrick, K. K. Tangi, D. M. Kral, and S. L. Hawkins, Eds. "Chemical Mobility and Reactivity in Soil Systems: Proceedings of a Symposium

Sponsored by the American Society of Agronomy and the Soil Science Society of America. American Society of Agronomy, the Soil Science Society of America, Madison, WI, 1983.

46. Maki, A. W., K. L. Dickson, and J. Cairns, Eds. *Biotransformation and Fate of Chemicals in Aquatic Environments.* American Society for Microbiology, Washington, DC, 1980.
47. MacKay, D., Wan Ying Shiu, A. Chau, J. Southwood, and C. I. Johnson. "Environmental Fate of Diesel Fuel Spills on Land," Department of Chemical Engineering and Applied Chemistry, University of Toronto. Report for the Association of American Railroads, Washington DC, 1986.
48. Menzer, R. E., and J. O. Nelson. "Water and Soil Pollutants," in J. Doull, D. C. Klaassen, and M. D. Amdur, *Toxicology,* (New York, NY: Macmillan Publishing Co., Inc., 1980).
49. Jaber, H. M., W. R. Mabey, A. T. Liu, T. W. Chou, H. L. Johnson, T. Mill, R. T. Podoll, and J. W. Winterle. "Data Acquisition for Environmental Transport and Fate Screening," EPA 600/6-84/009. U.S. Environmental Protection Agency, Office of Health and Environmental Assessment, Washington, DC, 1984.
50. Grain, C. F. "Vapor Pressure," in Lyman et al., *Handbook of Chemical Property Estimation Methods.* (New York, NY: McGraw-Hill Book Company, 1982).

CHAPTER 21

The Effect of Soil Type on Absorption of Toluene and Its Bioavailability

Rita M. Turkall,[*,**] **Gloria A. Skowronski,**[*] and **Mohamed S. Abdel-Rahman,**[*]
Department of Pharmacology, New Jersey Medical School,[*] and Clinical Laboratory Sciences Department, School of Health Related Professions,[**] University of Medicine and Dentistry of New Jersey, Newark

Exposures to soil contaminated by petrochemicals released during manufacture, use, storage, transport or disposal are widespread. Evaluation of the potential health risk to workers and communities exposed to petroleum contaminated soils is of growing concern, and, moreover, is a complex problem for which methods are being developed. The United States Environmental Protection Agency (EPA), the Center for Disease Control (CDC) and numerous independent scientists have shown that inhalation of polluted soils is minimal and is unlikely to cause significant health hazards.[1-4] On the other hand, ingestion and dermal contact represent significant routes of exposure. Children (ages 1.5 to 3.5 years) are especially susceptible to exposure by ingestion.[1] When all of the published information on soil ingestion is considered, the data indicates that the best estimate of soil ingestion by children is about 50 mg/day.[1] The EPA in its risk assessment of 2,3,7,8-tetra-chlorodibenzo-p-dioxin (TCDD) based its estimates of dermal exposures to TCDD polluted soils on actual field investigations.[5] The EPA cited the work of Roels et al.[6] which showed that about 0.5 mg of soil adheres per cm^2 of exposed skin.

The percentage of the total chemical in the soil which subsequently enters the body, i.e., bioavailability, can be influenced by the soil's composition, the chemical's characteristics, as well as the biological environment provided at the site of the body where exposure takes place. Previous studies from this laboratory have demonstrated that route of exposure, in addition to soil-chemical interactions, alter the bioavailability of chemicals from contaminated soils. Adsorption to either of two New Jersey soils increased the amount of benzene available to orally exposed rats, as well as produced changes in plasma kinetics and excretion patterns versus benzene in the absence of soil.[7,8] However, decreased bioavailability resulted when exposure to soil-adsorbed benzene occurred via the dermal route. Altered tissue distribution and excretion patterns were also seen.[8,9]

Similar studies were conducted on soil-adsorbed toluene to evaluate the influence of substituent addition to the benzene ring. The EPA in its 1985 Priority List for cleanup cited toluene as the third most frequent out of 465 substances recorded in 818 abandoned dump sites. Toluene together with benzene and xylene represent the major aromatic compounds of gasoline. Toluene is also used as a solvent for paints, plastics, varnishes, and resins, and in the manufacture of explosives, adhesives, drugs, perfumes, and other synthetic chemicals.[10] Although toluene lacks benzene's chronic hematopoietic effects, animal experiments indicate that toluene is more acutely toxic than benzene. High concentrations of toluene cause symptoms of central nervous system (CNS) depression such as headache, dizziness, weakness, and loss of coordination. Furthermore, chronic exposure to toluene may cause permanent CNS damage.[10-12]

Oral administration of ^{3}H-toluene in peanut oil to rats was followed by rapid absorption of radioactivity. Maximum blood levels of radioactivity were reached within 2 hr after gastric intubation.[13] Sato and Nakajima[14] showed that toluene is poorly absorbed through the skin while Dutkiewicz and Tyras[15] reported that the absorption rate of toluene through human skin is very high (14–23 mg/cm^2/hr). The aim of the present study is to apply pharmacokinetic techniques to assess the effect that soil adsorption has on the bioavailability of toluene by the oral and dermal routes.

MATERIALS AND METHODS

Radioisotopes

All studies were conducted using ring uniformly labeled ^{14}C-toluene with a specific activity of 16.4 mCi/mmole and radiochemical purity of 95% (Amersham Corp., Arlington Heights, IL). Prior to use, the radioisotope was diluted with unlabeled HPLC-grade toluene (Aldrich Chemical Co., Milwaukee, WI) to reduce specific activity to a workable range.

Soils

Two soils were utilized: (1) an Atsion sandy soil (90% sand, 2% clay, 4.4% organic matter) collected from an outcrop site of the Cohansey sand formation near Chatsworth in southcentral New Jersey; and (2) a Keyport clay soil (50% sand, 22% clay, 1.6% organic matter) collected from the Woodbury formation near Moorestown in southwestern New Jersey. The Atsion sandy soil is a deep, poorly drained soil formed in Atlantic coastal plain sediments of New Jersey and New York. Warmer thermic equivalent Leon soil occurs from Maryland to Florida.[16] The Keyport clay soil is a moderately well drained soil formed on clay bed marine deposits of the Atlantic inner-coast plain. In addition to New Jersey, the Keyport soil is found in Delaware, Maryland, and Virginia with similar soils occurring as far southwest as Texas.[17] Gas chromatography/mass spectrometry analysis of extracts of the soil did not detect contamination by toluene or halogenated hydrocarbons.

Animals

Male Sprague-Dawley rats (275–300 g) were purchased from Taconic Farms, Germantown, NY, and quarantined for at least one week prior to administration of the chemical. Animals were housed three per cage and were maintained on a 12 hr light/dark cycle at constant temperature (25°C) and humidity (50%). Ralston Purina rodent lab chow (St. Louis, MO) and tap water were provided ad libitum.

Toluene Administration

The oral administration of toluene was performed as follows: either 150 μL of ^{14}C-toluene solution (5 μCi) alone, or the same volume of radioactivity added to 0.5 g of soil, was combined with 2.85 mL of aqueous 5% gum acacia and a suspension formed by vortexing. This volume of toluene or toluene soil suspension was immediately administered by gavage to groups of rats which had been fasted overnight. Heparinized blood samples were collected at 5, 10, 15, 20, 30, 45, 60, 90, 120, and 180 minutes by cardiac puncture of lightly ether-anesthetized rats.

For the dermal application, a shallow glass cap (Q Glass Co., Towaco, NJ) circumscribing a 13 cm^2 area was tightly fixed with Lang's jet liquid acrylic and powder (Lang Dental Manufacturing Co., Inc., Chicago, IL) on the lightly-shaved right costo-abdominal region of each ether-anesthetized animal one-half hour prior to the administration of toluene. Then 225 μL of ^{14}C-toluene (30 μCi) alone or after the addition of 750 mg of soil was introduced by syringe through a small opening in the cap, which was immediately sealed. This volume of toluene coated the soil with no excess chemical remaining. Rats were rotated from side to side so that the soil-chemical mixture covered the entire circumscribed area.

Heparinized blood samples were collected by cardiac puncture under light ether anesthesia at 0.25, 0.5, 1, 2, 4, 5, 7, 9, 10, 11, 12, 24, 30, 36, and 48 hr. Samples from both routes of administration were processed and radioactivity was measured by liquid scintillation spectrometry as in previous studies.[7,9] Immediately after the collection of the 180 min blood sample in the oral study, rats were sacrificed by an overdose of ether; whole organs or samples of brain, thymus, thyroid, esophagus, stomach, duodenum, ileum, lung, pancreas, adrenal, testes, skin, fat, carcass, bone marrow, liver, kidney, spleen, and heart were collected and stored at $-75°C$. Thawed tissue samples of 300 mg or smaller were used to determine the distribution of radioactivity as previously reported.[7]

Excretion and Metabolism Studies

In the excretion studies, groups of six rats each were administered toluene or toluene adsorbed to the soil, as described above. Animals were housed in all-glass metabolism chambers (Bio Serve Inc., Frenchtown, NJ) for the collection of expired air, urine, and fecal samples. Expired air was passed through activated charcoal tubes (SKC Inc., Eighty-Four, PA) for the collection of ^{14}C-toluene, then bubbled through traps filled with ethanolamine: ethylene glycol monomethyl ether (1:2 v/v) for the collection of $^{14}CO_2$. Charcoal tubes and trap mixtures were collected at 1, 2, 6, 12, 24, and 48 hr after administration of compound. Urine samples were collected at 12, 24, and 48 hr, and fecal samples were collected at 24 and 48 hr. Samples were processed and radioactivity was measured as previously described.[7]

At the conclusion of the dermal excretion studies, rats were sacrificed by an overdose of ether. One to 1.2 mL of ethyl alcohol were introduced into the glass cap and the animals were rotated from side to side. Aliquots of ethanol wash (100 μL) were removed for counting to determine the percent of toluene dose remaining on the skin application sites. Then the glass caps were removed from the rats and tissue specimens were collected for the distribution determination.

Toluene metabolites were determined in n-butanol extracts ($>95\%$ efficiency) of urine, utilizing high performance liquid chromatography with a 3.8% phosphoric acid-acetonitrile mobile phase and a C-18 column.

Data Analysis

Exploratory data analysis was used to summarize replicate data in the plasma time course study.[18,19] The curve-fitting procedure which was utilized is called smoothing. For these studies, a "4235EH" smoother (a statistical procedure for treating the data) was utilized.[19] Each replicate was smoothed over all time points, a median value was calculated for all smoothed replicates at each time point, and a second smooth was applied to these median values. The final smoothed data was used to calculate rate constants, $t_{1/2}$ of absorption and elimination from plasma by regression analysis and the method of residuals[20] as well as to determine a maximum concentration and a time at which the maximum concentration was achieved. Plasma concentrations from 0 min to the time at which maximum concentration was achieved were used for absorption calculations.

For the calculation of toluene elimination from plasma, 60 through 180 min were used in oral route studies, while 30 through 48 hr were used in dermal route studies. Since the rate constants and half-lives were calculated from smoothed data, the standard errors (SE) of the rate constants were determined by the bootstrap method.[21,22] The area under the plasma concentration time curve (AUC) was calculated by the trapezoidal rule using individual replicate data and is reported as the mean ± standard error of the mean (SEM). Comparison of slopes were determined by analysis of covariance. Excretion, tissue distribution, and metabolite data are reported as a mean ± SEM. Statistical differences between treatment groups were determined by analysis of variance (ANOVA), F test, and Scheffe's multiple range test.

RESULTS

Data showing the absorption and elimination half-lives following oral and dermal administration of ^{14}C-toluene to male rats are presented in Table 1. No statistically significant differences in the half-life of absorption into plasma were determined in the presence of soil after oral treatment compared to toluene alone. However, adsorption to sandy soil produced a statistically significant decrease (two-fold) in $t_{1/2}$ of absorption after dermal exposure. Absorption half-lives were longer by about 45- to 70-fold after dermal administration compared to oral exposures. Clay soil significantly shortened the $t_{1/2}$ of ^{14}C elimination from plasma compared to control after oral treatment. Neither of the soils altered the half-life of elimination compared to pure toluene following dermal exposure. In the pure and sandy groups, the elimination half-lives after dermal treatment were decreased to approximately one-half of oral treatment, while in the clay group, the dermal elimination was about 2.5-fold of oral elimination. AUC calculations did not reveal any significant differences among the groups during the 3 hr oral period studied or the 48 hr dermal exposure period (Table 2). It is worth noting that there were 270- to 300-fold reductions in dermal AUC's compared to the oral values.

Excretion of radioactivity after oral or percutaneous absorption of ^{14}C-toluene occurs primarily through the kidneys (Table 3). After the oral treatments, radioactivity was excreted rapidly in urine (greater than 50% of the initial dose) for all groups in the first 12 hr of collection. An additional 10% to 20% of the dose was excreted during the 12 to 24 hr time interval. Only about 2% of the administered dose was excreted in urine during the 24–48 hr collection period.

Topical administration of toluene delayed the urinary excretion of radioisotope. About 20% of the applied dose was eliminated between 0–12 hr, while the majority of the dose (approximately 35%) was collected during the 12–24 hr time interval following application of the chemical. The total fractions of the initial dose eliminated in urine by the two routes of administration were comparable for the pure

Table 1. Absorption and Elimination Half-Lives of Radioactivity in Plasma Following Oral or Dermal Administration of ^{14}C-Toluene[a]

Treatment	$t_{1/2}$ (hr)			
	Absorption		Elimination	
	Oral	Dermal	Oral	Dermal
Pure[b]	0.14	6.4	18.1	10.3
Sandy[c]	—[e]	3.2[f]	23.2	10.8
Clay[d]	0.08	5.6	4.2[f]	10.8

[a]Values calculated from 4 to 7 rats per group.
[b]^{14}C-toluene alone.
[c]^{14}C-toluene adsorbed to sandy soil.
[d]^{14}C-toluene adsorbed to clay soil.
[e]Occurence of peak concentration in plasma at the first sampling point did not allow calculation of absorption $t_{1/2}$.
[f]Significantly different than treatment with ^{14}C-toluene alone ($p<0.05$).

Table 2. Area Under Plasma Concentration Time Curve (AUC) Following Oral or Dermal Administration of ^{14}C-Toluene[a]

Treatment	Percent Initial Dose/mL†min	
	Oral	Dermal
Pure[b]	1.13±0.12	0.0038±0.0017
Sandy[c]	1.00±0.07	0.0037±0.0017
Clay[d]	1.27±0.13	0.0042±0.0018

[a]Values calculated from 4 to 7 rats per group.
[b]^{14}C-toluene alone.
[c]^{14}C-toluene adsorbed to sandy soil.
[d]^{14}C-toluene adsorbed to clay soil.

Table 3. Urinary Recovery of Radioactivity Following Oral or Dermal Administration of ^{14}C-Toluene[a]

Time (hr)	Oral			Dermal		
	Pure	Sandy	Clay	Pure	Sandy	Clay
0–12	53.7±10.2	56.0±5.5	61.2±11.8	18.4± 2.8	16.5± 1.6	18.4±1.9
12–24	20.9± 7.3	10.1±1.7	18.6± 3.0	36.0± 7.5	37.0± 3.7	31.4±3.2
0–24	74.5± 2.9	66.1±5.4	79.9± 8.9	54.3±10.2	53.6± 2.5	49.8±4.7
24–48	2.2± 0.6	1.6±0.3	1.9± 0.3	21.2± 4.2	31.3± 8.1	30.2±4.4
0–48	76.8± 2.4	67.8±5.3	81.7± 8.7	75.5±14.2	84.8±10.4	80.1±1.2

[a]Values represent percentage of initial dose (mean±SEM) for six animals per group.

Table 4. Expired Air Recovery of Radioactivity Following Oral or Dermal Administration of ^{14}C-Toluene[a]

Time (hr)	Oral			Dermal		
	Pure	Sandy	Clay	Pure	Sandy	Clay
0–12	20.2±1.1	23.5±1.7	13.8±0.7[b]	2.7±1.4	2.6±0.3	0.2±0.1
12–24	0.2±0.0	0.4±0.1	0.5±0.1	0.8±0.3	1.3±0.4	0.5±0.3
0–24	22.4±1.1	23.9±1.8	14.3±0.8[b]	3.5±1.3	4.0±0.7	0.8±0.4
24–48	0.1±0.0	0.1±0.0	0.1±0.0	0.2±0.1	1.4±0.7	0.1±0.0
0–48	22.4±1.1	23.9±1.8	14.4±0.8[b]	3.8±1.2	5.4±1.3	0.9±0.4

[a]Values represent percentage of initial dose (mean±SEM) for six animals per group.
[b]Significantly different than treatment with ^{14}C-toluene alone ($p<0.01$).

and clay soil-adsorbed groups; however, total urinary recovery in the dermal sandy soil-adsorbed group (85%) exceeded the value of the oral sandy soil group (68%).

The excretion of radioactivity as parent compound in expired air was greater in the oral studies than in the dermal studies (Table 4). Most of the radioactivity (14–24%) was expired in the first 12 hr after oral treatment. Oral clay soil-adsorbed toluene produced significantly smaller amounts of total activity (14% of the initial dose versus 22% and 24%, respectively, for the pure and sandy soil-adsorbed chemicals). The amount of radioactivity eliminated by the lungs was minor after dermal application: 4%, 5%, and 1% for pure, sandy, and clay groups, respectively. Carbon dioxide comprised less than 2% of the total radioactivity in the expired air of all oral treatment groups (data not shown). None of the dose was expired as $^{14}CO_2$ from the dermal route of administration. During the 48-hr period, the radioactivity in feces after oral treatment was 0.8%, 1.2%, and 0.6% of initial dose in pure, sandy, and clay groups, respectively. Negligible amounts of radioactivity (less than 0.5%) were also recovered in all dermal groups during the same time period (data not shown).

The tissue distribution patterns of ^{14}C-activity 3 hr following oral treatment are presented in Table 5. Stomach and fat contained the highest tissue concentrations of radioactivity in the pure toluene group, followed by duodenum, pancreas, and kidney. Radioactivity in the sandy soil-adsorbed group was also greatest in the stomach, followed by the fat, and to a lesser extent, in pancreas and liver. In the clay soil-adsorbed group, fat and stomach contained larger amounts of radioactivity than duodenum, kidney, and ileum. No statistically significant differences in the tissue concentrations of radioactivity were observed between oral treatment groups.

Table 5. Tissue Distribution of Radioactivity in Male Rat Following Oral Administration of ^{14}C-Toluene[a]

Pure Toluene Treatment Group:
Stomach = Fat > Duodenum = Pancreas = Kidney
Sandy Treatment Group:
Stomach > Fat > Pancreas = Liver
Clay Treatment Group:
Fat = Stomach > Duodenum = Kidney = Ileum

[a]Data obtained from five rats per group, 3 hr following oral administration.

Although the sequence of ^{14}C-distribution is the same in all treatment groups after dermal exposure, soil-related differences were found in tissue distribution 48 hr post-administration (Table 6). Clay soil treatment significantly decreased radioactivity in treated skin compared to toluene alone. Statistically significant decreases were also found in the untreated skin of both soil groups compared to the control group. High concentrations of activity appeared in all areas of fat examined; namely, fat beneath the treated and untreated skin and gut fat. Ethanol washes of treated sites at necropsy established that only about 0.5% of the initial dose was loosely retained on the skin application sites.

Table 6. Tissue Distribution of Radioactivity in Male Rat Following Dermal Administration of ^{14}C-Toluene[a]

In All Treatment Groups:
Treated Skin[b] >> Untreated Fat = Treated Fat = Gut Fat > Untreated Skin[c]

[a]Data obtained from five rats per group, 48 hr following dermal administration.
[b]Significantly decreased in the clay group compared to the pure toluene group ($p<0.03$).
[c]Significantly decreased in the sandy and clay groups compared to the pure toluene group ($p<0.003$).

Table 7. Urinary Metabolites of ^{14}C-Toluene in the Male Rat[a]

Metabolite	Oral			Dermal		
	Pure	Sandy	Clay	Pure	Sandy	Clay
Hippuric Acid	71.4±6.9	63.5±3.6	72.9±2.0	71.8±4.6	64.5±2.5	63.3±2.3
Undetermined	22.8±4.7	23.6±2.7	24.8±2.2	19.4±1.6	25.4±0.5	27.4±1.2

[a]Values represent % of total radioactivity in the 0–12 hr collection period for six animals per group (mean ± SEM).

The urinary metabolites of ^{14}C-toluene in the male rat after oral and dermal treatments are given in Table 7. Hippuric acid was the major metabolite (>60%) detected in the 0–12 hr samples of all groups in both routes of administration. Smaller quantities of an undetermined metabolite were also found. The type and percentage of toluene metabolites were not significantly altered in the presence of soil after either route of exposure. Similar metabolite percentages were detected in the 12–24 hr urines of all treated groups (data not shown). The parent compound was not detected in the urine of any treatment group.

DISCUSSION

The results of this study indicate that the presence of either sandy or clay soil produced alterations in the bioavailability of toluene following oral or dermal treatments. The absorption of radioactivity into plasma was faster after oral administration of toluene than after topical application. Adsorption to sandy soil, however, significantly reduced the dermal absorption half-life compared to pure compound. Although soil significantly decreased the half-life of elimination of toluene after oral administration, the pure and sandy soil-adsorbed toluene half-lives were longer than by the dermal route. No change in the amount of chemical absorbed was produced by the soils, as evidenced by similar AUCs within the groups exposed by a particular route. Thus, the soils altered the time course of toluene absorption or elimination, but not the amount absorbed. The amount of radioisotope absorbed by the oral route, however, was much greater than by the dermal route.

The oral kinetic data in the current study agrees with the previous studies of Pyykko et al.[13] which showed that toluene is rapidly absorbed through the gastrointestinal tract. Approximately 80% of the toluene dose was absorbed at a rate of 1.2 mg/cm²/hr over 48 hr in the present dermal studies. Sato and Nakajima[14] reported low absorption of toluene through skin. Volatilization losses in their study were not determined. Volatilization losses from the application sites in the present experiment were minimized by employing the glass cap arrangement. The high percutaneous absorption rate found by Dutkiewicz and Tyras[15] is based on the dose of applied and remaining toluene on skin. Because no data are available on the quantity of parent compound and/or metabolites in the tissues and excreta of Dutkiewicz and Tyras'[15] study, an accurate comparison of results cannot be made with their work.

Urine as the primary route of excretion in all oral and dermal treatment groups is consistent with the metabolism of toluene to water soluble products which are eliminated via the kidney. The appearance of radioactivity in urine was delayed by the dermal treatments, while most of the oral urinary excretion occurred within 12 hr of exposure. Smaller percentages of toluene were excreted unmetabolized via the expired air by both routes. Furthermore, clay soil significantly decreased the amount of radioactivity excreted through the lungs by the oral route.

The tissue distribution pattern of oral toluene-derived radioactivity was unaffected by the presence of either soil. In all treatment groups, stomach (the site of administration) and fat exhibited the highest percentage of radioactivity. Pyykko et al.[13] also reported high concentrations of radioactivity in the same organs following gastric intubation of rats with ^{3}H-toluene. The high concentration of radioactivity in fat is most likely unmetabolized ^{14}C-toluene which has a high octanol/water partition coefficient. The higher a compound's octanol/water partition coefficient, the more readily it would be bioaccumulated into biological tissues.[23,24] Compared to the oral route, a smaller percentage of radioactivity distributed to fat when ^{14}C-toluene was administered by the dermal route. The difference may be a consequence of slower absorption of radioactivity into blood following dermal exposure, resulting in more extensive metabolism of toluene and less unmetabolized toluene in blood for distribution to tissues including fat.

The metabolic profile of toluene was not changed by the soils in either route of exposure. Hippuric acid appeared as the major urinary metabolite in all treatment groups during all collection periods. The oxidation of toluene to benzoic acid followed by conjugation with glycine is consistent with the findings of Bray et al.[25] in rabbits administered toluene orally. Pathiratne et al.[26] also reported an unidentified ^{14}C-metabolite after incubating ^{14}C-+toluene with rat liver microsomes in the presence of an NADPH-generating system.

The oral and dermal bioavailability of toluene in the presence of either soil is different from that reported for benzene. While the time course of each chemical in the body was altered by the soils, only oral benzene showed an increase in the amount absorbed. The relatively stronger adsorbance of benzene to clay soil resulted in a significantly decreased dermal penetration of chemical. On the other hand, the quantity of oral and dermal toluene absorbed was unaffected by the soils.

ACKNOWLEDGMENT

This research was supported as a project of the National Science Foundation/Industry/University Cooperative Center for Research in Hazardous and Toxic Substances at New Jersey Institute of Technology, an Advanced Technology Center of the New Jersey Commission on Science and Technology.

REFERENCES

1. Paustenbach, D. J. "A Methodology for Evaluating the Environmental and Public Health Risks of Contaminated Soil," in *Petroleum Contaminated Soils, Volume I: Remediation Techniques, Environmental Fate, Risk Assessment,* P. T. Kostecki and E. J. Calabrese, Eds. (Chelsea, MI: Lewis Publishers, Inc., 1989), pp. 225–261.
2. Kimbrough, R., H. Falk, P. Stehr, and G. Fries. "Health Implications of 2,3,7,8-tetrachlorodibenzo-p-dioxin (TCDD) Contamination of Residential Soil," *J. Toxicol. Environ. Health.* 14:47–93 (1984).
3. Paustenbach, D. J., H. P. Shu, and T. J. Murray. "A Critical Analysis of Risk Assessments of TCDD Contaminated Soil," *Regul. Toxicol. Pharmacol.* 6:284–304 (1986).
4. Eschenroeder, A., R. J. Jaeger, J. J. Ospital, and C. P. Doyle. "Health Risk Assessment of Human Exposures to Soil Amended with Sewage Sludge Contaminated with Polychlorinated Dibenzodioxins and Dibenzofurans," *Vet. Hum. Toxicol.* 28:435–442 (1986).
5. Schaum, J. "Risk Analysis of TCDD Contaminated Soil," (Washington, DC, Office of Health and Environmental Assessment, U.S. Environmental Protection Agency, 1983).
6. Roels, H., J. P. Buchet and R. R. Lauwerys. "Exposure to Lead by the Oral and Pulmonary Routes of Children Living in the Vicinity of a Primary Lead Smelter," *Environment.* 22:81–94 (1980).
7. Turkall, R. M., G. Skowronski, S. Gerges, S. Von Hagen, and M. S. Abdel-Rahman. "Soil Adsorption Alters Kinetics and Bioavailability of Benzene in Orally Exposed Male Rats," *Arch. Environ. Contam. Toxicol.* 17:159–164 (1988).
8. Abdel-Rahman, M. S., and R. M. Turkall, "Determination of Exposure of Oral and Dermal Benzene from Contaminated Soils," in *Petroleum Contaminated Soils, Volume I: Remediation Techniques, Environmental Fate, Risk Assessment,* P. T. Kostecki and E. J. Calabrese, Eds. (Chelsea, MI: Lewis Publishers, Inc., 1989) pp. 301–311.
9. Skowronski, G. A., R. M. Turkall, and M. S. Abdel-Rahman. "Soil Adsorption Alters Bioavailability of Benzene in Dermally Exposed Male Rats," *Am. Ind. Hyg. Assoc. J.* 49:506-511 (1988).
10. Von Burg, R. "Toxicology Update, Toluene," *J. Appl. Toxicol.* 1:140 (1981).
11. Bergman, K. "Whole-Body Autoradiography and Allied Tracer Techniques in Distribution and Elimination Studies of Some Organic Solvents," *Scand. J. Work Environ. Health.* 5 (suppl. 1): 1–263 (1979).
12. Sandmeyer, E. E., "Aromatic Hydrocarbons," in *Patty's Industrial Hygiene* and *Toxicology,* Vol. II B. G. D. Clayton and F. E. Clayton, Eds. (New York: John Wiley & Sons, 1981), pp. 3253–3431.
13. Pyykko, K., H. Tahti, and H. Vapaatalo. "Toluene Concentrations in Various Tissues of Rats after Inhalation and Oral Administration," *Arch. Toxicol.* 38:169–176 (1977).

14. Sato, A. and T. Nakajima. "Differences Following Skin or Inhalation Exposure in the Absorption and Excretion Kinetics of Trichloroethylene and Toluene," *Br. J. Ind. Med.* 35:43–49 (1978).
15. Dutkiewicz, T., and H. Tyras. "Skin Absorption of Toluene, Styrene, and Xylene by Man," *Br. J. Ind. Med.* 25:243 (1968).
16. National Cooperative Soil Survey: Official Series Description—Atsion Series. (Washington, DC: United States Department of Agriculture—Soil Conservation Service, 1977).
17. National Cooperative Soil Survey: Official Series Description—Keyport Series. (Washington, DC: United States Department of Agriculture—Soil Conservation Service, 1972).
18. Tukey, J. W. *Exploratory Data Analysis* (Reading, MA: Addison Wesley, 1977), pp. 205–235.
19. Velleman, P. F., and D. C. Hoaglin. *Applications, Basics and Computing* of *Exploratory Data Analysis* (*ABC's of EDA*) (Boston, MA: Duxbury Press, 1981), pp. 159–200.
20. Gibaldi, M., and D. Perrier. *Pharmacokinetics* (New York, Marcel Dekker, 1975), pp. 281–292.
21. Efron, B. *The Jacknife,* the *Bootstrap,* and *Other Resampling Plans* (Philadelphia, PA: Society of Industrial Applied Math, 1982).
22. Efron, B., and R. Tibshirani. "Bootstrap Method for Assessing Statistical Accuracy," Technical Report 101 (Stanford, CA: Division of Biostatistics, 1985).
23. Grisham, J. W., Ed. "Factors Influencing Human Exposure," in *Health Aspects of the Disposal of Waste Chemicals.* (New York: Pergamon Press, 1986), pp. 40–64.
24. Freed, V. H., C. T. Chiou, and R. Hague. "Chemodynamics: Transport and Behavior of Chemicals in the Environment—A Problem in Environmental Health," *Environ. Health Perspect.* 20:55–70 (1977).
25. Bray, H. G., W. V. Thrope, and K. White. "Kinetic Studies of the Metabolism of Foreign Compounds," *Biochem. J.* 48:88–96 (1951).
26. Pathiratne, A., R. L. Puyear, and J. D. Brammer. "Activation of ^{14}C-toluene to Covalently Binding Metabolites by Rat Liver Microsomes," *Drug Metab. Dispos.* 14:386–391 (1986).

CHAPTER 22

Estimates for Hydrocarbon Vapor Emissions Resulting from Service Station Remediations and Buried Gasoline-Contaminated Soils

Paul C. Johnson, Marvin B. Hertz, and Dallas L. Byers, Shell Development, Westhollow Research Center, Houston, TX

I. INTRODUCTION

Soils become contaminated at service stations as the result primarily of leaking underground storage tanks, leaking transport lines, or spills that occur during storage tank filing. Upon detection of a spill, a site investigation is conducted and a remediation plan is formulated. While the specific remediation plan for any site depends on the level of contamination, location of contaminated soil, soil stratigraphy, and other site-specific factors, a typical service station remediation will involve soil excavation, pumping and treating of contaminated groundwater and free-liquid residual gasoline, and in-situ treatment (soil venting or enhanced biodegradation) of the unsaturated zone. Because some hydrocarbon vapors are released to the atmosphere during each stage, it is important to know the range of possible emission levels in order to evaluate the health risk that they may pose to a nearby community.

This chapter is divided into three main sections. In the first section models are developed for computing conservative emissions estimates for each stage of a hypothetical service station cleanup, which consists of tank excavation and replacement, a pump-and-treat operation that removes contaminated groundwater and free-liquid residual gasoline, and in-situ soil venting of the remaining contaminated soil. A model that estimates the emissions for the case in which gasoline-contaminated soils are left in place is also presented. In the second main section these models are used to estimate the benzene emissions associated with this hypothetical remediation, and with leaving the soils in-place (no treatment). The vapor fluxes are used as air dispersion model inputs in the third main section, and ambient air concentrations are calculated for a nearby community. For comparison, the ambient air hydrocarbon concentrations due to hydrocarbon vapor emissions from undisturbed underground gasoline-contaminated soils are also computed.

II. MODEL DEVELOPMENT

II.1 Vapor Equilibrium Models

An integral part of any vapor transport model is the calculation of vapor concentrations at the source based on measured residual soil contamination levels, contaminant composition, soil properties (organic carbon content, soil moisture content), and environmental factors (temperature). Two main approaches are used in vapor transport models, but rarely is their use justified by the authors. Before presenting vapor emissions models, therefore, it is useful to briefly review the various methods for calculating vapor concentrations and justify the approach used in this work.

The influence of soil type, moisture content, chemical type, temperature, and residual soil contaminant levels has been the focus of studies by Chiou and Shoup,[1] Spencer,[2] Poe et al.,[3] and Valsaraj and Thibodeaux.[4] In each study the effect of the parameters listed above on the equilibrium vapor concentration above a soil matrix was studied for a single component. Briefly, changes in the moisture content significantly influence the vapor concentration when the soils are "dry"; that is, the moisture content is less than that required to provide a complete monolayer coverage of water molecules on the soil particle surfaces. This corresponds roughly to the "wilting point" of a soil, and for sandy soil types is in the 0.02 to 0.05 g-H_2O/g-soil moisture content range. It has been observed that the sorptive capacity of soil is greatly increased when the soil is dry.[1,2] When the contaminant concentration is low enough that free adsorptive sites are available on the soil ($\approx < 100$ mg-contaminant/kg-soil), the adsorbed contaminant/vapor equilibrium can be modeled by a modified Brauner-Emmett-Teller (BET) equation.[5] If the moisture content is great enough that there is more than a monlayer of water molecules adhering to the soil surface, then the vapor equilibrium appears to be governed by the partitioning between four phases: vapor, dissolved in the soil moisture, sorbed to the soil particles, and free-residual (when concentrations are great enough).[2] More often than not, the moisture content of soils buried more than a foot below ground surface will be greater than the wilting point, so we will focus on modeling the partitioning of contaminants in this moisture content regime.

As stated above, components in the residual contaminant partition between vapor, adsorbed, soluble (dissolved in soil moisture), and free-liquid (or solid) residual phases. Mathematically, this can be described for any component i:

$$\frac{M_i}{M_{soil}} = y_i\left[\frac{\alpha_i P_i^v \epsilon_A}{RT\rho_{soil}} + \alpha_i \frac{M^{HC}}{M_{soil}} + \frac{M^{H_2O}}{M_{soil}} + \frac{k_i}{M_{w,H_2O}}\right] \quad (1)$$

where:

M_i = total moles of i in soil matrix
y_i = mole fraction of i in soil moisture phase
α_i = activity coefficient for i in water
k_i = sorption coefficient for i [(mass-i/mass-soil)/(mass-i/mass-H_2O)]
P_i^v = pure component vapor pressure of i
ϵ_A = vapor-filled void fraction in soil matrix
ρ_{soil} = soil matrix density
R = gas constant (=82.1 cm^3-atm/mole-K)
T = absolute temperature
M^{HC} = total moles of free-liquid residual contaminant
M^{H_2O} = total moles in soil moisture phase
M_{soil} = mass of soil matrix
M_{W,H_2O} = molecular weight of water

The first term on the right-hand side of Equation 1 represents the number of moles of i in the vapor phase, the second represents the number of moles of i in the free-liquid residual phase, the third term is the number of moles of i dissolved in the soil moisture, and the last term is the number of moles of i sorbed to the soil particles. In writing Equation 1 we assume equilibrium between an ideal gas vapor phase, an ideal mixture free-liquid hydrocarbon phase, and a nonideal soil moisture phase. When contaminant levels are great enough that a free-liquid (or solid) residual phase is present, then Equation 1 must be solved iteratively, subject to the condition that $\Sigma\alpha_i y_i = 1$.

Once Equation 1 is solved, the vapor concentration in equilibrium with the contaminant/soil matrix, $C_{i,v}^{eq}$, [mass-i/volume-vapor] is obtained from:

$$C_{i,v}^{eq} = \frac{\alpha_i y_i M_{w,i} P_i^v}{RT} \quad (2)$$

where $M_{w,i}$ denotes the molecular weight of component i. In the limits of low and high residual contaminant soil concentrations Equation 1 reduces to forms that do not require iterative solutions. In the low concentration limit (i.e., no free-liquid or solid precipitate phase present), Equation 2 becomes:

$$C_{i,v}^{eq} = \frac{H\ C_{i,soil}}{[(H\epsilon_A/\rho_{soil}) + \theta_M + k_i]} \tag{3}$$

where:

- H = Henry's law constant ($=\alpha_i P_i^v M_{w,H_2O}/RT$)
- $C_{i,soil}$ = residual contamination level of i [mass-i/mass-soil]
- θ_M = soil moisture content [mass-H_2O/mass-soil]

In the high residual contaminant concentration limit Equation 2 becomes:

$$C_{i,v}^{eq} = \frac{x_i P_i^v M_{w,i}}{RT} \tag{4}$$

where x_i is the mole fraction of component i in the free-liquid residual phase. For mixtures composed of compounds with similar molecular weights, x_i is roughly equal to the mass fraction of compound i.

Equations 3 and 4 are the two most commonly incorporated in vapor transport models. Note that Equation 3 predicts vapor concentrations that are proportional to the residual soil concentration of each species and are independent of the relative concentrations of each chemical species in the contaminant, while the vapor concentrations of each chemical species in the contaminant, while the vapor concentrations predicted by Equation 4 are independent of residual soil concentration levels and depend only on the relative concentrations of species. Due to its mathematical characteristics, rather than any model validation, transient transport models[6,7] most often incorporate Equation 3. Steady-state landfill emission models[8] often utilize Equation 4. It is important to recognize that these models are only valid for specific limiting conditions, and generalization to other concentration ranges can produce very misleading results. For example, Equation 3 predicts that vapor concentrations always increase with increasing residual contaminant levels, but realistically the equilibrium vapor concentration of any compound can not exceed its saturated vapor concentration ($=P_i^v M_{w,i}/RT$).

Figure 1 compares vapor concentrations predicted by Equations 1, 3, and 4 for the regular gasoline defined by Table 1. The required chemical parameters (vapor pressures, octanol-water partition coefficients, water solubility values) can be found in Johnson et al.[9] Example model parameters for a sandy soil are:

- f_{oc} = organic carbon fraction = 0.002
- θ_M = soil moisture content = 5%
- ϵ_T = total void fraction = 0.35
- ρ_{soil} = soil bulk density = 1.60 g/cm^3

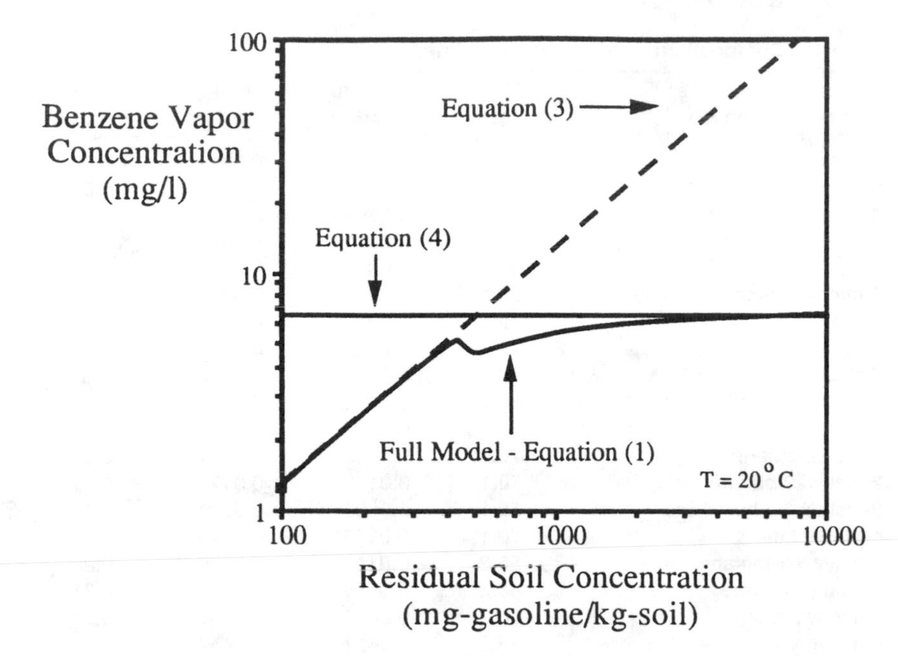

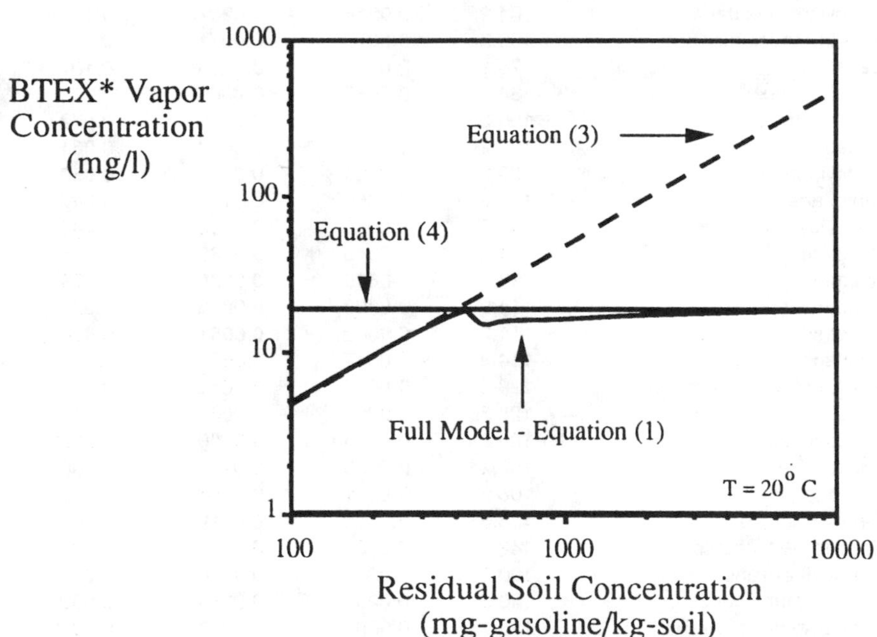

***denotes the sum of benzene, toluene, ethylbenzene, and xylenes vapor concentrations**

Figure 1. Comparison of vapor concentration prediction models.

Table 1. Composition of a Regular Gasoline

Compound Name	Mw (g)	Weight Fraction	Mole Fraction	P_i^v(20°C) (atm)
propane	44.1	0.0005	0.0001	8.50
isobutane	58.1	0.0085	0.0137	2.93
n-butane	58.1	0.0259	0.0415	2.11
trans-2-butene	56.1	0.0019	0.0032	1.97
cis-2-butene	56.1	0.0018	0.0030	1.79
3-methyl-1-butene	70.1	0.0010	0.0013	0.96
isopentane	72.2	0.0916	0.1181	0.78
1-pentene	70.1	0.0032	0.0042	0.70
2-methyl-1-butene	70.1	0.0068	0.0090	0.67
2-methyl-1,3-butadiene	68.1	0.0068	0.0092	0.65
n-pentane	72.2	0.0628	0.0810	0.57
trans-2-pentene	70.1	0.0138	0.0184	0.53
2-methyl-2-butene	70.1	0.0129	0.0171	0.51
3-methyl-1,2-butadiene	68.1	0.0003	0.0004	0.46
cyclopentane	70.1	0.0185	0.0245	0.35
2,3-dimethylbutane	86.2	0.0111	0.0120	0.26
2-methylpentane	86.2	0.0515	0.0556	0.21
3-methylpentane	86.2	0.0314	0.0340	0.20
n-hexane	86.2	0.0411	0.0444	0.16
methylcyclopentane	84.2	0.0214	0.0237	0.15
2,2-dimethylpentane	100.2	0.0077	0.0071	0.11
benzene	78.1	0.0172	0.0205	0.10
cyclohexane	84.2	0.0059	0.0065	0.10
2,3-dimethylpentane	100.2	0.0063	0.0058	0.072
3-methylhexane	100.2	0.0099	0.0092	0.064
3-ethylpentane	100.2	0.0168	0.0156	0.060
n-heptane	100.2	0.0356	0.0331	0.046
methylcyclohexane	98.2	0.0055	0.0052	0.048
2,2-dimethylhexane	114.2	0.0046	0.0038	0.035
toluene	92.1	0.0899	0.0908	0.029
2-methylheptane	114.2	0.0028	0.0023	0.021
3-methylheptane	114.2	0.0062	0.0051	0.020
n-octane	114.2	0.0647	0.0528	0.014
2,4,4-trimethylhexane	128.3	0.0015	0.0011	0.013
2,2-dimethylheptane	128.3	0.0003	0.0002	0.011
ethylbenzene	106.2	0.0205	0.0180	0.0092
p-xylene	106.2	0.0153	0.0134	0.0086
o-xylene	106.2	0.0221	0.0194	0.0066
n-nonane	128.3	0.0155	0.0112	0.0042
3,3,5-trimethylheptane	142.3	0.0033	0.0022	0.0037
n-propylbenzene	120.2	0.0346	0.0268	0.0033
1,3,5-trimethylbenzene	120.2	0.0201	0.0156	0.0024
1,2,4-trimethylbenzene	120.2	0.0061	0.0047	0.0019
n-decane	142.3	0.0343	0.0224	0.0036
methylpropylbenzene	134.2	0.0210	0.0146	0.0010
dimethylethylbenzene	134.2	0.0173	0.0120	0.00070
n-undecane	156.3	0.0078	0.0046	0.00060

Table 1. Continued

Compound Name	Mw (g)	Weight Fraction	Mole Fraction	P_i^v(20 : C) (atm)
1,2,4,5-tetramethylbenzene	134.2	0.0511	0.0354	0.00046
1,2,3,4-tetramethylbenzene	134.2	0.0053	0.0037	0.00033
1,2,4-trimethyl-5-ethylbenzene	148.2	0.0191	0.0120	0.00029
n-dodecane	170.3	0.0050	0.0027	0.00040
naphthalene	128.2	0.0041	0.0030	0.00014
methylnaphthalene	142.2	0.0061	0.0040	0.00005
Total		0.996	0.999	

For these values, Figure 1 indicates that Equation 3 is applicable below a residual soil contamination level of about 500 mg-gasoline/kg-soil. Above this residual concentration level, however, Equation 3 predicts increasing vapor concentrations with increasing residual levels, while the complete model predicts that vapor concentrations become independent of the residual concentration level. This limiting behavior is predicted by Equation 4. One must be very careful when using transport models based on a "three-phase model," such as Equation 3, because they will overpredict vapor concentrations and emission rates for many situations. Usually there are no internal checking procedures in these models to ensure that unrealistic vapor concentrations are not being predicted.

Throughout this chapter, Equation 4 is used to predict equilibrium vapor concentrations because it is most applicable for the residual concentration levels encountered at typical service station spill sites.

II.2 Vapor Flux Models

In the following analysis, models are presented for computing conservative estimates for emissions resulting from processes associated with typical service station remediations. Specifically, we will consider emissions associated with tank excavation, in-situ soil venting, and the pumping of groundwater and free-liquid hydrocarbons.

II.2a Emissions Associated with Tank or Soil Excavation

During the excavation of leaky storage tanks, tank backfill material (usually pea gravel) is removed from the tank area and placed in a pile. In some states the pile of contaminated soil can remain uncovered as long as the excavation does not cease for more than an hour (a "one-hour working pile"). Figure 2 illustrates this operation. Both the excavated soil pile and the empty tank pit are potential sources of hydrocarbon emissions. In addition, during the excavation process fine particulate material may be released to the atmosphere. In this study transport of particulate materials is neglected.

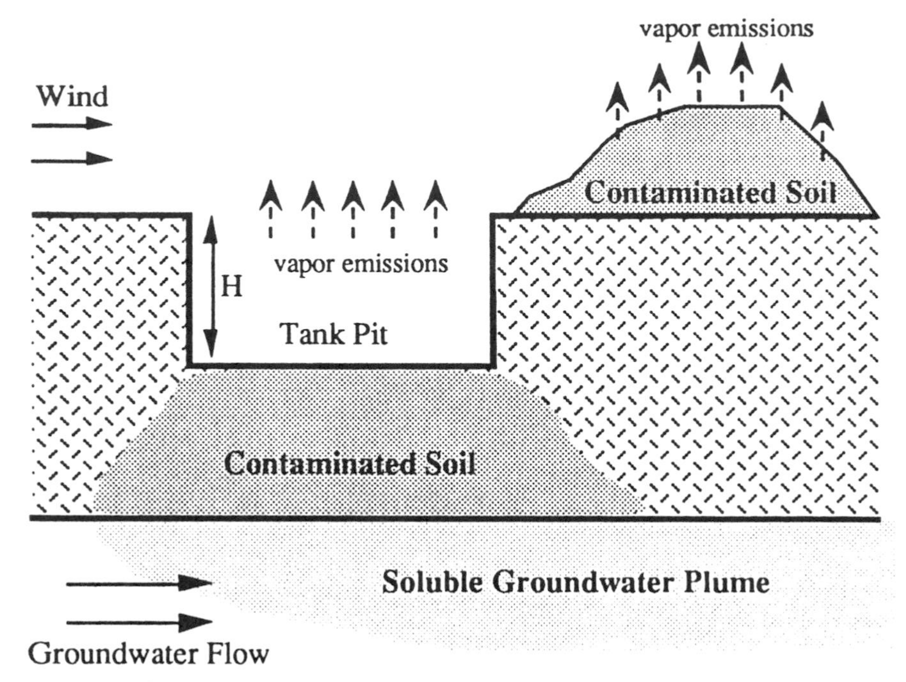

Figure 2. Sources of emissions during excavation.

We can estimate emissions from an excavated pit by assuming that vapor transport is limited by diffusion upwards from the pit bottom through a quasi-stagnant layer of air. At ground surface the vapors are swept away by the wind. For this situation, the vapor flux in the pit can be obtained as a solution to the diffusion equation:

$$\frac{\partial C_{i,v}}{\partial t} = D_i^o \frac{\partial^2 C_{i,v}}{\partial z^2} \tag{5}$$

subject to the following boundary and initial conditions:

$$\begin{aligned} C_{i,v} &= 0; & t &= 0 \\ C_{i,v} &= 0 & z &= H \\ C_{i,v} &= C_{i,v}^{eq} & z &= 0 \end{aligned} \tag{6}$$

where:

$C_{i,v}$ = vapor phase concentration of species i
t = time
z = distance above the pit bottom
H = depth of excavation
D_i^o = vapor phase molecular diffusion coefficient in air

and $C_{i,v}^{eq}$ is the vapor concentration of species i in equilibrium with the contaminant/soil matrix, and as discussed is given by Equation 4. Experimentally measured D_i^o values are available in the literature for many compounds of interest; they can also be predicted reliably from kinetic theory formulas.[10] The solution to Equation 5 is:

$$C_{i,v}(t,z) = C_{i,v}^{eq} \left\langle \left(1 - \frac{z}{H}\right) - \sum_{n=1}^{n=\infty} \frac{2}{n\pi} \exp\left(-\frac{n^2\pi^2 D_i^o t}{H^2}\right) \sin(n\pi z/H) \right\rangle \quad (7)$$

and the corresponding vapor flux, $\mathfrak{F}_{pit}$, from the pit is:

$$\mathfrak{F}_{pit} = -D_i^o \frac{\partial C_{i,v}}{\partial z}_{z=0} = \frac{D_i^o C_{i,v}^{eq}}{H} \left\langle 1 + \sum_{n=1}^{n=\infty} 2(-1)^n \exp\left(-\frac{n^2\pi^2 D_i^o t}{H^2}\right) \right\rangle \quad (8)$$

The maximum vapor flux, $\mathfrak{F}_{pit,max}$, occurs at steady state ($t \to \infty$):

$$\mathfrak{F}_{pit,max} = \frac{D_i^o C_{i,v}^{eq}}{H} \quad (9)$$

Note that the assumptions used in this model (nondiminishing source, zero concentration boundary condition) lead to conservative emissions estimates. Emissions during excavation may be higher than those predicted by Equation 9 due to the mixing within the pit induced by digging machinery. For most cases, however, Equation 9 should provide a good estimate of the average emissions during the excavation and tank replacement stage.

II.2b Emissions Associated with Exposed Piles of Contaminated Soil

Estimating emissions from an excavated pile of contaminated soil requires a more complex analysis. When fresh contaminated soil is placed on the pile, volatilization occurs rapidly from the soil layers on the outside of the pile. As layers of soil near the surface dry out, the vapor emission rate decreases because vapors must diffuse through a longer path to reach the atmosphere. For this process:

$$\frac{\partial \epsilon_A C_{i,v}}{\partial t} + \frac{\partial \rho_b C_{i,s}}{\partial t} = \frac{\partial}{\partial y} D_{i,v}^{eff} \frac{\partial C_{i,v}}{\partial y} \quad (10)$$

where:

ϵ_A = air-filled void fraction in soil
ρ_b = soil bulk density
t = time
$D_{i,v}^{eff}$ = effective porous media vapor diffusion coefficient for species i

$C_{i,s}$ = residual concentration of species i in soil [mass-i/mass-soil]
y = depth into contaminated soil pile
$C_{i,v}$ = concentration of species i in vapor phase

The air-filled void fraction ϵ_A, is a function of the soil moisture content θ_M, the total residual level of hydrocarbons in soil $C_{T,s}(=\Sigma C_{i,s}$ [mass-i/mass-soil]), and the total void fraction ϵ_T:

$$\epsilon_A = \epsilon_T - \frac{\theta_M \rho_b}{\rho_{H_2O}} - \frac{C_{T,s}\rho_b}{\rho_{HC}} \tag{11}$$

where ρ_{H_2O} and ρ_{HC} denote the liquid densities of water and the hydrocarbon mixture. The effective porous media vapor diffusion coefficient $D_{i,v}^{eff}$ is generally calculated by the Millington-Quirk expression:[11,12]

$$D_{i,v}^{eff} = D_i^o \frac{\epsilon_A^{3.33}}{\epsilon_T^2} \tag{12}$$

Again, D_i^o denotes the molecular vapor diffusion coefficient for species i in air.
Without simplifying assumptions, Equation 10 must be solved numerically because $C_{i,v}$ is dependent on composition, not just $C_{i,s}$. In addition ϵ_A and $D_{i,v}^{eff}$ will change with time due to the drying process. Fortunately, for our purposes (emissions from gasoline-contaminated soils) we are typically interested in estimating emissions of benzene, which happens to be one of the more volatile compounds in gasoline (see Table 1). As a result, benzene volatilizes at a much greater rate than the majority of gasoline components. Based on this observation, we can model this situation as the volatilization of a volatile compound from a relatively nonvolatile mixture. Therefore, we assume that:

ϵ_A = constant
$D_{i,v}^{eff}$ = constant
$C_{T,s}$ = constant
$C_{i,v} = (C_{i,s}\ M_{w,T}/C_{T,s}\ M_{w,i})\ M_{w,i}\ P_i^v/RT$ (Equation 4)

where $M_{w,T}$ and $M_{w,i}$ denote the molecular weights of the hydrocarbon mixture and component i, respectively. We also adopt the following initial and boundary conditions:

$$\begin{aligned} C_{T,s} &= C_{T,s}^o & t &= 0 \\ (C_{i,s}/C_{T,s}) &= x_i^o & t &= 0 \\ C_{i,s} &= 0 & y &= 0 \\ C_{i,s} &= C_{i,s}^o & y &= \infty \end{aligned} \tag{13}$$

The solution to Equation 10, subject to the assumptions discussed above, yields the following expression for the flux, $\mathfrak{F}_{soil}$, of volatile species i:

$$\mathfrak{F}_{soil} = D_{i,v}^{eff}\, C_{i,v}^{eq} \frac{1}{\sqrt{\pi \alpha t}} \tag{14}$$

where:

$$\alpha = \frac{D_{i,v}^{eff}}{\epsilon_A + \dfrac{\rho_b RT(C_{T,s}/M_{w,T})}{P_i^v}} \tag{15}$$

As expected, Equation 15 predicts that the emission rate decreases with time. The average flux, $\mathfrak{F}_{soil,avg}$, betweent $t=0$ and $t=\tau$ is:

$$\mathfrak{F}_{soil,avg} = D_{i,v}^{eff}\, C_{i,v}^{eq} \frac{2}{\sqrt{\pi \alpha \tau}} \tag{16}$$

Equation 16 is expected to provide good emissions estimates for volatile compounds, until significant depletion of the less volatile gasoline components occur. In §III.1c emission rates measured during laboratory experiments are compared with predictions from Equation 16.

II.2c Emissions from a Soil Venting Operation

Figure 3 depicts a typical soil venting operation. Vapors are removed from the soil at a volumetric flowrate Q_{vent}, and then are treated by a vapor treatment unit, which may consist of a vapor incinerator, catalytic oxidizer, carbon bed, or diffuser stack. Of these four options, the greatest emission rate of any compound i occurs when the vapors are untreated and discharged through a diffuser stack at a rate $\mathfrak{F}_{i,untreated}$ equal to:

$$\mathcal{E}_{i,untreated} = Q_{vent}\, C_{i,vent} \tag{17}$$

where $C_{i,vent}$ denotes the vapor concentration of species i in the extraction well. The greatest vapor concentration that can be obtained at any time during venting is the equilibrium concentration, $C_{i,v}^{eq}$, defined by Equation 4. When the vapors are treated by a process with a destruction/removal efficiency η, then the emission rate $\mathfrak{F}_{i,treated}$, will be reduced to:

$$\mathcal{E}_{i,treated} = (1-\eta)\, Q_{vent}\, C_{i,vent} \tag{18}$$

Typical gasoline-range hydrocarbon destruction efficiencies for incinerators and catalytic oxidizers are >0.95.

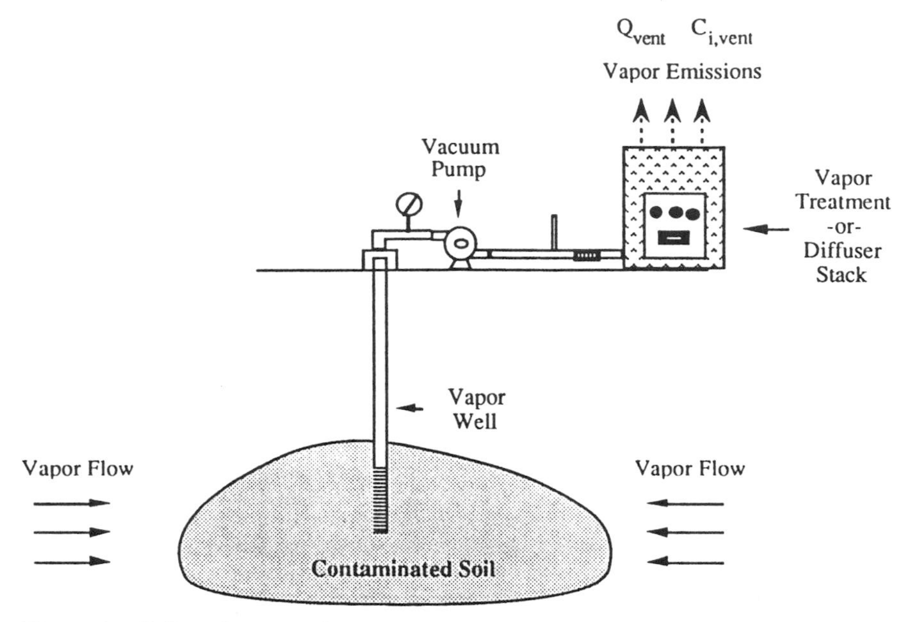

Figure 3. Soil venting operation.

As mentioned above, the most conservative emissions estimates for venting operations are obtained by using Equation 17 for the emission rate and Equation 4 for the vapor concentrations. While this approach might provide good estimates of the emission rate at the start of venting, vapor concentrations decrease with time during venting due to changes in the composition of the residual and due to mass-transfer resistances.[9] To account for this behavior, we calculate a time-averaged emission rate, $\mathcal{E}_{i,avg}$, and average vapor concentration, $C_{i,avg}$, based on the time period for remediation, τ_{vent}, and the mass of compound i removed during this period, m_i:

$$\begin{aligned} \mathcal{E}_{i,avg} &= m_i/\tau_{vent} \\ C_{i,avg} &= m_i/Q_{vent}\tau \end{aligned} \tag{19}$$

Typical venting vapor flowrates are $10 < Q_{vent} < 200$ ft^3/min. For gasoline spill remediations, typical total hydrocarbon vapor concentrations can be as great as 300 mg/L at start-up, but then usually decrease to <50 mg/L.

II.2d Emissions from Groundwater Pump and Treat Operations and Free-Liquid Product Recovery

A pumping operation is pictured in Figure 4, where groundwater and free-liquid product are being removed. If submersible electric pumps are used, then emissions will be minimal. Often, however, injector pumps are used. These work

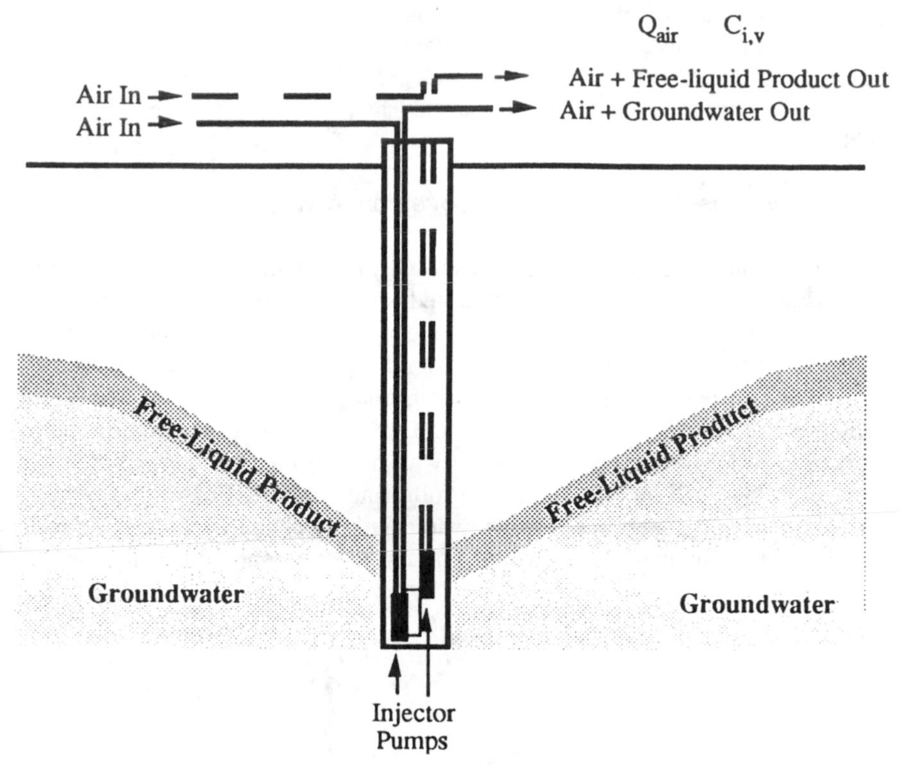

Figure 4. Pump and treat operation.

by displacing fluid with air, and as the fluid is driven to a collection and treatment system aboveground, the air is allowed to escape untreated to the atmosphere. The air flowrate (at 1 atm), Q_{air}, required to obtain a given fluid pumping rate, Q_f, is:

$$Q_{air} = [1 + \frac{H[m]}{10.3 \text{ m}}] \, Q_f \text{ (pumping groundwater)}$$

$$Q_{air} = [1 + \frac{H[m]}{12.9 \text{ m}}] \, Q_f \text{ (pumping free-liquid hydrocarbon)} \qquad (20)$$

where H is the depth to groundwater expressed in m. The maximum vapor emissions, $\mathcal{E}_{i,pump}$, from the pumps will then be:

$$\mathcal{E}_{i,pump} = Q_{air} \, C_{i,v} \qquad (21)$$

where $C_{i,v}$ again denotes the hydrocarbon vapor concentrations in the air. These conservative emission estimates are obtained by assuming that vapor and liquid phases are in equilibrium. In this case $C_{i,v}$ is given by Equation 4 for free-liquid product pumping, and Equation 2 for contaminated groundwater pumping. Soluble levels of hydrocarbons in groundwater are typically an order of magnitude less than saturation, so more realistic levels (such as 1 ppm benzene) might also be used.

II.2e Emissions Associated with Aboveground Water Treatment

Aboveground water treatment systems usually consist of carbon beds, air strippers, or aerobic biotreaters. If a carbon bed is used, emissions will be insignificant because the effluent water must be cleaned to a discharge limit (often ≈ 1 ppb) and all contaminant is transferred from the groundwater to the carbon bed. Biotreaters utilize air spargers to maintain high dissolved oxygen levels, and hence they produce emissions during operation. Those emission levels are difficult to estimate, but at worst, they could be only as great as the emissions from an air stripper, which will be modeled in the following analysis. Figure 5 presents a basic air stripping operation. Groundwater with a soluble hydrocarbon level, $C_{i,soluble}$, enters the unit at a flowrate, Q_f, and is then contacted with clean air

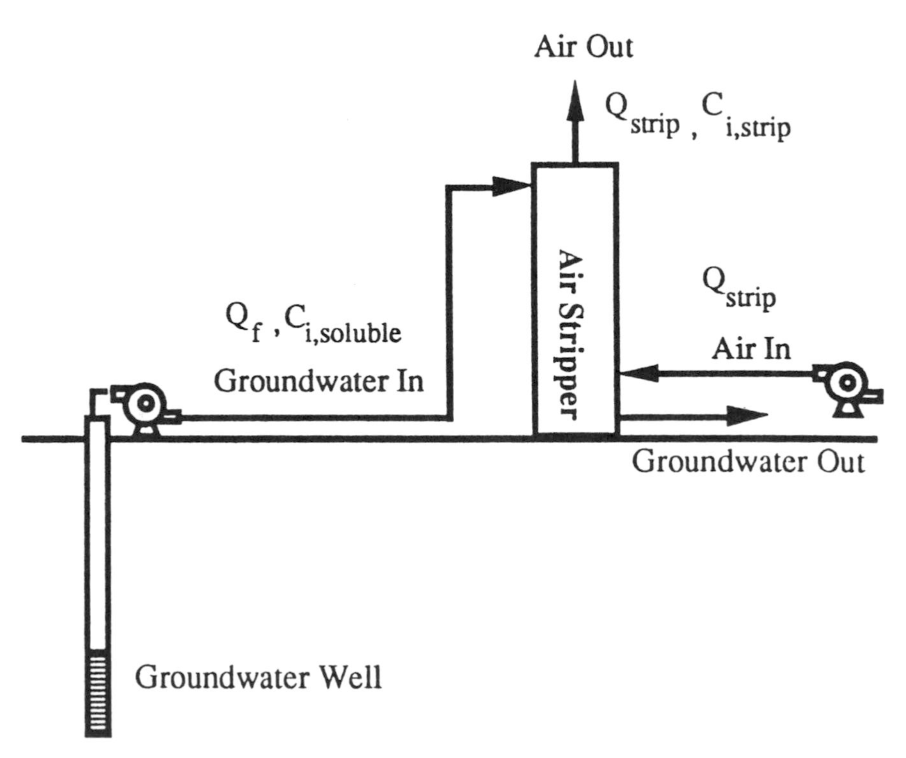

Figure 5. Air stripping operation.

entering at a flowrate, Q_{strip}. Maximum emissions rates occur when the air-stripping efficiency is 100%, and there is no vapor treatment system connected to the air stripper. In this case the emissions rate, $\mathcal{E}_{i,strip}$, and exit vapor concentration, $C_{i,strip}$, are:

$$\begin{aligned} \mathcal{E}_{i,strip} &= Q_f \, C_{i,soluble} \\ C_{i,strip} &= Q_f \, C_{i,soluble}/Q_{strip} \end{aligned} \tag{22}$$

II.2f Emissions from Buried Gasoline-Contaminated Soils

For comparison, we estimate the emissions emanating from gasoline-contaminated soils that are left in place. For the case pictured in Figure 6, where there is a nondiminishing vapor source located a distance H below ground

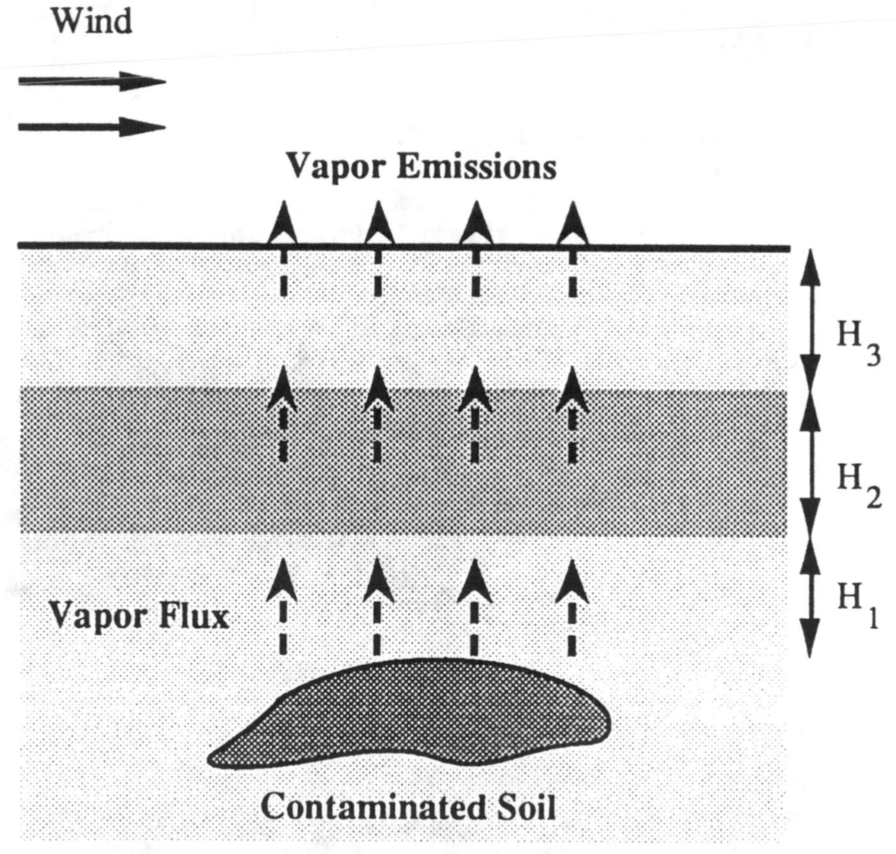

Figure 6. Emissions from buried contaminated soils.

surface and no biodegradation of the vapors (conservative assumptions), the steady-state one-dimensional solution to the governing diffusion equation yields the following expression for the vapor flux, $\mathcal{F}_{i,\text{undisturbed}}$:

$$\mathcal{F}_{i,\text{undisturbed}} = C_{i,v}^{eq} \frac{1}{\sum_n \frac{H_n}{(D_{i,v}^{eff})_n}} \tag{23}$$

where $C_{i,v}^{eq}$ is given by Equation 4, and H_n and $(D_{i,v}^{eff})_n$ are the thickness and effective porous media vapor diffusion coefficient (see Equation 12) for each distinct soil layer above the source. Equation 23 is a generalized form of an equation that is often used for estimating diffusive vapor fluxes from landfills.[8,13] In writing Equation 23 we assume that the vapor concentration is zero at ground level (this yields the maximum emissions rate), and any contribution from diffusion through the soil moisture is insignificant. The latter is a valid assumption for volatile hydrocarbons, but is not necessarily applicable for very nonvolatile compounds.[7]

III. EMISSION RATE CALCULATIONS

To illustrate the use of the models presented in §II.2, benzene vapor emissions estimates are calculated below for a hypothetical service station cleanup consisting of a tank excavation and replacement, soil venting, free-liquid product pumping, and a groundwater pump and treat operation. It will be assumed that gasoline containing 1 mole % ($\approx 1\%$ by weight) benzene is the source of soil and groundwater contamination. "Fresh" gasolines usually contain between 0.5% to 3% by weight benzene. Table 2 summarizes the results presented in this section; these are inputs to the air dispersion models discussed in §IV.1.

Table 2. Summary of Benzene Vapor Emissions Predictions[a]

Source	Flux (g/cm^2-d)	Emission Rate (g/d)	Vapor Conc. at Source (g/cm^3)
Empty Tank Pit[b]			
—maximum steady-state	7.7×10^{-5}	58	NC
—three-day average	2.9×10^{-5}	22	NC
Soil Pile[c]			
—1 hr average for 8 hrs/d	3.0×10^{-3}	370 (24 h average)	NC
Soil Venting[d]			
—untreated vapors, constant source	NC	6532	3.2×10^{-6}
—treated vapors, constant source ($\eta = 0.95$)	NC	327	1.6×10^{-7}
—untreated vapors, 500 gal spill	NC	84	4.2×10^{-8}
—treated vapors, 500 gal spill ($\eta = 0.95$)	NC	4.7	2.1×10^{-9}

Table 2. Continued

Source	Flux (g/cm^2-d)	Emission Rate (g/d)	Vapor Conc. at Source (g/cm^3)
Pumping[e]			
—gasoline-saturated groundwater	NC	560	3.2×10^{-6}
—groundwater with 1 ppm benzene	NC	32	1.8×10^{-7}
—free-liquid gasoline	NC	0.26	3.2×10^{-6}
Groundwater Treatment[f]			
—carbon beds with 1 ppb discharge	NC	0.1	NC
—air-stripper with gas.-sat. water	NC	1965	2.4×10^{-7}
—air-stripper with 1 ppm benzene water	NC	109	1.3×10^{-8}
Leaving Soils In Place[g]	5.9×10^{-6}	2.19	NC

NC = not calculated, or not applicable.
[a]These values correspond to specific conditions (see below).
[b]T = 20°C, 1 mole % benzene, 3 m deep excavation, 6.1 m x 12.2 m area.
[c]T = 20°C, 1 mole % benzene, 10000 mg/kg TPH initial residual gasoline, 6.1 m x 6.1 m area.
[d]T = 20°C, 1 mole % benzene, 1400 1/min (50 SCFM) vapor extraction rate.
[e]T = 20°C, 1 mole % benzene, 18 mg/l saturated conc. in groundwater, 76 l/min (20 gpm) groundwater pumping rate, 0.038 l/min (0.01 gpm) free-liquid product pumping rate.
[f]T = 20°C, 1 mole % benzene, 18 mg/l saturated conc. in groundwater, 76 l/min (20 gpm) groundwater pumping rate, 5660 l/min (200 SCFM) air flowrate.
[g]T = 20°C, 1 mole % benzene, 5 m depth to contamination, no surface seal, 6.1 m x 6.1 m area.

II.1a Emissions Associated with Excavated Pits

Emissions from the empty pit are estimated by assuming vapors diffuse upward from the pit bottom through a stagnant layer of air and are swept away by the wind at ground level. Equations 4 and 9 were used to generate the benzene emissions estimates that appear in Figure 7 for a range of excavation depths and benzene mole fractions. For the air dispersion analyses appearing in §IV.2 the emissions rate corresponds to the parameter values:

x_i	= mole fraction of benzene in the hydrocarbon spill	= 0.01
T	= absolute temperature	= 293 K (20°C)
P_i^o	= vapor pressure of pure benzene	= 0.10 atm
$M_{w,i}$	= molecular weight of benzene	= 78 g/mole
R	= Universal Gas Constant	= 82.1 cm^3-atm/mole-°K
D_i^o	= vapor phase diffusion coefficient in air [cm^2/d]	= 7270 cm^2/d
H	= depth of excavation [cm]	= 305 cm (10 ft)

The result is:

$$\mathfrak{F}_{pit,max} = 7.7 \times 10^{-5} \ g/cm^2\text{-d}$$

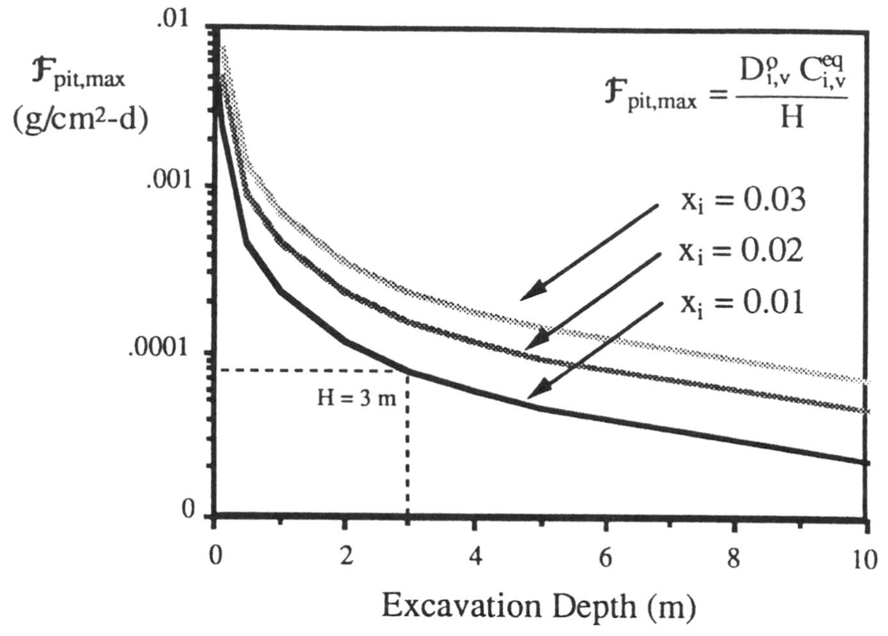

Figure 7. Benzene emission estimates for excavated pit.

For a 6.1 m by 12.2 m (20 ft by 40 ft) pit area, this corresponds to a maximum benzene emission rate $\mathcal{E}_{pit,max}$:

$$\mathcal{E}_{pit,max} = 58 \text{ g/d}$$

A less conservative, but more realistic estimate can be obtained from Equation 8 which models the transient period before a steady-state flux is established. The three-day average flux $<\mathcal{F}_{pit}>_{3\text{-day}}$, and average emission rate $<\mathcal{E}_{pit}>_{3\text{-day}}$ are:

$$<\mathcal{F}_{pit}>_{3\text{-day}} = 2.9 \times 10^{-5} \text{ g/cm}^2\text{-d} \qquad <\mathcal{E}_{pit}>_{3\text{-day}} = 22 \text{ g/d}$$

III.1b Emissions from Exposed Piles of Contaminated Soil

Figure 8 presents emission rates predicted by Equations 16, 15, 11, and 5 for the following parameter values:

ϵ_T	= total void fraction in soil	= 0.38
ρ_b	= soil bulk density	= 1.6 g/cm³
θ_M	= soil moisture content	= 0.05 g-water/g-soil
ρ_{HC}	= hydrocarbon liquid bulk density	= 0.80 g/cm³
$M_{w,T}$	= equivalent molecular weight of gasoline	= 100 g/mole

ρ_{H_2O}	= bulk density of water	= 1.00 g/cm³
$C_{T,s}$	= total hydrocarbon concentration in soil	= 0.01 g-gasoline/g-soil
x_i	= benzene mole fraction in gasoline	= 0.01

and the values given above in §III.1a for P_i^o, $M_{w,i}$, R, T, and D_i^o. The one-hour average flux $<\mathcal{F}_{soil}>_{1\text{-hour}}$ for these values is:

$$<\mathcal{F}_{soil}>_{1\text{-hour}} = 3.0 \times 10^{-3} \text{ g/cm}^2\text{-d}$$

For a pile with a 37 m² (a 20 ft × 20 ft pile) surface area, that is only uncovered for the duration of an 8 h shift each day, the daily average emission rate $<\mathcal{E}_{soil}>$ is:

$$<\mathcal{E}_{soil}> = 370 \text{ g/d}$$

This is calculated by multiplying the one-hour average by eight.

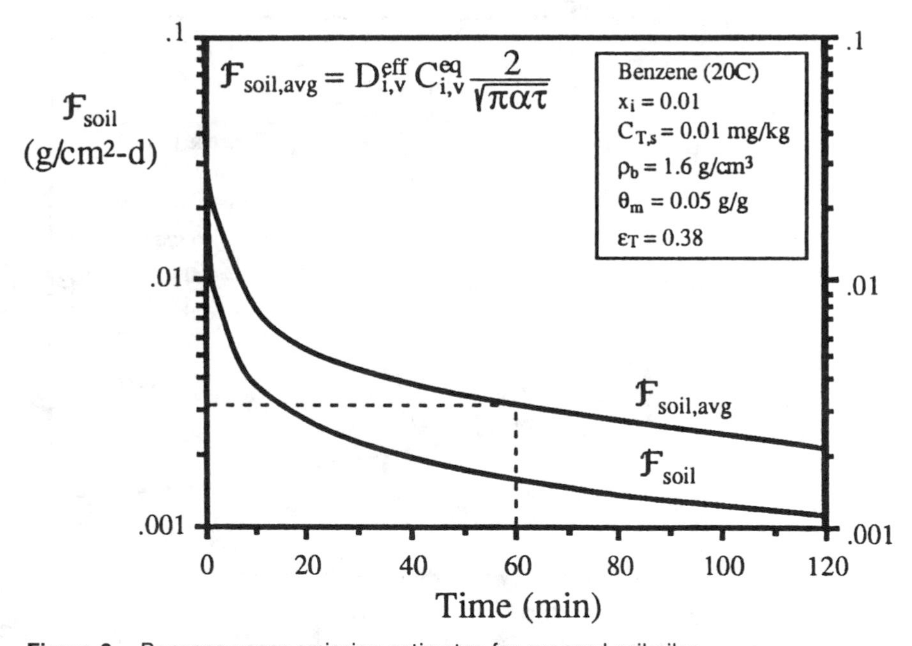

Figure 8. Benzene vapor emission estimates for exposed soil piles.

III.1c Soil Pile Emissions Experiments

Soil pile emissions experiments were conducted with the apparatus pictured in Figure 9. To a coarse (1.2 mm diameter) sand was added enough water and

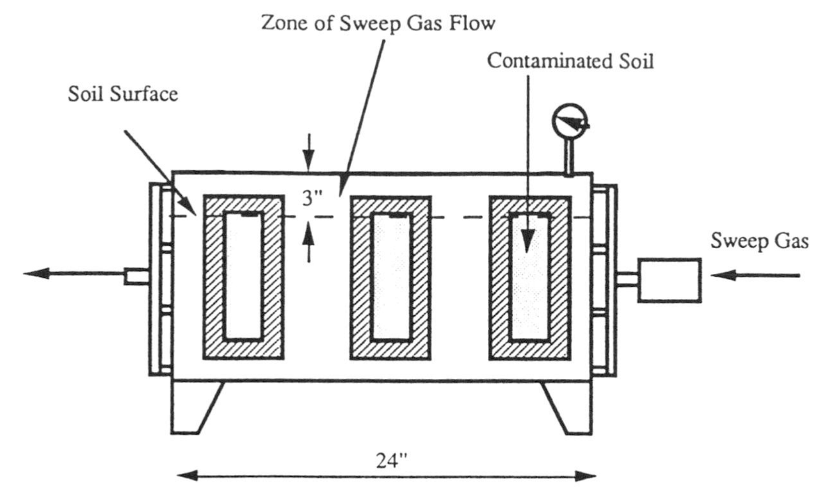

Figure 9. Gasoline emissions experiment apparatus.

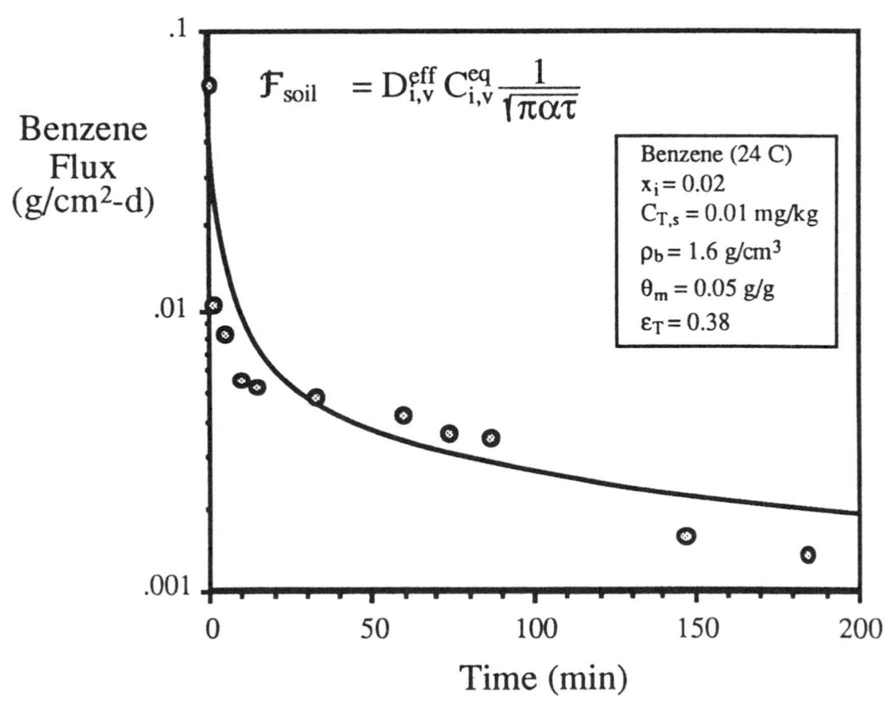

Figure 10. Comparison of measured and predicted benzene fluxes from gasoline contaminated soil.

gasoline to create a 5% moisture content, 10,000 mg-gasoline/kg-soil gasoline-contaminated soil. This soil was placed in the tank, a N_2 sweep gas flow started, and effluent benzene vapor concentrations were monitored with time. The vapor concentration data, sweep gas flowrate, and tank dimensions were used to calculate the vapor flux rates presented in Figure 10. The sweep gas velocity was 3.4 cm/s (0.08 mph), which was determined experimentally to be great enough to ensure that the flux was not dependent on the sweep gas flowrate. An approximate analysis of the gasoline is contained in Table 1, where the initial mole fraction of benzene is 0.0205. The sweep gas temperature was 24°C.

The measured benzene vapor emissions from this experiment are compared with the predictions from Equation 14 in Figure 10. For these conditions, the model reasonably predicts the observed emissions, at least to the accuracy that is desired for emissions estimation purposes.

III.1d Emissions from Soil Venting Operation

Figure 11 presents emission rates for a range of vapor flowrates (Q_{vent}) and vapor concentrations ($C_{i,vent}$). In §IV.2 we use the emission rates corresponding to $Q_{vent} = 1400$ L/min (50 ft^3/min), and the maximum benzene vapor concentration for soils contaminated with gasoline containing 1 mole percent benzene (3.24 mg/L). The worst case is the situation in which vapors are discharged untreated, and there is a nondiminishing benzene source. Equations 17 and 4 predict that the emission rate $\mathcal{E}_{untreated}$, and vapor concentration in the exit gas $C_{v,untreated}$, will be:

$$\mathcal{E}_{untreated} = 6532 \text{ g/d}$$

$$C_{v,untreated} = 3.2 \times 10^{-6} \text{ g/cm}^3$$

If the vapors are treated by a vapor treatment unit with a destruction efficiency $\eta = 0.95$, then the benzene emission rate $\mathcal{E}_{treated}$, and vapor concentration in the exit gas $C_{v,treated}$ are:

$$\mathcal{E}_{treated} = 327 \text{ g/d}$$

$$C_{v,treated} = 1.6 \times 10^{-7} \text{ g/cm}^3$$

It is typically observed that vacuum-well hydrocarbon vapor concentrations decrease during venting, so the values presented above should be regarded as worst-case estimates. To calculate the worst-case average vapor emissions during a soil venting project, we use Equation 19. Suppose that venting removed 100% of the benzene from a 500 gal gasoline spill, which contained 1% by weight benzene, over a six-month period. Then the average emission rate $\mathcal{E}_{avg}$, for the venting operation is calculated with Equation 19 to be:

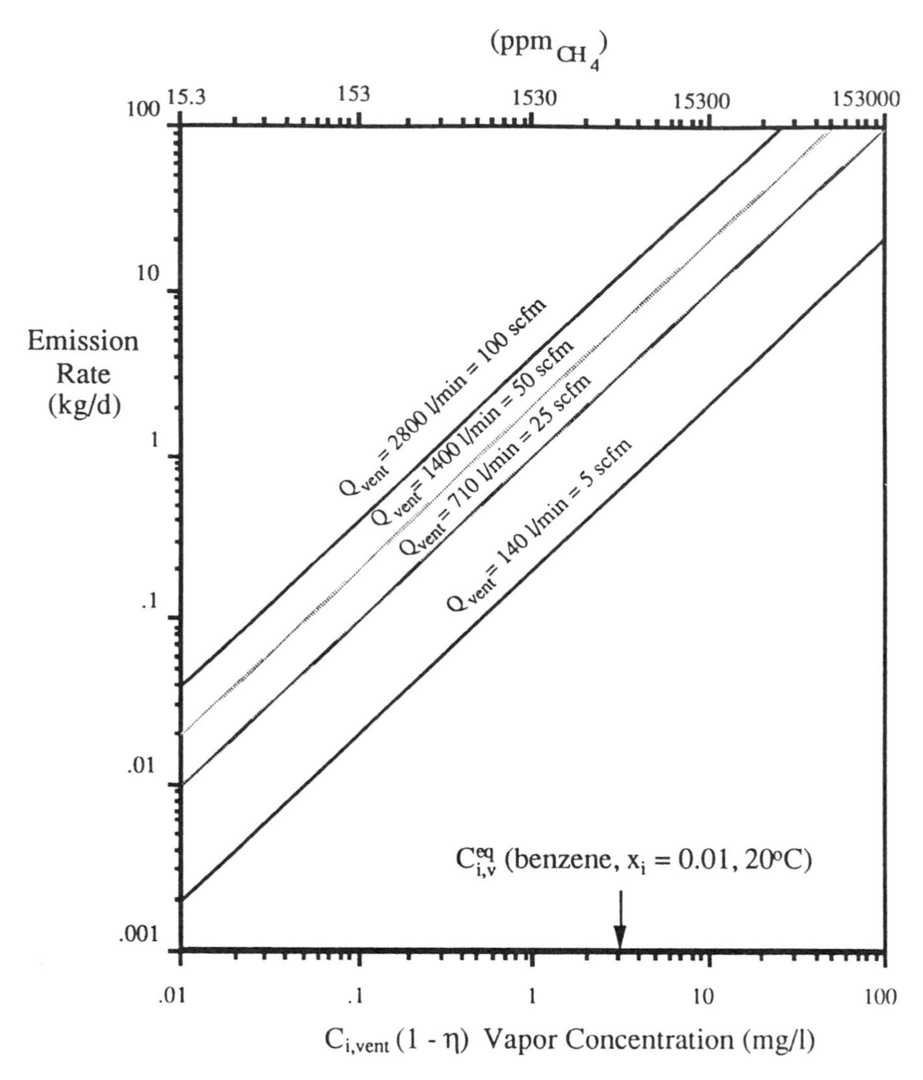

Figure 11. Soil venting emissions estimates.

$$\mathcal{E}_{avg} = 84 \text{ g/d (if untreated vapors are emitted)}$$

$$\mathcal{E}_{avg} = 4.7 \text{ g/d (for a vapor treatment efficiency } \eta = 0.95)$$

Note that these numbers are several orders of magnitude less than the conservative estimates predicted by Equation 17.

III.1e Emissions from Groundwater Pump & Treat Operations and Free-Liquid Product Recovery

Equations 3, 4, and 21 predict conservative emission rate estimates for groundwater and free-liquid product pumping systems that use injector (air displacement) pumps. Calculations were performed for three different cases: (a) pumping gasoline saturated groundwater with a benzene concentration of 18 ppm (equilibrium value for groundwater in contact with gasoline containing 0.01 mole fraction benzene), (b) pumping gasoline-contaminated groundwater with a more realistic benzene level of 1 ppm, and (c) pumping free-liquid product gasoline containing 1 mole percent benzene. At service stations, groundwater pumping rates are generally on the order of 76 L/min (20 gal/min), and average free-liquid product pumping rates might be 0.01 gal/min (860 gal over a two-month period). The results for these pumping rates, and a 6 m (20 ft) depth to the water table are:

(*a*) *gasoline-saturated groundwater,* $Q_f = 20$ *gpm*

$$\mathcal{E}_{pump} = 560 \text{ g/d} \qquad C_v = 3.2 \times 10^{-6} \text{ g/cm}^3$$

(*b*) 1 *ppm benzene in groundwater,* $Q_f = 20$ *gpm*

$$\mathcal{E}_{pump} = 32 \text{ g/d} \qquad C_v = 1.8 \times 10^{-7} \text{ g/cm}^3$$

(*c*) *free-liquid gasoline pumping,* $Q_f = 0.01$ *gpm*

$$\mathcal{E}_{pump} = 0.26 \text{ g/d} \qquad C_v = 3.2 \times 10^{-6} \text{ g/cm}^3$$

III.1f Emissions Due to Aboveground Water Treatment

Aboveground water treatment systems may be composed of carbon beds, biotreaters, or air-strippers. Of these options, a carbon-bed system will produce the lowest vapor emissions, because discharge water must meet a minimum water quality standard, which is often ≈ 1 ppb benzene. Even if all the benzene volatilizes from the water as it leaves the carbon beds, the emission rate for a 76 L/min (20 gal/min) operation with effluent water containing 1 ppb benzene calculated from Equation 22 is:

$$\mathcal{E}_{carbon\ bed} = 0.1 \text{ g/d} \qquad C_{v,max} = 1.8 \times 10^{-10} \text{ g/cm}^3$$

The maximum benzene vapor emission rate for the aboveground treatment systems comes from an air-stripper (with no subsequent vapor treatment). Equation 22 predicts the emission rates for a 100% efficient air-stripper. Emissions predictions for a wide range of flowrates and concentrations appear in Figure 12a. The relationship between air flowrate, water flowrate, and the effluent vapor

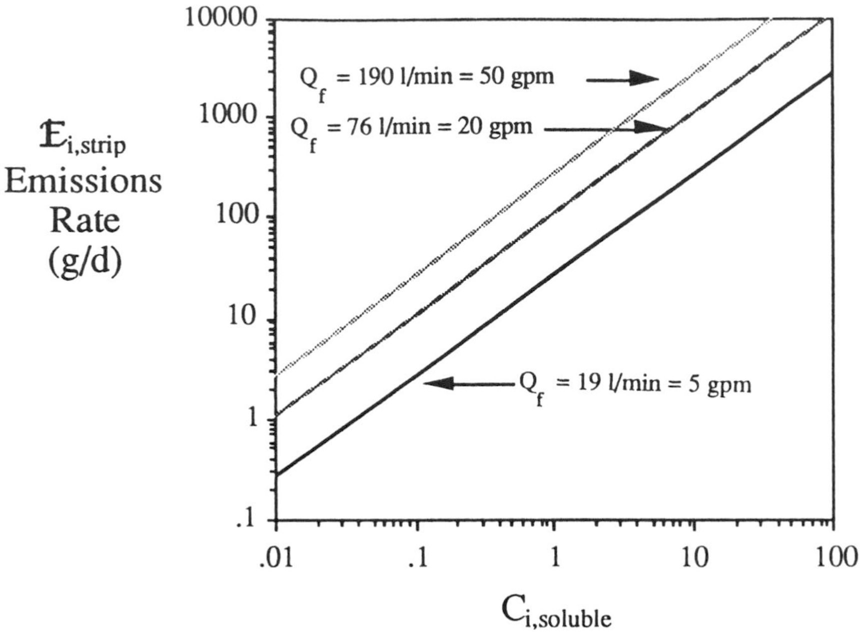

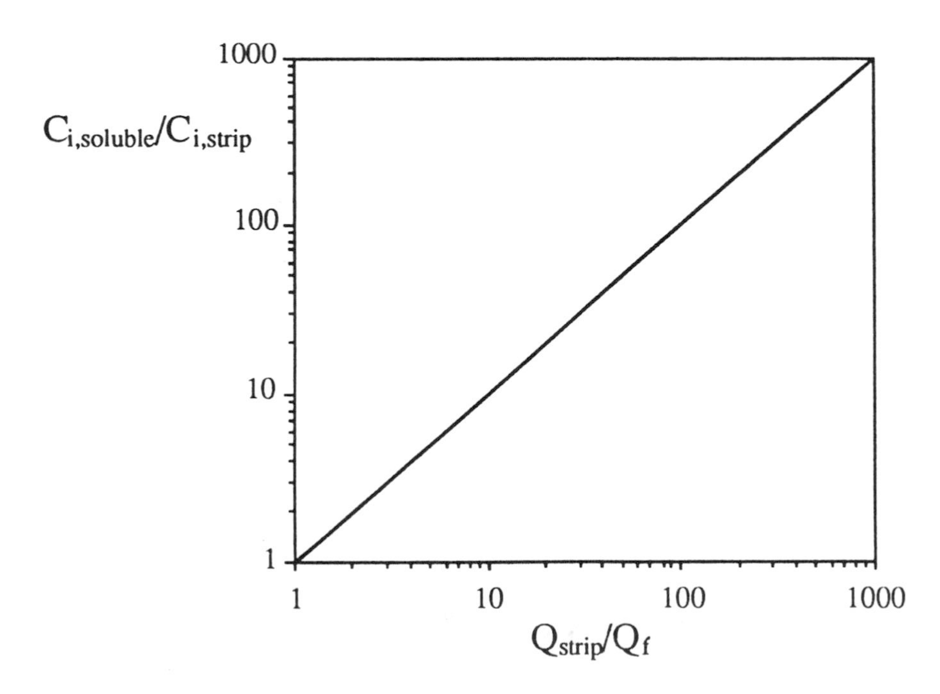

Figure 12. Air stripper emissions.

concentrations shown in Figure 12b. The emission rates from a 76 L/min (20 gal/min) treatment of gasoline-saturated groundwater (18 ppm benzene), and groundwater containing a more realistic 1 ppm benzene, are predicted by Equation 22 to be:

$$\mathcal{E}_{\text{sat gw}} = 1965 \text{ g/d} \qquad C_v = 2.41 \times 10^{-7} \text{ g/cm}^3$$

$$\mathcal{E}_{\text{1 ppm gw}} = 109 \text{ g/d} \qquad C_v = 1.33 \times 10^{-8} \text{ g/cm}^3$$

It has been assumed that the air flowrate into the stripper is 5660 L/min (200 ft³/min, $Q_{strip}/Q_f = 75$).

III.1g Emissions from Buried Gasoline-Contaminated Soils

Figure 13 presents benzene emission estimates for gasoline-contaminated soils left in-place as predicted by Equation 23 for the following conditions:

$\epsilon_T = 0.38$
$\theta_M = 0.05$
$D_{i,v}{}^o = 7270 \text{ cm}^2/\text{d}$
$T = 20°C$

These conservative estimates are for the situation in which relatively homogeneous soils lie above the gasoline-contaminated soil, and there is no low permeability surface cover (i.e., paving, clay liner).

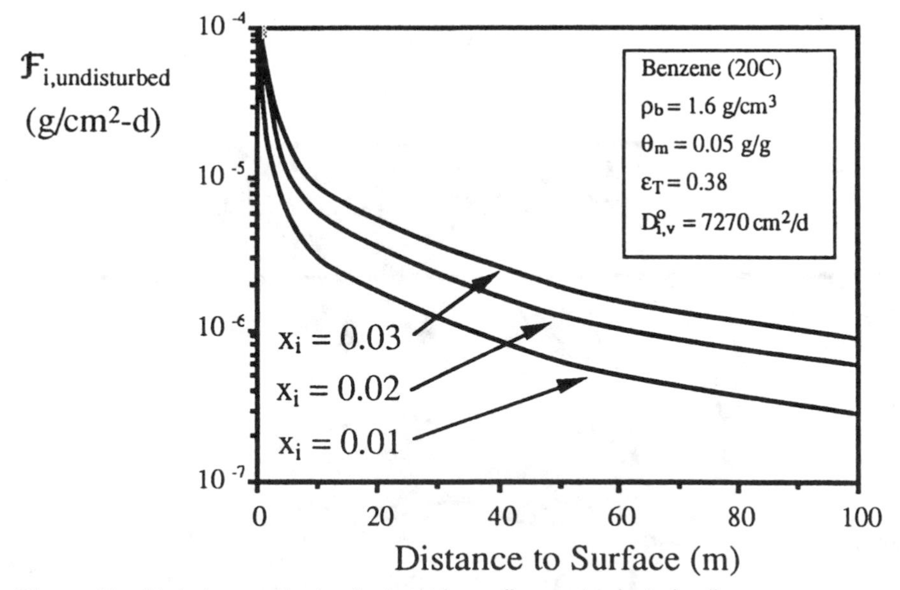

Figure 13. Emissions estimates for buried gasoline-contaminated soils.

IV. AIR DISPERSION MODELING

IV.1 Model Description

The EPA-recommended Industrial Source Complex (ISC) Model, which can be used to predict ambient concentrations from both stack and area sources for receptors located at least 100 m from a source, was used to perform the air dispersion modeling. For this analysis, a polar-coordinate grid was generated with receptors located on concentric rings at radii of 100 m, 300 m, 500 m, 700 m, and 1000 m from each source. Thirty-six receptors, at 10 degree intervals, were located on each ring, making a total of 180 receptors. The ISC model employs standard gaussian dispersion algorithms to calculate the hourly concentration at each receptor. If an entire year of meteorological data is input to the model, then a full year of hourly concentrations can be calculated for each receptor. These hourly values can then be averaged to give an annual average for each receptor, or they can be sorted and ranked to give maximum 1-, 8-, and 24-h averages, For this modeling, meteorological data from Long Beach, California for the year 1964 was used. Other required meteorological parameters include wind speed and direction, ambient temperature, mixing height, and stability class. For area sources, the size and height of the release as well as the emission flux are required input parameters. For stack releases, the stack height and diameter and exit gas velocity and temperature are needed. Modeling options selected for this

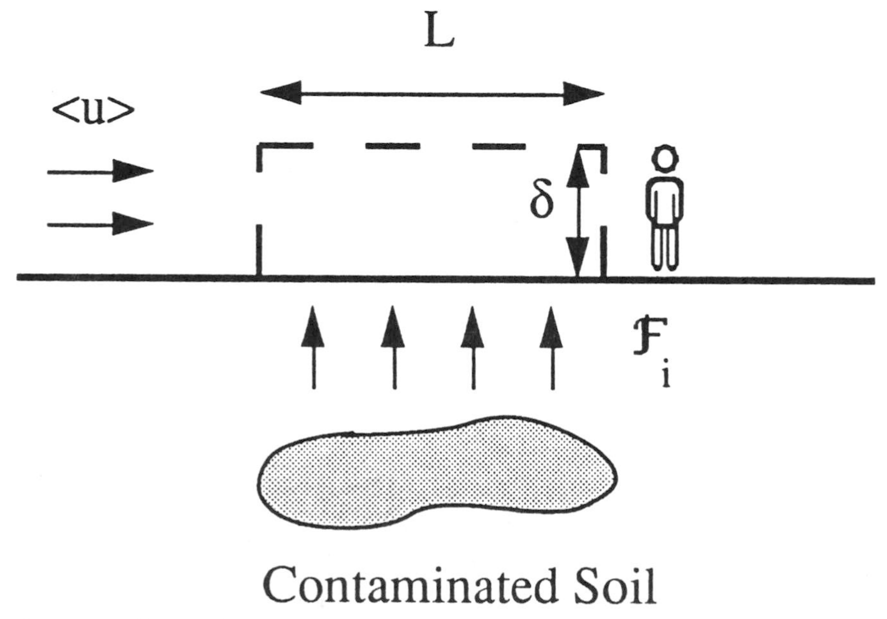

Figure 14. Box model schematic.

study included gradual plume rise, buoyancy-induced dispersion, and stack downwash. The source was assumed to be located in an urban environment.

The "box model" illustrated in Figure 14 was used to predict conservative ambient concentration estimates for receptors located on the downwind edge of the area sources considered in this chapter. It was assumed that vapors within the box were "well-mixed," and the height of the box, δ, was set to be equal to a human breathing zone height – 2 m. Given the wind speed $<u>$, breathing zone height δ, length of emission source parallel to the wind direction L, and emissions flux $\mathfrak{F}_i$, then the ambient concentration $C_{i,amb}$, is:

$$C_{i,amb} = \frac{\mathfrak{F}_i L}{\delta <u>} \tag{24}$$

IV.2 ISC Model Predictions

ISC model predictions are summarized in Table 3. Each scenario modeled is described breifly below.

Table 3. Air Dispersion Modeling Results

		Benzene Vapor Concentrations (μg/m³) Radial Distance					
Scenario	Duration	100 m	300 m	500 m	700 m	1000 m	Box Model
1. excavation[a]		0.39	0.07	0.04	0.03	0.02	24
(70 y average)	3 d	<0.0001	<0.0001	<0.0001	<0.0001	<0.0001	<0.01
2. excavation[a]		0.15	0.03	0.02	0.01	0.01	9.2
(70 y average)	3 d	<0.0001	<0.0001	<0.0001	<0.0001	<0.0001	<0.01
3. soil pile[b]		7.5	1.3	0.78	0.57	0.41	480
(70 y average)	3×8 h	<0.001	<0.0001	<0.0001	<0.0001	<0.0001	<0.01
4. venting, untreated vapors[c]		6.5	0.9	0.36	0.2	0.11	NA
(70 y average)	180 d	0.05	<0.01	<0.01	<0.01	<0.001	NA
5. venting, untreated vapors[c]		0.08	0.01	<0.01	<0.01	<0.01	NA
(70 y average)	180 d	<0.0001	<0.0001	<0.0001	<0.0001	<0.0001	NA
6. venting, treated vapors[c]		0.12	0.04	0.02	0.01	<0.01	NA
(70 y average)	180 d	<0.001	<0.001	<0.001	<0.0001	<0.0001	NA
7. venting, treated vapors[c]		<0.01	<0.001	<0.001	<0.001	<0.001	NA
(70 y average)	180 d	<0.0001	<0.00001	<0.00001	<0.00001	<0.00001	NA
8. free-liquid pumping[c]		<0.001	<0.001	<0.001	<0.001	<0.001	NA
(70 y average)	60 d	<0.00001	<0.00001	<0.00001	<0.00001	<0.00001	NA
9. gasoline-saturated GW pump[c]		0.64	0.08	0.03	0.02	<0.01	NA
(70 y average)	5 y	0.07	<0.01	<0.01	<0.01	<0.001	NA
10. 1 ppm benzene GW pump[c]		0.04	<0.01	<0.01	<0.001	<0.001	NA
(70 y average)	5 y	<0.01	<0.001	<0.001	<0.0001	<0.0001	NA
11. air-stripper gas-sat. GW[c]		1.2	0.25	0.10	0.06	0.03	NA
(70 y average)	5 y	0.09	0.02	<0.01	<0.01	<0.01	NA
12. air-stripper 1 ppm benzene GW[c]		0.07	0.01	<0.01	<0.01	<0.01	NA
(70 y average)	5 y	<0.01	<0.001	<0.001	<0.001	<0.001	NA
13. buried soils left in-place		0.044	<0.01	<0.01	<0.01	<0.01	0.94
(70 y average)	70 y	0.044	<0.01	<0.01	<0.01	<0.01	0.94

[a]peak 24 h concentration.
[b]peak 8 h concentration.
[c]annual average concentration.
[d]nondiminishing source located 5 m below uncovered ground surface.

IV2.a Emissions Associated with Excavated Pits

Scenario 1: The maximum estimated emission rate from the empty pit was estimated to be 58 g/d from a 6.1 m by 12.2 m pit area at a temperature of 20°C (see §III.1a). Since these emissions are hypothesized to last for three days, the peak 24-h concentration was assumed to approximate the highest exposure that might result during the three days. A ground level emission height was assumed.

Scenario 2: A more realistic average emission rate estimate from the pit was calculated to be 22 g/d (see §III.1a). As for Scenario 1, the pit was assumed to be 6.1 m by 12.2 m, the temperature to be 20°C, and the emission height to be ground level. As in Scenario 1, the 24-h peak exposure was determined.

IV.2b Emissions from Exposed Piles of Gasoline-Contaminated Soils

Scenario 3: The estimated one-hour average flux rate at ground level from the 6.1 m by 6.1 m excavated soil pile was calculated to be 3.0×10^{-3} g/cm^2-d in §III.1b, for a temperature of 20°C. It was assumed that the pile would remain exposed for an 8 h shift, and then be covered at night. Therefore, an 8-h peak concentration was calculated to represent the exposure from the pile.

IV.2c Emissions from Soil Venting Operations

In a typical soil venting operation, vapors are removed from the soil by a vacuum pump and then treated by a vapor treatment unit before being released to the atmosphere. In areas with strict emissions regulations, vapor incinerators and catalytic oxidizers are often used. In some areas, diffuser stacks are still allowed. In §III.1d four emission rates were estimated for the venting operations: (a) a worst-case situation in which it was assumed that there was a nondiminishing gasoline source and the vapors were released without any treatment; (b) a more realistic case in which it was assumed that 100% of the benzene from a 1900 L (500 gal) gasoline spill was removed over a six-month period, again with no vapor treatment; (c) emissions from case (a), except that they are treated by a unit with a vapor destruction efficiency $\eta = 0.95$; and (d) same as case (b), except the vapors are treated by a unit with a vapor destruction efficiency $\eta = 0.95$. It should be noted that vapor incinerator and catalytic oxidizer units are quite capable of achieving destruction efficiencies $<95\%$, so the 95% value used in these calculations should be regarded as a worst-case (very conservative) estimate.

All of these scenarios were modeled assuming that the vapor was released from a 4.6 m (15 ft) stack with a 0.6 m (2 ft) diameter. The total vapor flowrate was assumed to be 1400 L/min (50 SCFM) for all cases.

Scenario 4: A worst-case untreated emission rate of 6532 g/d, and 20°C release temperature were assumed.

Scenario 5: The more realistic six-month average untreated emission rate of 84 g/d, and 20°C release temperature.

Scenario 6: A worst-case treated (95% treatment efficiency) emission rate of 327 g/d, and a release temperature of 370°C.

Scenario 7: A more realistic, treated (95% treatment efficiency) benzene emission rate of 4.7 g/d, and a release temperature of 370°C.

IV.2d Emissions from Groundwater Pump & Treat Operations and Free-Liquid Product Recovery

Groundwater and free-liquid product recovery systems utilize various pumping hardware, including submergible air-injected pumps. These pumps work by displacing fluid with air, and while the fluid is collected aboveground, the air is generally emitted untreated. Some volatilization may occur during the period that air and fluid are in contact, so the released air will be a source of emissions. The worst case (maximum emission rate) occurs when the air and fluid phases equilibrate during pumping. Scenario 8 represents the case in which the discharged air is in equilibrium with free-liquid product containing 1 mole % benzene (at 20°C), whereas Scenarios 9 and 10 describe the cases of pumping gasoline-saturated groundwater (18 ppm benzene for 1 mole % benzene in gasoline) and groundwater containing a more realistic 1 ppm benzene, respectively. All scenarios assume that the vapor is released from a 0.64-cm diameter stack at a height of 5 cm above ground. Since pump and treat operations are expected to last for a few years, the annual average was used to estimate exposure levels. In summary, the following scenarios were analyzed:

Scenario 8: An average benzene emission rate of 0.26 g/d (see §III.1e) for the pumping of free-liquid gasoline at a rate of 0.038 L/min (860 gal over a two-month period) and at 20°C temperature.

Scenario 9: Pumping gasoline-saturated groundwater containing 18 ppm benzene at a rate of 76 L/min (20 gpm), which results in a maximum emission rate of 560 g/d at 20°C (see §III.1e).

Scenario 10: Pumping groundwater containing 1 ppm benzene at a rate of 76 L/min (20 gpm), which results in a maximum emission rate of 32 g/d at 20°C (see §III.1e).

IV.2e Emissions Due to Aboveground Water Treatment

The maximum vapor emissions for aboveground water treatment systems occur in air-stripping systems. In §III.1f benzene emissions for air-strippers were calculated and presented in Figure 12. For the air dispersion modeling, the following situations were selected:

Scenario 11: A benzene emission rate of 1965 g/d for an air-stripper treating gasoline-saturated groundwater (18 ppm benzene for 1 mole % benzene in gasoline) at a throughput rate of 76 L/min (20 gpm). The air flowrate through the stripper is 5600 L/min (200 SCFM).

Scenario 12: A benzene emission rate of 109 g/d for an air-stripper treating groundwater containing 1 ppm benzene at a throughput rate of 76 L/min (20 gpm). The air flowrate through the stripper is 5660 L/min (200 SCFM).

IV.2f Emissions from Buried Gasoline-Contaminated Soils

In §III.1g benzene emission estimates for gasoline-contaminated soils left in-place were calculated and presented in Figure 13. For the air dispersion modeling, the following situation was selected:

Scenario 13: The case of a nondiminishing gasoline source containing 1 mole % benzene at 20°C located 5 m below ground surface. The soil is homogeneous and there is no low permeability surface cover (i.e., paving, clay liner).

IV.3 "Box Model" Predictions

As mentioned above, the box model was used to calculate ambient vapor concentrations close to the source. Maximum vapor concentrations will be encountered on the downwind edge of an area source, and for the purpose of generating very conservative exposures, we will assume that a receptor is located at this downwind edge. For the sample calculations a 2.2 m/s (5 mph) wind speed and a 2 m box height were used. The length, L, of the source in the windward direction was taken to be the greatest "edge" length of the source. For example, L = 12.2 m for the excavated pit, and L = 6.1 m for the contaminated soil pile. Values were not computed for the stack sources (air-stripper, soil venting), because we feel that this model is not appropriate for such sources. The "box model" predictions are listed in Table 3 for comparison with the results from the ISC model. As expected, the box model predictions are always greater than the ISC model predictions listed in Table 3.

V. EXPOSURE CALCULATIONS AND CONCLUSIONS

The vapor concentrations listed in Table 3 can be combined with a breathing rate, body weight, duration of exposure, and potency factor to predict the additional human health risk due to each of the sources described in this chapter. Given the current debate over the proper choice of poetency factors, breathing rates, and demographics, however, calculated risks are not presented in this chapter. For comparison though, the IRIS[14] potency factor for benzene (0.029 kg-d/mg), which is based on a 70 y exposure for a 70 kg person, yields a $>1.0 \times 10^{-6}$ risk for average lifetime benzene vapor concentrations $>0.12\ \mu g/m^3$, for a 20 m^3/d breathing rate. In order to compare this value with the predicted concentrations presented in Table 3, one must consider that the exposure duration during any of these processes is much less than 70 y. For example, the maximum duration of exposure from soil pile emissions is only 0.066 y (3×8 h). In Table 3, therefore, equivalent 70-y average vapor concentrations are also presented for each scenario. Based on these values, the only 70-y average vapor concentration >0.12 $\mu g/m^3$ is due to the emissions from contaminated soils left in place (0.94 $\mu g/m^3$). It is important to note, however, that these predictions correspond to the most conservative assumptions (uncovered, permeable soils, nondiminishing source), and a specific depth-to-contamination. It should be clear that for other soil types, surface covers, and contamination depths, the predicted aboveground vapor concentrations will be low enough so as to not pose a significant health risk. For this reason, future cleanup criteria should be influenced by site-specific environmental risk assessments, which evaluate all significant migration pathways and the potential impact on people and the environment.

In summary, models have been presented for estimating the hydrocarbon vapor emissions for typical service station remediation operations. A range of example scenarios was presented in the text, emission rates were calculated, and air-dispersion models were used to calculate the corresponding vapor concentrations in nearby communities. It is important to remember that these models incorporate conservative assumptions and are intended for use as screening tools to determine whether the emission rates or local vapor concentrations are potentially "large" or "small" as compared to values specified in regulatory requirements.

REFERENCES

1. Chiou, G. T., and T. D. Shoup. "Soil Sorption of Organic Vapors and Effects of Humidity on Sorptive Mechanism and Capacity," *Environ. Sci. Technol.*, 19, 1196–1200, 1985.
2. Spencer, W. F. "Distribution of Pesticides Between Soil, Water, and Air, Pesticides in the Soil: Ecology, Degradation, and Movement," Michigan State University, East Lansing, Michigan, 1970.
3. Poe, S. H., K. T. Valsaraj, L. J. Thibodeaux, and C. Springer. "Equilibrium Vapor Phase Adsorption of Volatile Organic Chemicals on Dry Soils," *J. Hazard. Mat.*, 19, 17–32, 1988.
4. Valsaraj, K. T., and L. J. Thibodeaux. "Equilibrium Adsorption of Chemical Vapors on Surface Soils, Landfills, and Landfarms—A Review," *J. Hazard. Mat.*, 19, 79–99, 1988.
5. Brauner, S., P. H. Emmett, and E. Teller. *J. Am. Chem. Soc.*, 60, 309, 1938.
6. Jury, W. A., W. F. Spencer, and W. J. Farmer. "Behavior Assessment Model for Trace Organics in Soil: I. Model Description," *J. Environ. Qual.*, 12, 558–564, 1983.
7. Jury, W. A., W. F. Spencer, and W. J. Farmer. "Behavior Assessment Model for Trace Organics in Soil: II. Application of Screening Model," *J. Environ. Qual.*, 13, 573–579, 1984.
8. Shen, T. "Estimating Hazardous Air Emissions from Disposal Sites," *Pollution Engineering*, 13(8), 31–34, 1981.
9. Johnson, P. C., M. W. Kemblowski, and J. D. Colthart. "Practical Screening Models for Soil Venting Applications," in *Proceedings of the NWWA/API Conference on Petroleum Hydrocarbons and Organic Chemicals in Groundwater: Prevention, Detection, and Restoration,* November 9–11, Houston, TX, 1988.

10. Bird, R. B., W. E. Stewart, and E. N. Lightfoot. *Transport Phenomena,* (New York, NY: John Wiley & Sons, 1960), pp. 511–512.
11. Millington, R. J., and J. M. Quirk. "Permeability of Porous Solids," *Trans. Faraday Soc.,* 57:1200–1207, 1961.
12. Bruell, C. J., and G. E. Hoag. "The Diffusion of Gasoline-Range Hydrocarbon Vapors in Porous Media, Experimental Methodologies," in *Proceedings of the NWWA/API Conference on Petroleum Hydrocarbons and Organic Chemicals in Groundwater: Prevention, Detection, and Restoration,* November 12–14, Houston, TX, 1986.
13. "Superfund Exposure Manual," U.S. Environmental Protection Agency, Office of Remedial Response, Washington, D.C., EPA/540/1-88/001, 1988, p. 16.
14. IRIS (Integrated Risk Information System), Office of Health and Environmental Assessment, Environmental Protection Agency, "Benzene," last revised 3/1/88.

CHAPTER 23

A Toxicological Assessment of Unleaded Gasoline Contamination of Drinking Water

William R. Hartley, Tulane University School of Public Health and Tropical Medicine, Department of Environmental Health Sciences, New Orleans, Louisiana
Edward V. Ohanian, United States Environmental Protection Agency, Office of Drinking Water, Washington, D.C.

The purpose of this chapter is to present toxicological approaches for determining water quality criteria for drinking water and/or groundwater contaminated with unleaded gasoline from leaking underground storage tanks. Of approximately 12,444 release incidents of petroleum products occurring between 1970 and 1984, more than 70% were due to gasoline alone.[1] There are a reported[2] 1.4 million underground storage tank systems with mostly petroleum products in the United States. Although the actual quantity of gasoline migrating into groundwater is unknown, it is clear that a need exists to develop a strategy for evaluating the health risks from unleaded gasoline contaminated drinking water.

UNLEADED GASOLINE COMPOSITION

Unleaded gasoline is a complex mixture of hydrocarbons consisting of aliphatic hydrocarbons, cycloalkanes, alkenes, aromatic hydrocarbons, and polynuclear aromatic hydrocarbons (PAHs). It frequently contains additives or octane enhancers such as methyl tertiary butyl ether (MTBE). It is difficult, if not impossible, to define an exact chemical profile for unleaded gasoline since there are numerous blends to meet seasonal and performance requirements. Performance characteristics include freedom from engine knock (octane quality), quick starting both summer and winter (full boiling range volatility), and many others.[3] Some typical profiles of the major chemical classes of hydrocarbons in unleaded gasoline, excluding additives, are listed in Table 1. For the purposes of this risk assessment, unleaded gasoline may be defined as 56% alkanes, 34% aromatics, 10% alkenes, and <1% PAHs.

In consideration of a mixtures approach to unleaded gasoline risk assessment, selection of components of concern should be based upon the characteristics of solubility, toxicity, and occurrence in groundwater. Dietrich[5] evaluated this issue and selected nine chemicals as representative of unleaded gasoline in drinking water. These components were benzene, toluene, o-xylene, m-xylene, p-xylene, ethylbenzene, and n-hexane, and two additives (MTBE and methanol). Page[6] proposed a similar group of chemicals, with the exception of n-hexane and ethylbenzene. These compounds were selected based upon their chemical, physical, and biological properties/toxicity. The selection of chemical components representative of unleaded gasoline impacts upon implementation of the Safe Drinking Water Act and Amendments of 1986 as well as monitoring requirements. Considering that unleaded gasoline is a frequently increasing contaminant of groundwater and consists of over 100 chemicals, the speed and economy of monitoring and risk management would be enhanced by recognizing only those chemicals of toxicological concern that are likely to migrate to groundwater.

The solubilities of the components most likely to migrate to groundwater were evaluated in a study by the American Petroleum Institute.[7] Components in the aqueous solution resulting from gasoline-water equilibrium were compared to the solubilities of the components in pure water. The results for components of toxicological concern are presented in Table 2. The API concluded that neither the solubilities of the components in pure water nor the gasoline composition solely influence the composition of the aqueous solution. Mihelcick[8] determined through modeling that additives such as MTBE and methanol had a significant effect on enhancing hydrocarbon solubility at levels of 5% or greater by volume cosolvent. After a review of environmental fate studies, Page[6] proposed that the properties of gasoline components that increase the likelihood of reaching groundwater are high water solubility, low volatility, low soil sorption, low biodegradability, stability in water, and specific gravity more than 1. Based upon high solubility and low biodegradation, Page[6] concluded the gasoline components will concentrate in groundwater in the order: MTBE, benzene, toluene, and xylenes. Methanol is no longer used as an additive for unleaded gasoline and is unlikely to be used again in the future.[9]

Using eight components (benzene, toluene, ethylbenzene, [o-, m-, p-] xylenes, n-hexane, and MTBE) of a 100 or greater component mixture as representative of toxic potential in drinking water is supported by monitoring data, migration models, and the toxicity of other gasoline components. The aliphatic hydrocarbons[1] are insoluble to slightly soluble as a chemical group and are unlikely to migrate

Table 1. Approximate Percent Volume Composition of the Main Constituents of Unleaded Gasoline

Hydrocarbon Class	Typical Data a	b	c	Average
Aliphatic hydrocarbons				
n-Paraffins (C_3–C_{10})	11.40			
Isoparaffins (C_4–C_{13})	46.55			
Cycloalkanes				
Cycloparaffins (C_5–C_{13}) (Naphthenes)	4.68			
Total Paraffins	62.63	58.23	46.47	55.78
Alkenes				
mono-Olefins (C_2–C_{12})	8.89	10.54	9.33	9.58
Aromatic hydrocarbons (C_6–C_{12})	26.08	31.23	44.20	33.84
Polynuclear aromatics (Naphthalenes)	<1			<1
Totals	99	100	100	

[a]Domask et al.[3]
[b]Amoco 76 no-lead regular.[4]
[c]Amoco premium no-lead.[4]

Table 2. Solubilities of Selected Hydrocarbons[a]

Compound	Aqueous Solution mg/L	Pure Compound in Water mg/L	Percent Gasoline by Weight
Benzene	58.7	1740–1860	1.94
Toluene	33.4	500–627	4.73
Ethylbenzene	4.3	131–208	2.00
o-Xylene	6.9	167–213	2.27
m-Xylene	11.1	134–196	5.66
n-Hexane[b]		9	0.83–1.32

[a]Adopted from Reference #7, except where noted.
[b]Reference #4.

to or solubilize in groundwater to any significant extent. MacFarland,[10] in reviewing the toxicity of this chemical group, noted that n-hexane is metabolized by oxidation at the 2 and 5 carbon atoms to the hydroxyl group and then to the keto group, resulting in the intermediate metabolite methyl n-butyl ketone and the active neurotoxin 2,5-hexanedione. Other aliphatic hydrocarbons in gasoline are extremely volatile (methane to butane) or remain liquids at room temperature (pentane to hexadecane), but represent low percent composition of unleaded gasoline and are relatively insoluble. The remaining alkane or paraffin fraction, containing primarily aliphatic carbons from C_3 to C_8, has relatively low toxicity with narcotic properties following inhalation.[11] Although n-hexane is only approximately 1% by weight of total unleaded gasoline and of low solubility (approximately 9 mg/L), it should be included in the mixtures model for unleaded gasoline contamination of groundwater due to toxicity.

The cycloalkanes (cycloparaffins) are similar to open-chain aliphatic hydrocarbons, are central nervous system (CNS) depressants and general anesthetics upon inhalation, and are skin irritants. The alkenes (mono-olefins) are CNS depressants with weak anesthetic properties by inhalation, and produce skin and mucous membrane irritation.[10,11] The alkenes are a small fraction of unleaded gasoline and are generally insoluble with the exception of propylene, which is less than 1% by volume of gasoline. The smallest fraction of unleaded gasoline, the PAHs, are insoluble and associated with high boiling fractions such as heating oils.[10]

By far the most water soluble, and second largest fraction of unleaded gasoline, are the aromatic hydrocarbons, including benzene and alkylbenzenes. These chemicals, including benzene, ethylbenzene, m-,o-,p-xylenes, and toluene have been previously evaluated as drinking water contaminants. MTBE is currently under evaluation by the USEPA.

TOXICITY OF WHOLE UNLEADED GASOLINE AND SELECTED COMPONENTS

Unleaded Gasoline

The most direct way to evaluate the health effects of drinking water contaminated with unleaded gasoline would be to conduct toxicity tests on drinking water spiked with whole unleaded gasoline. This type of data is not available and would be of questionable use for risk assessment, considering that leaking underground storage tanks are the main source of unleaded gasoline contamination. Most of the components in a spiked sample would not be expected to migrate through soil. In a fatal case of gasoline ingestion,[12] there was general systemic distribution of gasoline with the major components concentrated in the liver, gastric wall, and lungs. The estimated lethal oral dose of gasoline is approximately 7.5 g/kg for humans[13] and produced initial symptoms of central nervous system depression of increasing severity. There are no oral subchronic or chronic toxicity tests with whole unleaded gasoline. MacFarland et al.[14] exposed mice and rats (both sexes) to the vapor of unleaded gasoline in a two-year chronic toxicity study at concentrations of 57, 292, and 2,056 ppm. There was reduction of body weight at the 2,056 ppm level. There was also an increased incidence of hepatocellular tumors in the high-dose female mice. Renal tumors occurred in the male rats at all doses. Using the USEPA[15] guidelines for carcinogen risk assessment, whole unleaded gasoline by the inhalation route of exposure meets the criteria to be classified in Group B, probable human carcinogen, or Group C, possible human carcinogen, depending on the significance ascribed to male rat kidney tumors. The relevance of male rat kidney tumors to human cancer evaluation is currently under review by the USEPA. Halder et al.[16] reported that male rats, upon inhalation of full-range alkylate naphtha (98% normal and branched alkanes), exhibited severe nephrotoxicity. Heavy catalytic-reformed naphtha (93% aromatics and 7% alkanes) produced no nephrotoxicity in inhalation studies with male rats. Both these test substances had minimal amounts of alkenes. The findings suggested that exposure to alkanes is the source of nephrotoxicity and tumors in the male rat kidney rather than the aromatic fraction.

We believe that inhalation toxicological studies have little relevance to assessment of the toxicity of unleaded gasoline-contaminated drinking water. The prime water soluble component likely to migrate to groundwater is benzene, which is classified as USEPA Group A, human carcinogen. Hence, this chapter will not attempt to review the extensive epidemiological and toxicological literature on the inhalation of gasoline, but concentrate on the assessment of the oral toxicity of components likely to migrate to drinking water sources.

Benzene

The USEPA[17] classified benzene as a Group A, human carcinogen, based on several studies of increased incidence of nonlymphocytic leukemia from occupational exposure, increased incidence of neoplasia in rats and mice exposed by inhalation and gavage, and some additional supporting data. The International Agency for Research on Cancer[18] placed benzene in Group 1 (carcinogenic to humans). The USEPA developed a quantitative cancer risk assessment for pooled human inhalation exposure and leukemia incidence data from Rinsky et al.,[19] Ott et al.,[20] and Wong et al.[21] Using a one hit or modified linearized multistage model, the USEPA[17] calculated an oral slope factor (also termed potency factor or q_1^*) of 2.9 E-2 per mg/kg bw/day. This slope factor was based on a human respiratory rate of 20 m^3/day, 100% absorption, and an air concentration of benzene such that 1 ppm equals 3.25 mg/m^3. This slope factor can be used in the following generalized equations to determine the cancer risks from consuming drinking water contaminated with benzene.

$$C=(35,000/q_1^*)(R) \text{ or } UR=q_1^*/35,000$$

where:

q_1^* = (mg/kg bw/day)$^{-1}$
R = risk (e.g., 0.0001 or 1 excess tumor per 10,000 population)
C = concentration of the chemical in drinking water (μg/L)
35,000 = conversion factor for mg to μg and exposure assumptions of a 70-kg body weight adult drinking 2 liters of water per day
UR = unit risk for exposure to 1 μg/L of the chemical for a 70-year lifetime

Using the above equations in the case of benzene results in a unit risk of 8.3E-7/μg benzene/L (i.e., 8.3 excess tumors per 10 million population drinking 2 L/day of water containing 1 μg/L of benzene for a lifetime of 70 years) or the risk range shown in Table 3 may be calculated.

Table 3. Benzene Concentration in Water in μg/L

Cancer Risk	Actual	Rounded
10^{-4}	120	100
10^{-5}	12	10
10^{-6}	1.2	1

The USEPA has established the Maximum Contaminant Level Goal or MCLG (nonenforceable health goal) of zero since benzene is a USEPA Group A carcinogen. However, the enforceable standard for benzene in drinking water as a single contaminant is the Maximum Contaminant Level (MCL) of 5 μg/L, which falls in the 10^{-5} to 10^{-6} cancer risk range for lifetime exposure to this chemical.

Toluene

The USEPA[22] proposed both an MCLG and MCL for toluene of 2.0 mg/L based upon a study by CIIT[23] in which Fischer 344 rats were exposed to toluene via inhalation at 0, 113, 337, or 1,130 mg/m^3 for 6 hours/day, 5 days/week for two years. There were no significant treatment-related effects in this study.[24] Therefore, using the standard inhalation physiological parameters and human exposure assumptions, a Lifetime Health Advisory for toluene of 2.42 mg/L was determined. The MCLG and MCL are the same as the Lifetime Health Advisory rounded to one significant figure. Shorter-term exposure of humans to toluene vapors usually results in central nervous system related effects including fatigue, headache, nausea, muscular weakness, confusion, and incoordination. Toluene is only moderately toxic upon ingestion, with no appropriate oral chronic toxicity studies. Since, following oral or inhalation exposure in both animals and humans, toluene is excreted rapidly as the unchanged compound in expired air, and as hippuric acid in urine, it is reasonable to assume that the toxicity endpoint of concern via the oral route is CNS effects similar to the inhalation route. Toluene may be classified in Group D, not classifiable as to human carcinogenicity.[24]

Xylenes (m-, o-, and p-)

The USEPA proposed an MCLG of 0.4 mg/L for total xylenes in November, 1985, based upon the study by Jenkins et al.[25] in which o-xylene was administered by inhalation to rats, guinea pigs, monkeys, and dogs for 30 repeated exposures at 3,358 mg/m^3, 8 hr/day and 5 day/wk or 90 days continuous exposure at 337 mg/m^3. The high concentration resulted in mortalities and CNS effects including tremors. Thus, the low concentration was considered a no-observed-adverse-effect level and, using standard inhalation parameters and exposure assumptions, the resultant Lifetime Health Advisory of 0.4 mg/L was determined.[26] The USEPA[22] proposed a new MCLG and MCL for total xylenes of 10 mg/L based on new data. The new proposal is based upon a National Toxicology Program[27] study in which a test substance consisting of m-xylene (60.2%), p-xylene (13.6%), o-xylene (9.1%), and ethylbenzene (17%) was administered by gavage to rats and mice for two years. There were no increases in the incidence of neoplasia related to xylenes exposure. However, there were significant decreases in the body weights of the high-dose (500 mg/kg) group. Using the 250 mg/kg dose as the no-observed-adverse-effect level, the new MCLG of 12 mg/L (rounded to 10 mg/L) was determined, using standard dose conversion and exposure assumptions. However, inhalation studies at high concentrations indicate that both humans and animals exhibit CNS disturbances. Xylenes may be classified in Group D, not classifiable as to human carcinogenicity.[26]

Ethylbenzene

The USEPA[22] proposed both an MCLG and MCL of 0.7 mg/L for ethylbenzene based upon a study by Wolf et al.[28] in which rats were administered, by gavage, doses of 13.6, 136, 408, or 680 mg/kg/day ethylbenzene in olive oil for 130 days of the 182-day test period. The two highest doses resulted in kidney and liver weight increases as well as minor swelling of hepatocytes and renal tubular epithelium. Using the 136 mg/kg/day as the no-observed- adverse-effect level, the Lifetime Health Advisory[29] of 0.68 mg/L was calculated using standard dose conversion and exposure assumptions. The reproposed MCLG and MCL values are derived from the Lifetime Health Advisory rounded to one significant figure. Humans exposed to high levels of ethylbenzene by inhalation result in the CNS effects typical of other aromatic hydrocarbons. Ethylbenzene may be classified in Group D, not classifiable as to human carcinogenicity.[29]

n-Hexane

The USEPA[30] developed a Longer-term (up to seven years exposure) Health Advisory of 14.3 mg/L for n-hexane. This health advisory value was based upon the study by Krasavage et al.[31] in which the neurotoxicity of n-hexane was evaluated in Charles River male rats. The rats were given oral doses of 570 mg/kg for 5 days/week for a total of 90 days or 1,140 and 4,000 mg/kg for 120 days. No clinical or histopathological signs of neuropathy were observed at the two lowest dose levels, but body weights were reduced at all three dose levels. At the 4,000 mg/kg dose level there were clinical and histopathological signs of neuropathy (hind-limb weakness and axonal lesions) at 101 days. The Longer-term Health Advisory was determined using the 570 mg/kg/day dose as a lowest-observed-adverse-effect level, along with standard dose conversion factors and exposure assumptions resulting in the Longer-term Health Advisory value. As an interim value, the Longer-term Health Advisory could be modified to estimate a provisional lifetime advisory by using an additional uncertainty factor to account for the less than lifetime duration of the study and a 20% relative source contribution (RSC) factor in the absence of chemical-specific data. The resultant value would be as follows:

Lifetime Health Advisory = [(Longer-term HA)/UF][RSC]
(provisional)
= [(14.3)/10][0.2]
= 0.29 (rounded to 0.3 mg/L)

where:

Longer-term HA = 14.3 mg/L
UF = uncertainty factor of 10 for extrapolation to lifetime exposure
RSC = relative source contribution of 20% (USEPA Office of Drinking Water policy)

This provisional Lifetime Health Advisory should be protective of known adverse effects due to the metabolism of the chemical to active neurotoxins such as 2,5-hexanedione. The n-hexane may be classified in Group D, not classifiable as to human carcinogenicity.[30]

Methyl-tertiary-butyl-ether (MTBE)

The USEPA currently does not have a Lifetime Health Advisory for MTBE but is in the process of considering the data. There are no adequate chronic studies of MTBE upon which to base a Lifetime Health Advisory value. The only studies concerning the possible health effects in humans are associated with the use of MTBE as a cholelitholytic agent.[32,33] In this procedure, MTBE is perfused through a catheter introduced into the gallbladder to dissolve gallstones. Adverse health effects were not reported. The oral LD_{50} of MTBE in rats (strain not reported) is 3.9 g/kg, and toxic effects included CNS depression and respiratory problems.[34] In the absence of subchronic and chronic oral toxicity studies, the Greenough et al.[35] study was used to develop a provisional Lifetime Health Advisory value. In this study Sprague-Dawley rats of both sexes were exposed by inhalation to MTBE vapor for 6 hrs/day and 5 days/wk for 13 weeks. The concentrations of MTBE were 250, 500, or 1,000 ppm. There were no exposure-related effects reported except a transient increase in hemoglobin in the high-dose male rats. However, increasing depth of anesthesia was observed with increasing vapor concentration. There was no further explanation of this observation and it was considered to be an unwanted/adverse effect. Taking the low dose of 250 ppm MTBE as a lowest-observed-adverse-effect level for anesthesia, a provisional Lifetime Health Advisory is proposed as follows:

TAD = [(MTBE conc)(exposure duration)(rat tidal vol)(absorption)]/rat bw

TAD = [(901 mg/m^3)(6 hr/24 hr)(5 day/7 day)(0.237 m^3/day)(0.5)]/.336 kg

TAD = 56.7 mg/kg/day

where:

TAD = Total Absorbed Dose
901 mg/m^3 = low dose of 250 ppm (3.6 mg/m^3 per 1 ppm MTBE)
6 hr/24 hr = factor for exposure duration
5 day/7 day = factor for exposure duration
0.237 m^3/day = rat tidal volume calculated using the equation proposed by Guyton (1947): Respiratory Volume (mL/min) = 2.10 (body weight in gm)$^{0.75}$
0.5 = 50% absorption
0.336 kg = mean body weight for male and female rats at 250 ppm

RfD = TAD/UF

RfD = (56.7 mg/kg/day)/10,000

RfD = 0.00567 mg/kg/day

where:

RfD = Reference Dose with an adequate margin of safety to protect sensitive populations
UF = Uncertainty Factor (10× for intraindividual variability, 10× for interspecies extrapolation, 10× for use of a less-than-lifetime study and 10× for use of a study with a lowest-observed-adverse-effect level and lack of oral toxicity data

DWEL = [RfD (human body weight)]/water consumption

DWEL = [(0.00567 mg/kg/day)(70 kg)]/(2 L/day)

DWEL = 0.198

where:

70 kg = assumed body weight of adult
2 L/day = assumed water consumption for an adult
DWEL = Drinking Water Equivalent Level assuming 100% drinking water contribution

Lifetime HA = DWEL (RSC)
(provisional)
= 0.198 (0.2)
= 0.0396 (rounded to 0.04 mg/L or 40 μg/L)

where:

HA = Health Advisory
RSC = Relative Source Contribution, assumed to be 20%.

This provisional Lifetime Health Advisory for MTBE of 40 μg/L seems prudent considering that MTBE has very limited subchronic, reproductive, and genotoxicity studies and almost no information on long- or short-term effects from oral exposure.

A summary of the USEPA MCLs, proposed MCLs, and provisional Lifetime Health Advisories developed in this chapter, along with the general health effects, is presented in Table 4.

Table 4. Major Soluble Components for Consideration in a Mixtures Model

Component	Critical Effect	MCL	Proposed MCL	Provisional HA
Benzene	CNS, cancer	5 μg/L		
Toluene	CNS, blood		2 mg/L	
Xylenes	CNS, liver, body weight		10 mg/L	
Ethylbenzene	CNS, liver, kidney		0.7 mg/L	
n-Hexane	CNS, body weight			0.3 mg/L
MTBE	CNS			40 μg/L

A MIXTURES APPROACH

Most of the existing studies on unleaded gasoline were conducted via the inhalation route. The actual exposure of humans to components of unleaded gasoline from drinking water or groundwater is probably vastly different, both quantitatively and qualitatively, from the inhalation route. Gasoline vapors consist mainly of alkanes, particularly short chained n-paraffins and isoparaffins.[16,37,38] Aromatic compounds will predominate in groundwater/drinking water.[5-7] The hydrocarbon components with high water solubility are most likely to reach drinking water sources. The major source of contamination is leaking underground storage tanks, where the soluble components migrate quickly to groundwater. In the unusual occurrence where unleaded gasoline is spilled directly into drinking water, the largely insoluble portion would float on the top and render the water unpalatable. Therefore, considering the overview of the toxicology of the components of concern, the following approach for managing the risks from unleaded gasoline contamination of groundwater or drinking water is proposed.

The USEPA has provided guidelines for the risk assessment of chemical mixtures.[39] Using these guidelines, the approach would include conducting a complete analysis of the suspect contaminated water sample. The analysis should include, at a minimum, the water soluble components of unleaded gasoline and the additives detailed in this chapter. For systemic (i.e., noncarcinogenic) components, the Hazard Index (H.I.) should be employed for components that have the same general toxicological effects. In the case of unleaded gasoline, these components include (exclusive of additives) the CNS depressants (toluene, xylenes, ethylbenzene, and n-hexane). Benzene is not included since the critical toxicological endpoint is cancer. Assuming an additive toxicity model, in the absence of evidence of synergism, the following generalized H.I. is applicable:

$$H.I. = E_1/AL_1 + E_2/AL_2 + \ldots\ldots + E_i/AL_i$$

where:

E_i = exposure concentration of the i-th toxicant in drinking water
AL_i = maximum contaminant level or maximum acceptable level for the i-th toxicant in drinking water

therefore:

$$H.I. = [\text{toluene}]/2\text{ mg/L} + [\text{xylenes}]/10\text{ mg/L} + [\text{ethylbenzene}]/0.7\text{ mg/L} + [\text{n-hexane}]/0.3\text{ mg/L}$$

where:

[ ] = concentration of the component in mg/L
AL values = proposed MCLs or provisional HA values from Table 4.

If the H.I. is less than or equal to 1.0, there is not a significant risk of CNS effects.

An argument might be made for including benzene and MTBE in the H.I. for systemic toxicants, since they are also CNS depressants. We recommend excluding benzene, since the guidelines[39] do not recommend mixing different toxicological endpoints in an H.I. Also, significant cancer risk from benzene occurs from exposure to relatively low lifetime doses. Following the mixture guidelines, chemicals with carcinogenic potential are usually included in a separate H.I. as follows:

$$H.I. = E_1/DR_1 + E_2/DR_2 + \ldots\ldots + E_n/DR_n$$

where:

E_n = exposure level or drinking water concentration of the n-th component
DR_n = concentration of the n-th component at a selected cancer risk level

Since benzene is the only component in unleaded gasoline-contaminated drinking water of known carcinogenic potential, the MCL of 5.0 μg/L should not be exceeded. If other carcinogens are identified in future research, then the H.I. detailed above for carcinogens should be employed.

In the case of additives such as MTBE, we do not recommend including them in the systemic toxicity H.I., since they are only occasionally present in unleaded gasoline. We recommend that additives be evaluated separately. In the case of MTBE, the provisional Lifetime Health Advisory of 40 μg/L should not be exceeded.

CONCLUSIONS

- A large portion of the total health risks from exposure to unleaded gasoline in drinking water is due to cancer risk from the benzene component.
- Not exceeding the 5 μg/L MCL for benzene will probably be adequate to protect against toxicities due to other water soluble components, since the MCL for benzene is orders of magnitude lower than the proposed MCLs and provisional HAs of the other components.
- The health effects of exposure to ethylbenzene, toluene, xylenes, and n-hexane should be evaluated using an H.I. approach.
- The health effects of exposure to occasional additives should be evaluated individually.
- A continued research effort is necessary to ensure that all toxic components of unleaded gasoline with potential to migrate to drinking water are identified. New components may be incorporated into the mixtures model proposed in this chapter.

ACKNOWLEDGMENTS

The views expressed are those of the authors and do not necessarily represent those of the United States Environmental Protection Agency. The authors are indebted to the staff of the Health Effects Branch, EPA, Office of Drinking Water, for many useful discussions and review of this document.

REFERENCES

1. "Summary of State Reports on Releases from Underground Storage Tanks," U.S. Environmental Protection Agency, Office of Underground Storage Tanks, Washington, DC. EPA/600/M-86/020, 1986a.
2. *Federal Register.* "Underground Storage Tanks; Technical Requirements," U.S. Environmental Protection Agency, Proposed Rule, 74 FR, April 17, 1987, pp. 12662–12663.
3. Domask, W. G. "Introduction to Petroleum Hydrocarbons, Chemistry and Composition in Relation to Petroleum-Derived Fuels and Solvents," in *Advances in Modern Environmental Toxicology,* Vol. VII, Renal Effects of Petroleum Hydrocarbons,

M. A. Mehlman, G. P. Hemstreet, J. J. Thorpe and N. K. Weaver, Eds. (Princeton, NJ: Princeton Scientific Publishers, Inc., 1984), pp. 1–24.

4. Sigsby, J. E., Jr., S. Tejada, and W. Ray. "Volatile Organic Compound Emissions from 46 In-Use Passenger Cars," *Environ. Sci. Technol.* 24(5): 466–475 (1987)
5. Dietrich, A. M. "An Evaluation of the Composition and Toxicity of Unleaded Gasoline in Drinking Water," in *Reports,* 1988 *Environmental Science and Engineering Fellows Program,* American Association for the Advancement of Science, Washington, DC, 1988, pp. 2–5.
6. Page, N. P. "Gasoline Leaking from Underground Storage Tanks: Impact on Drinking Water Quality," in *Trace Substances in Environmental Health* XXII, D. D. Hemphill, Ed. (Columbia, MO: University of Missouri Press, 1988), pp. 233–245.
7. "Laboratory Study on Solubilities of Petroleum Hydrocarbons in Groundwater," American Petroleum Institute Publication No. 4395, August, 1985.
8. Mihelcick, J. R. "Corrective Action Considerations Associated with Leaking Underground Storage Tanks: Additives and Weathered Spills," in *Reports,* 1988 *Environmental Science and Engineering Fellows Program,* American Association for the Advancement of Science, Washington, DC, 1988, pp. 19–23.
9. Personal Communication. U.S. Environmental Protection Agency, Office of Mobile Sources, Mr. Lester Wyborny, Ann Arbor, MI., September 5, 1989.
10. MacFarland, H. N. "Toxicology of Petroleum Hydrocarbons," *Occupational Medicine: State of the Art Reviews,* 3 (3): pp. 445–454 (1988).
11. "Toxicity of Inorganic Contaminants in Drinking Water," in *Drinking Water and Health,* 4:251–264 (1982), National Research Council.
12. Carnevale, A., M. Chianotti, and N. DeGiovanni. "Accidental Death by Gasoline Ingestion," *Am. J. Forensic Med. Pathol.* 4(2):153–157 (1983).
13. Machle, W. "Gas Intoxication," *J. Am. Med. Assoc.* 117:1965–1971 (1941).
14. MacFarland, H. M., C. E. Ulrich, C. E. Holdsworth, D. N. Kitchen, W. H. Halliwell, and S. C. Blum. "A Chronic Inhalation Study with Unleaded Gasoline Vapor," *J. Am. Coll. Toxicol.* 3:231–248 (1984).
15. "Guidelines for Carcinogen Risk Assessment," U.S. Environmental Protection Agency, 51 *Federal Register* 33992–34003, 1986b.
16. Halder, C. A., T. M. Warne, and N. S. Hatoun. "Renal Toxicity of Gasoline and Related Petroleum Naphthas in Male Rats," in *Advances in Modern Environmental Toxicology,* Vol. VII. Renal Effects of Petroleum Hydrocarbons. M. A. Mehlman, Ed. (Princeton, NJ: Princeton Scientific Publishers, 1984), pp. 73–88.
17. "Integrated Risk Information System (IRIS). Carcinogenicity Assessment for Lifetime Exposure to Benzene," U.S. Environmental Protection Agency, 1987a. Revised: Verification date 10/09/87.
18. "IARC Monographs on the Evaluation of the Carcinogenic Risk of Chemicals," *Industrial Processes and Industries Associated with Cancer in Humans.* (Lyon, France: IARC, Supplement 4, 1982), p. 56.
19. Rinsky, R. A., R. J. Young, and A. B. Smith. "Leukemia in Benzene Workers," *Am. J. Ind. Med.* 2:217–245 (1981).
20. Ott, M. G., J. C. Townsend, W. A. Fishbeck, and R. A. Langnar. "Mortality Among Individuals Occupationally Exposed to Benzene," *Arch. Environ. Health* 33:3–10 (1978).

21. Wong, O., R. W. Morgan, and M. D. Whorton. "Comments on the NIOSH Study of Leukemia in Benzene Workers," *Environmental Health Associates report for Gulf Canada Ltd.* August 31, 1983.
22. National Primary and Secondary Drinking Water Regulations; Proposed Rule, U.S. Environmental Protection Agency, *Federal Register,* Vol. 54, No. 97, Monday, May 22, 1989, pp. 22062–22160.
23. "A Twenty-Four Month Inhalation Toxicology Study in Fischer 344 Rats Exposed to Atmospheric Toluene," Chemical Industry Institute of Toxicology. Executive Summary and Data Tables. October 15, 1980.
24. "Health Advisory for Toluene," U.S. Environmental Protection Agency, Office of Drinking Water, Washington, DC, 1987b, 17 pp.
25. Jenkins, L. J., R. A. Jones, and J. Siegel. "Long-Term Inhalation Screening Studies of Benzene, Toluene, o-Xylene and Cumene on Experimental Animals," *Toxicol. Appl. Pharmacol.* 16:818–823 (1970).
26. "Health Advisory for Xylenes," U.S. Environmental Protection Agency, Office of Drinking Water, Washington, DC, 1987c, 15 pp.
27. "Toxicology and Carcinogenesis Studies of Xylenes," NTP Technical Report No. 327, 1986.
28. Wolf, M. A., V. K. Rowe, D. D. McCollister, R. L. Hollingsworth, and F. Oyen. "Toxicological Studies of Certain Alkylated Benzenes and Benzene," *Arch. Ind. Health.* 14:387–398 (1956).
29. "Health Advisory for Ethylbenzene," U.S. Environmental Protection Agency, Office of Drinking Water, Washington, DC, 1987d, 13 pp.
30. "Health Advisory for n-Hexane," U.S. Environmental Protection Agency, Office of Drinking Water, Washington, DC, 1987e, 14 pp.
31. Krasavage, W. J., J. L. O'Donoghue, G. D. DiVincenzo, and C. J. Terhaar. "The Relative Neurotoxicity of Methyl-n-butyl-ketone, n-Hexane and Their Metabolites," *Toxicol. Appl. Pharmacol.* 52:433–441 (1980).
32. Allen, M. J., T. F. Borody, G. R. Bugliosi, G. R. May, N. F. LaRusso, and J. L. Thistle. "Cholelitholysis Using Methyl Tertiary Butyl Ether," *Gastroenterol.* 88:122–125 (1985a).
33. Allen, M. J., T. F. Borody, G. R. Bugliosi, G. R. May, N. F. LaRusso, and J. L. Thistle. "Rapid Dissolution of Gallstones by Methyl Tertiary Butyl Ether," *N. Eng. J. Med.* 312:217–220 (1985b).
34. "Methyl Tertiary Butyl Ether: Acute Toxicological Studies," Arco Chemical Company. Unpublished study for Arco Research and Development, Glenolden, PA, 1980.
35. Greenough, R. J., P. McDonald, P. Robinson, J. R. Cowie, W. Maule, F. Macnaughtan, and A. Rushton. "Methyl Tertiary Butyl Ether (DRIVERON) Three Month Inhalation Toxicity in Rats," Unpublished report by Inveresk Research International for Chemische Werke Hols AG., West Germany, 1980, 227 pp.
36. Guyton, A. C. "Measurement of the Respiratory Volumes of Laboratory Animals," *Am. J. Physiol.* 150:70–77 (1947).
37. McDermott, H., and S. E. Killiany. "Quest for Gasoline TLV," *Am. Ind. Hyg. Assoc. J.* 39:110–117 (1978).
38. Runion, H. E. "Benzene in Gasoline," *Am. Ind. Hyg. Assoc. J.* 36:338–350 (1975).
39. "Guidelines for the Health Risk Assessment of Chemical Mixtures," 51 *Federal Register* 34014–34025 (1986c).

CHAPTER 24

Choosing a Best Estimate of Children's Daily Soil Ingestion

Edward J. Stanek III, Edward J. Calabrese, and **Charles E. Gilbert,** School of Public Health, University of Massachusetts, Amherst

Soil ingestion estimates were made by Calabrese et al.[1] based on a mass balance approach for 64 preschool age children in Western Massachusetts. The estimates were based on a study design that collected duplicate food samples 3 days a week for 2 weeks, fecal and urine for 4 days a week for 2 weeks, and element concentrations in soil and house dust. Estimates of soil ingestion for each subject were made by subtracting the daily amount of an element ingested from the average daily amount of an element excreted, and dividing by the concentration of the element in soil and/or house dust. Eight elements (Al, Ba, Mn, Si, Ti, V, Y, and Zr) were used in the study, but two elements (Ba and Mn) were judged unsatisfactory for soil ingestion estimation.

Nine estimates of soil ingestion were presented by Calabrese et al.[1] The estimates differed in how the amount excreted was calculated, whether or not food ingestion was accounted for, and which estimate—soil, house dust, or a combination of the two—was used to represent the element concentration in soil. The "best" estimate was judged to be based on calculating fecal and urine output by dividing the total output by 8 days, subtracting food ingestion, and using a weighted average (using time indoors as a weighting factor) of soil and dust exposure. The estimates were made for 64 subjects, including one subject with pica.

Since output and food intake vary from day to day, and there is not a simple 1 day transit time for food, estimates of the difference between output and food reflected differences in transit time as well as soil ingestion. These differences in transit time lead to some overestimates, and some underestimates of soil ingestion. On average, the overestimates and underestimates should balance out, so that the resulting estimator of soil ingestion is unbiased. However, for individual subjects, some estimates of soil ingestion may actually be negative, due to transit time variability. These negative estimates are certainly due to transit time variability, and a better estimate for these subjects is zero soil ingestion. However, the similar overestimates in soil ingestion cannot be identified. As a result, estimates of soil ingestion based on setting subjects negative estimates to ''zero'' will overestimate actual soil ingestion. It is for this reason that negative estimates were not converted to ''zero'' in Calabrese et al.[1] Nevertheless, converting negative estimates to ''zero'' will produce a conservative (high) estimate of soil ingestion that may be of interest for some purposes.

The distribution of soil ingestion is not normally distributed among subjects, but skewed to the right. In forming confidence intervals for soil ingestion, better coverage can be obtained by forming confidence intervals based on the geometric mean (using log base e), rather than the simple arithmetic mean. To form such estimates, we take the log (base e) of soil ingestion estimates for each subject, and then calculate the mean and standard deviation (SD) of these log (e) estimates. A confidence interval can be formed by adding and subtracting a multiple of the SD to the log(e) mean. For example, for a 95% confidence interval, we would use:

$$U = \text{mean (log soil ingestion)} + 1.96\ [\text{SD(log soil ingestion)}]$$
$$L = \text{mean (log soil ingestion)} - 1.96\ [\text{SD(log soil ingestion)}]$$

The upper and lower limits of the confidence interval would be formed by exponentiating U and L.

Since the logarithm is a monotonic transformation, the resulting estimated distribution of soil ingestion based on this transformation will not differ from the distribution of untransformed data (Table 1).

CURRENT RESULTS

The current results augment the results presented by Calabrese et al.[1] by providing estimates of soil ingestion under different assumptions. Estimates are provided for six elements (Tables 2, 3, and 4). All estimates are based on an 8 day fecal output average, subtraction of food intake, and using soil concentrations. Three sets of estimates are provided, with the estimates differing in terms of the number of subjects included and criteria for exclusion. The first set of estimates uses all 64 subjects (Table 2); the second set excludes one subject (ID=851) with suspected pica (Table 3), and the third set excludes ID=851 and 4 subjects in which one or more laboratory data value for a particular day are unusually variable and inconsistent with other data (847, 841, 854, 856) (Table 4). For these

Table 1. Median and Geometric Mean for the 3 Elements Considered to Most Accurately Reflect Soil Ingestion in the UMass Study under Different Assumptions

Estimates in mg/day	All Subjects			Excluding 1 Pica Subj.			Excluding 5 Subj.		
	Element			Element			Element		
Soil Ingestion Criteria	Al	Si	Y	Al	Si	Y	Al	Si	Y
Based on MEDIAN									
Soil and dust									
Including Negatives	30	49	11	29	48	8	27	48	15
Negatives set to 1	30	49	11	29	48	8	27	48	15
Soil only									
Including Negatives	29	40	9	28	39	8	24	39	10
Negatives set to 1	29	40	9	28	39	8	24	39	10
Based on GEOMETRIC MEAN									
Soil and dust									
Negatives set to 1	26	29	9	24	26	8	22	25	9
Soil only									
Negatives set to 1	22	23	7	20	21	7	18	20	7

Table 2. Soil Estimates (mg/day)

Name	N	Mean	Median	Std	Min	P5th	P10th	P90th	P95th	Max
Based on Simple Averages (8 Day Fecal-Food)/(Soil) (All Subjects)										
Al	64	153	29	852	−75	−23	−3	138	223	6837
Si	64	154	40	693	−53	−29	−17	219	276	5549
Ti	64	218	55	1150	−3069	−601	−381	702	1432	6707
V	62	459	96	1037	−650	−148	1	1366	1903	5676
Y	62	85	9	890	−1733	−84	−56	91	106	6736
Zr	62	21	16	209	−597	−70	−35	67	110	1391
Based on Simple Averages (8 Day Fecal-Food)/(Soil) with All Negative Intakes Set to Zero Intake (All Subjects)										
Al	64	156	29	851	0	0	0	138	223	6837
Si	64	158	40	692	0	0	0	219	276	5549
Ti	64	343	55	1026	0	0	0	702	1432	6707
V	62	482	96	1021	0	0	1	1366	1903	5676
Y	62	133	9	853	0	0	0	91	106	6736
Zr	62	47	16	177	0	0	0	67	110	1391
Based on Simple Averages (8 Day Fecal-Food)/(Soil) with All Negative Intakes Set to 1 mg/day Intake (All Subjects)										
Al	64	156	29	851	1	1	1	138	223	6837
Si	64	159	40	692	1	1	1	219	276	5549
Ti	64	344	55	1026	1	1	1	702	1432	6707
V	62	482	96	1021	1	1	1	1366	1903	5676
Y	62	133	9	853	1	1	1	91	106	6736
Zr	62	47	16	177	1	1	1	67	110	1391

continued

Table 2. Continued

Name	N	Mean	Median	Std	Min	P5th	P10th	P90th	P95th	Max
Based on Geometric (log e) Averages (8 Day Fecal-Food)/(Soil) with All Negative or Zero Intakes Set to 1 mg/day										
Al	64	22	29	5900	1	1	1	138	223	6837
Si	64	23	40	8272	1	1	1	219	276	5549
Ti	64	29	55	13532	1	1	1	702	1432	6707
V	62	87	96	9118	1	1	1	1366	1903	5676
Y	62	7	9	7640	1	1	1	91	106	6736
Zr	62	9	15	6300	1	1	1	67	110	1391

Table 3. Soil Estimates (mg/day)

Name	N	Mean	Median	Std	Min	P5th	P10th	P90th	P95th	Max
Based on Simple Averages (8 Day Fecal-Food)/(Soil) (All Subjects Except 1 Pica Subject)										
Al	63	47	28	70	−75	−23	−3	122	178	359
Si	63	68	39	104	−53	−29	−17	193	234	572
Ti	63	115	52	809	−3069	−601	−381	419	791	4469
V	61	389	89	888	−650	−148	1	1294	1831	5676
Y	61	−24	8	237	−1733	−84	−56	78	106	132
Zr	61	−2	14	113	−597	−70	−35	66	109	152
Based on Simple Averages (8 Day Fecal-Food)/(Soil) with All Negative Intakes Set to Zero Intake and Excluding 1 Pica Subject										
Al	63	50	28	67	0	0	0	122	178	359
Si	63	73	39	100	0	0	0	193	234	572
Ti	63	242	52	637	0	0	0	419	791	4469
V	61	413	89	871	0	0	1	1294	1831	5676
Y	61	25	8	35	0	0	0	78	106	132
Zr	61	25	14	34	0	0	0	66	109	152
Based on Simple Averages (8 Day Fecal-Food)/(Soil) with All Negative Intakes Set to 1 mg/day Intake and Excluding 1 Pica Subject										
Al	63	50	28	67	1	1	1	122	178	359
Si	63	73	39	100	1	1	1	193	234	572
Ti	63	243	52	637	1	1	1	419	791	4469
V	61	413	89	871	1	1	1	1294	1831	5676
Y	61	25	8	35	1	1	1	78	106	132
Zr	61	25	14	34	1	1	1	66	109	152
Based on Geometric (log e) Averages (8 Day Fecal-Food)/(Soil) with All Negative or Zero Intakes Set to 1 mg/day and Excluding 1 Pica Subject										
Al	63	20	28	5109	1	1	1	122	178	359
Si	63	21	39	7464	1	1	1	193	234	572
Ti	63	27	52	12582	1	1	1	419	791	4469
V	61	82	89	8735	1	1	1	1294	1831	5676
Y	61	7	8	6348	1	1	1	78	106	132
Zr	61	8	14	5669	1	1	1	66	109	152

Table 4. Soil Estimates (mg/day)

Name	N	Mean	Median	Std	Min	P5th	P10th	P90th	P95th	Max
Based on Simple Averages (8 Day Fecal-Food)/(Soil) (Excluding 5 Subjects with Some Questionable Data)										
Al	59	43	24	71	−75	−46	−7	122	223	359
Si	59	60	39	80	−49	−29	−17	181	234	287
Ti	59	68	52	370	−1212	−601	−381	402	714	1432
V	57	400	89	893	−203	−96	5	1294	1903	5676
Y	57	13	10	50	−93	−79	−55	91	106	132
Zr	57	17	17	44	−72	−66	−30	66	110	152
Based on Simple Averages (8 Day Fecal-Food)/(Soil) with All Negative Intakes Set to Zero Intake (Excluding 5 Subjects with Some Questionable Data)										
Al	59	47	24	67	0	0	0	122	223	359
Si	59	64	39	76	0	0	0	181	234	287
Ti	59	150	52	253	0	0	0	402	714	1432
V	57	408	89	889	0	0	5	1294	1903	5676
Y	57	26	10	36	0	0	0	91	106	132
Zr	57	25	17	34	0	0	0	66	110	152
Based on Simple Averages (8 Day Fecal-Food)/(Soil) with All Negative Intakes Set to 1 mg/day Intake (Excluding 5 Subjects with Some Questionable Data)										
Al	59	47	24	67	1	1	1	122	223	359
Si	59	64	39	76	1	1	1	181	234	287
Ti	59	150	52	253	1	1	1	402	714	1432
V	57	408	89	889	1	1	5	1294	1903	5676
Y	57	26	10	36	1	1	1	91	106	132
Zr	57	25	17	34	1	1	1	66	110	152
Based on Geometric (log e) Averages (8 Day Fecal-Food)/(Soil) with All Negative or Zero Intakes Set to 1 mg/day (Excluding 5 Subjects with Some Questionable Data)										
Al	59	18	24	5126	1	1	1	122	223	359
Si	59	20	39	7145	1	1	1	181	234	287
Ti	59	26	52	10911	1	1	1	402	714	1432
V	57	88	89	7625	1	1	5	1294	1903	5676
Y	57	7	10	6485	1	1	1	91	106	132
Zr	57	9	17	5498	1	1	1	66	110	152

subjects, one of the element results is unusually low or high and does not correspond to the estimated soil intake as indicated by other values for that child. Other element values for a set of estimates consists of four panels of estimates. The first panel contains estimates comparable to those presented in Calabrese et al.[1] These estimates agree with published estimates for the 64 subjects, and give comparable results for smaller subject groups.

The second panel of estimates is based on setting all negative soil ingestion estimates to zero. These estimates are once again based on simple arithmetic means and distributions of soil ingestion.

The third panel of estimates is based on setting all negative or zero soil ingestion estimates to .001 g/day. These estimates are useful in comparison with estimates based on log transformed data. Since the log of numbers less than or equal to zero is undefined, it is necessary to replace zero with a small positive value to calculate the log.

The final panel of estimates is based on the geometric mean, calculated by taking log(e) of individual soil ingestion estimates, estimating statistics, and then exponentiating the results. All negative and zero soil ingestion estimates are set to 0.001 g/day for these calculations. Tables 1–4 contain these soil ingestion estimates.

DISCUSSION

A basic conclusion of the tables is that the median soil ingestion estimate is insensitive to the various assumptions made in forming the different estimates. As such, the median is the most robust, and best, estimate of soil ingestion.

For regulatory purposes, some persons may recommend using a mean estimate rather than a median. The mean has the advantage of being simpler and more easily understood. We feel that this is a weak argument for using a mean, since the mean will be influenced strongly by extreme values. A percentile (such as the 75% percentile) would be a sounder measure if some estimate larger than the median were desired. If some estimate of the mean is insisted upon, a better estimate will be the geometric mean, since this estimate will account for the skewness in ingestion among children. However, the use of this statistic argues against simplicity. Presented in Table 1 is a summary of the Median and Geometric Mean for the 3 elements considered to most accurately reflect soil ingestion in the University of Massachusetts study under different assumptions.

If it were necessary to select a single estimate for soil ingestion, we would base our estimate on a combined soil and dust element concentration, and consider the best estimate to be based on the median. Among the three most valid tracers (Al, Si, Y) in the adult recovery study, yttrium would be eliminated because of the high number of negative values in the children's study. The present analysis would not be able to distinguish whether Al or Si would be the most preferred tracer. Since it is not possible to distinguish between these two tracers, we would select the Si value since it was the higher of the two, and therefore most conservative from a public health perspective. We would further round the value up from 49 mg to 50 mg for practical reasons. It is important to realize that the values obtained when excluding the one pica subject or excluding the five subjects with the questionable data would not affect the rounding up to 50 mg since both values were 48 mg.

During the development of a value for soil ingestion in children there has been some discussion concerning use of the arithmetic mean as a measure of central tendency and as the best estimate for exposure estimation purposes. We believe that the levels should have some meaning in the assessed population. With variables that have skewed distributions, the mean does not represent any particular benchmark, in terms of ranking a percentile, in the population distribution. Some of these concerns are illustrated below.

1. If a population has the following values, 0.1, 0.3, 0.7, 0.8, and 16.2, then 80% of the population falls below the mean.
2. If a population has the following values, 0.1, 14, 14.7, 15, and 17, then 80% of the population is above the mean.

In both of these examples the mean does not represent the central tendency of the population, whereas the median does reflect the central tendency of the population. It makes sense to pick a percentile of the variable's distribution (e.g., the 75th percentile) as an exposure, knowing that a certain percentage of the population is expected to exceed that standard.

If it were necessary to select a single "mean" estimate we would consider the best estimate to be derived from the geometric mean because it is also less sensitive to the various assumptions made in forming the various estimates. As in the case of the median, the value from Si would be used for the soil and dust from all subject groups. This value of 29 mg would be rounded up to 30 mg. The values obtained when excluding the one pica subject or excluding the subjects with the questionable data would not markedly affect the selected value since their respective values were 26 mg and 25 mg. Of course, these are estimates that are meant as best guesses of population ingestion. The estimates must certainly be qualified by limitations of the University of Massachusetts study, and the somewhat arbitrary choice between the different element estimators.

REFERENCES

1. Calabrese, E. J., R. Barnes, E. J. Stanek, H. Pastides, C. E. Gilbert, P. Veneman, and P. T. Kostecki. "How Much Soil Young Children Ingest: An Epidemiologic Study," *Regulatory Toxicology and Pharmacology*. In press.

CHAPTER 25

Adult Soil Ingestion Estimates

Edward J. Calabrese, Edward J. Stanek III, and **Charles E. Gilbert,** School of Public Health, University of Massachusetts, Amherst
Ramon M. Barnes, Department of Chemistry, University of Massachusetts, Amherst

Since groundwater is a significant drinking water source, it has long been recognized that contaminated soil may present a potential public health concern because of groundwater contamination. Recently, regulatory and public health agencies have become concerned that consumption of contaminated soil by children may be a significant public health problem. For example, soil levels of lead that range from 1500 to 7500 μg/g in certain sections of Boston[1] are suspected of contributing to elevated blood lead levels as a result of soil ingestion. Also widely discussed has been the dioxin contamination of Times Beach, Missouri. The Centers for Disease Control (CDC) derived a theoretical cancer risk associated with levels of dioxin in soil.[2] A major concern was the assumed consumption by children of soil containing dioxin, a contaminant that is relatively tightly bound to soil.

While the issue of soil contamination has correctly focused on young children, it is important to note that the U.S. Environmental Protection Agency (EPA) has developed guidance recommendations for how much soil adults ingest. In a 1989 communication[3] EPA concluded that 100 mg of soil per day is assumed to be ingested by adults. To our knowledge no paper has yet been published concerning how much soil adults ingest, although several extensive studies exist on quantifying how much soil children eat.[4–6] In the course of conducting a large-scale study of soil ingestion by children,[4] part of the analytical methodology was validated via exposure of a limited number of adults to known quantities of a well characterized soil. While this investigation was not designed to estimate the amount of soil normally ingested by adults, it did offer potential and limited insights as to how much soil these adults ingested. In light of the new initiative by EPA[3] to consider adult soil ingestion in risk assessment and the paucity of data on this topic, it is felt that these findings deserve broader dissemination.

METHODOLOGY

Participants and Soil Capsules

Participants in the study were six healthy adults; 3 males, 3 females, 25–41 years old, who did not have chronic illness such as diabetes, heart disease, or gastrointestinal disorders. Each of the volunteers ingested: (1) one empty gelatin capsule at breakfast and one with dinner Monday, Tuesday, and Wednesday during the first week of the study; (2) fifty (50) milligrams (mg) of sterilized soil within a gelatin capsule at breakfast and dinner for a total of 100 milligrams of sterilized soil per day for the 3 days during week two; (3) two hundred and fifty (250) milligrams of sterilized soil in a gelatin capsule at breakfast and dinner for a total of 500 mg of soil per day over the 3 days during week three.

Soil Preparation and Security

Soil for the adult ingestion study was selected from a soil library maintained by the Department of Plant and Soil Sciences on the University of Massachusetts, Amherst campus. The soil was selected because it was previously characterized as noncontaminated[4] and the tracer elements, aluminum, barium, manganese, silicon, titanium, vanadium, yttrium, and zirconium were of sufficient concentration to be detected during analysis of the volunteer's excretory samples. A formal security system was employed following the selection of this soil to prevent tampering with the soil and soil-filled capsules.

The soil was sterilized by autoclaving, oven dried, and evaluated microbiology to detect the presence of anaerobic and aerobic organisms and fungi.

Human subjects review required that the soil be chemically analyzed for metal lead and the U.S. EPA's extractable priority pollutants, consisting of approximately 100 compounds, including chlorinated hydrocarbon insecticides, PCBs, numerous polynuclear aromatic hyrocarbons, and phenolic compounds. The concentrations of these compounds in the soil were less than the detection limits: all compounds except polychlorinated biphenyls, one microgram per gram; polychlorinated biphenyls, five micrograms per gram.

Duplicate Meal

Duplicate meal samples, food and beverage, collected from the six adults, included all food ingested from breakfast Monday through the evening meal

Wednesday during each of the three weeks to reflect total dietary intake. All medications and vitamins ingested by the adults were included in the duplicate meals or were addressed during chemical analysis. The adults prepared two meals each containing the same types and quantities of food. One meal was eaten by the volunteer; the second meal was used as the duplicate meal. All uneaten food on the volunteer's plate was consolidated and compared to the duplicate meal. The food portions in the duplicate meal were separated to represent ingested food and uneaten food. The estimated amount of eaten food was placed in a polyethylene storage container for study analysis. The polyethylene containers were enclosed in plastic bags and transported in an insulated cooler bag with a reusable refrigerant.

Excretory Output

Total excretory output, feces and urine, was collected from Monday noon through Friday midnight (4 days) over three consecutive weeks. Urine and stool samples were collected in polyethylene commodes. Toilet paper or other wipes were not included in the commode and not collected by the research team. Used commodes were enclosed in plastic bags and placed into insulated bags containing portable refrigerants.

Results

Complete data were recorded on all six subjects for the entire three-week period of the study. The results are described for food ingestion, soil capsule ingestion, and fecal output for the adults, and these results are used to estimate the amount of soil ingested by the adults (methodology previously described).[4]

Food Ingestion

Daily freeze-dried weights of food samples varied from 97 g to 913 g. There was large variability in food intake between study subjects, with less variability between days of the week or weeks (Table 1).

The mean amount of food ingested per subject per day was calculated for each week. These amounts were then averaged per week, and overall. The average amounts ingested per week and overall are given in Table 2.

Table 1. Variance Components for Freeze-Dried Weight of Food Ingested by 6 Subjects over 3 Days for 3 Weeks

Source	Variance Component	Percent of Variance
Subject	63071 g sq/day	64%
Week	17071 g sq/day	17%
Day	17826 g sq/day	18%

Table 2. Mean Daily Tracers in Food Ingested Per Day for 3 Days for 6 Adults for Three Consecutive Weeks and Overall

Week	Freeze-Dried wt g	Al mg	Ba μg	Mn mg	Si mg	Ti mg	V μg	Y μg	Zr μg
1	553	2.06	759	3.11	34.1	0.94	9.97	1.59	2.51
2	461	3.66	547	1.77	34.3	2.46	65.09	1.52	6.97
3	504	1.32	631	2.59	28.8	4.05	8.66	0.77	3.01
Mean	506	2.35	646	2.49	32.4	2.48	27.91	1.29	4.16

Table 3. Components of Variance (and %) for Amount of Tracer Elements Ingested in Food for 6 Adults

Element	Unit (Squared)	Subject	Week	Day
Al	mg	0.1 (0.4%)	0.8 (3.6%)	21.8 (96.0%)
Ba	μg	13427 (9.2%)	6576 (9.2%)	51358 (72.0%)
Mn	mg	0.04 (2.1%)	0.8 (40.4%)	1.1 (57.5%)
Si	mg	0 (0.0%)	0 (0.0%)	993.0 (100.0%)
Ti	mg	1.3 (4.7%)	15.5 (55.7%)	11.0 (39.7%)
V	μg	2078 (16.3%)	2548 (20.0%)	8116 (63.7%)
Y	μg	0.04 (2.5%)	0.2 (12.9%)	1.3 (84.6%)
Zr	μg	18.9 (27.3%)	2.0 (2.8%)	48.3 (69.8%)

Table 3 summarizes components of variability in food ingestion for each tracer element. Variance components are given between subjects, between weeks, and between days of the week. The variance attributed to each component is also indicated.

The components of variance indicate that most variability in amount of elements ingested occurred from day to day, or week to week. The magnitude of variability from subject to subject in amount of tracer element ingested was low relative to these other sources.

Soil Concentrations

Soil was homogenized and the concentrations of each of the eight study elements were analyzed prior to filling the soil capsules. A total of 300 mg of soil was ingested in the second week of the study, and 1.5 g of soil ingested in the third week of the study. The total amount of each element ingested by week from soil capsules is given in Table 4.

Since fecal samples were collected for four days in each week, we also express weight in mg or μg, of each element ingested per day over a four-day period for comparison with the fecal sample results (Table 5). The average daily capsule ingestions reported in Table 5 for each element can be compared with the ingestion from food (Table 2) to quantify the experimental dose relative to background intake. For Al, Si, Y, and Zr, maximal daily capsule dose was more than four times average daily ingestion from food.

Table 4. Total Soil Tracer Amount in Capsule Ingested by Adults in Each Week

Week	Al mg	Ba μg	Mn mg	Si mg	Ti mg	V μg	Y μg	Zr μg
1	0.00	0	0.00	0.0	0.00	0	0.0	0
2	24.36	177	0.25	83.4	1.56	42	7.2	54
3	121.80	885	1.23	417.0	7.80	210	36.0	270

Table 5. Daily (over 4 days) Tracer Amount in Capsule Ingested by Adults by Week

Week	Al mg	Ba μg	Mn mg	Si mg	Ti mg	V μg	Y μg	Zr μg
1	0.00	0.00	0.00	0.00	0.00	0.0	0.0	0.0
2	6.09	44.25	0.06	20.85	0.39	10.5	1.8	13.5
3	30.45	221.25	0.31	104.25	1.95	52.5	9.0	67.5

Excretory Concentrations

Fecal and urine samples were collected from the Monday first morning void through Friday midnight of each week. The number of samples per day ranged from one to four for subjects. The most common number of excretory samples per day was one or two, and the average number of excretory samples per day for each week was two.

Samples were pooled on a given day, freeze-dried, and weighed. Concentrations of each element were then measured for each pooled sample. Daily excretory freeze-dried weight varied from 28.1 g to 141.4 g. There was no apparent association between excretory weight and number of excretory samples. There was marked variability in freeze-dried weight between subjects; 73% of the variability in freeze-dried weight was attributable to differences between subjects, with the remaining variability attributable to week to week differences. The mean freeze-dried excretory weight per day for the six subjects was 58.4 g for the first week, 66.0 g for the second week, and 71.2 g for the third week.

Concentrations for each element in the excretory samples were evaluated and used to estimate the amount of each element excreted per day in each of the study weeks. The average amount excreted by week is given in Table 6.

Table 6. Average Daily Amount of Tracers Excreted in 4 Days for 6 Subjects by Week

Week	Al mg	Ba μg	Mn mg	Si mg	Ti mg	V μg	Y μg	Zr μg
1	7.98	660.35	3.24	43.5	4.21	100.4	3.08	27.8
2	15.18	1341.40	2.79	59.6	2.94	78.4	3.82	31.2
3	33.30	919.67	3.48	125.8	9.07	95.6	11.34	56.6

INGESTION ESTIMATES

Table 7 provides the mean and median values of soil ingestion for each element by week, while Table 8 provides daily soil ingestion estimates by subject averaged over the three weeks of the study. These resultant values in Tables 7 and 8 have had tracer quantities in ingested food and capsules subtracted from amounts excreted. Table 9 provides the estimated median and mean soil ingestion values for the six subjects over the three weeks of the study. The median values for the eight tracers ranged from 28 mg (Si) to 778 mg (mn). The mean values ranged from 5 mg (Si) to 836 mg (Mn). Based on simple percent recovery values, Al, Si, and Y were considered the most valid tracers.[4] The median values of these three tracers are: Al-60 mg, Si-28 mg, and Y-44 mg. The mean values are Al-76mg, Si-5mg, and Y-49mg.

Table 7. Adult Daily Soil Ingestion Estimates (grams) by Week and Tracer Element After Subtracting Food and Capsule Ingestion, Based on Median Amherst Soil Concentrations: Means and Medians over Subjects

Week	Al	Ba	Mn	Si	Ti	V	Y	Zr
Means								
1	0.110	−0.232	0.330	0.030	0.071	1.288	0.063	0.134
2	0.098	2.265	1.306	0.014	0.025	0.043	0.021	0.058
3	0.028	0.201	0.790	−0.023	0.896	0.532	0.067	−0.074
Medians								
1	0.060	−0.071	0.388	0.031	0.102	1.192	0.044	0.124
2	0.085	0.597	1.368	0.015	0.112	0.150	0.035	0.065
3	0.066	0.386	0.831	−0.027	0.156	0.047	0.060	−0.144

Table 8. Adult Daily Soil Ingestion Estimates (grams) by Subject and Tracer Element, After Subtracting Food and Capsule Ingestion, Based on Median Amherst Soil Concentrations Averaged Over Weeks

Week	Al	Ba	Mn	Si	Ti	V	Y	Zr
1	0.068	0.218	0.681	−0.006	0.908	−1.824	0.111	−0.172
2	0.046	0.097	0.626	−0.070	−0.126	1.474	0.075	−0.025
3	0.037	3.968	0.826	−0.016	0.317	0.027	−0.035	0.148
4	0.001	−0.229	0.542	0.007	−0.054	0.712	0.054	−0.052
5	0.173	0.617	1.411	0.014	0.104	0.118	0.027	0.017
6	0.134	−0.053	0.883	0.099	1.113	2.183	0.084	0.216
Mean	0.077	0.769	0.828	0.005	0.377	0.448	0.053	0.022
Std Dev	0.065	1.592	0.312	0.055	0.517	1.348	0.051	0.140
Median	0.052	0.157	0.753	0.005	0.211	0.415	0.645	0.004

Table 9. Comparison of Soil Ingestion Values in Children and Adults

Tracer Elements	Adult		Children		Children/Adult Ratio	
	Mean	Median	Mean	Median	Mean	Median
Al	77 mg	52 mg	153 mg	29 mg	2.0	0.6
Si	5 mg	0.5 mg	154 mg	40 mg	30.8	80.0
Y	53 mg	65 mg	85 mg	9 mg	1.6	0.1

[a]Data from Calabrese et al.[4]

DISCUSSION

The results of this pilot study on adult soil ingestion indicate that soil ingestion values based on both mean and median estimates for the three most reliable tracers (Al, Si, Y) are less than the EPA guidance figure of 100 mg/day and more closely approximate a value near 50 mg/day. The preliminary nature of these findings should be emphasized, given the small sample size (n=6) and the fact that the adult data were originally designed to validate the children's ingestion study[4] and not to determine adult soil ingestion. Nonetheless, these findings provide initial estimates of adult soil ingestion, and should provide assistance in the development of methodological approaches, including study design, tracer selection, mass-balance consideration, analytical processes, and statistical evaluation.

Despite the limited nature of the adult soil ingestion data present here, comparison with the children ingestion data is warranted in light of EPA regulatory guidance procedures which address both children and adults.[3] While the factors which contribute to enhancing the occurrence of human soil ingestion remain to be more comprehensively and quantitatively assessed, it is generally believed that the greater hand to mouth behavior observed in young children as compared with older children and adults would contribute to their displaying greater soil ingestion than older age groups. The EPA guidance values appear to reflect this conceptual difference, with the agency assuming that adults would ingest about one-half as much soil as young children. In the absence of data it is certainly reasonable to assume that adults should ingest less soil than young children. However, the issue is how much less. The present data reveal the mean adult values for soil ingestion were lower than the childrens' values by a factor of from 1.6 (Y) to 30.8 (Si). In contrast, the comparison of adult versus children soil ingestion values using the median revealed no consistent relationship among the three tracers (Table 9). Conclusions based on these observations must be tempered by the fact that the adult study was of a much more limited nature (six subjects) than the children's study, which involved 64 children, and was not designed to determine adult soil ingestion.

The data also indicate that estimates of mean and median values in adults for Al and Y are similar, while the children display considerably different mean versus median estimates. This observation is consistent with the idea that mean values of childrens' data would be expected to be skewed upward, due to possible pica-like behavior for soil in some young children.

ACKNOWLEDGEMENT

This research was supported in part by Syntex Agribusiness, Incorporated and the Gradient Corporation. An earlier version of this study is published in *Regulatory Toxicology and Pharmacology,* Volume 12, 1990.

REFERENCES

1. Spittler, T. Minnesota Pollution Control Agency, Roseville, MN. Technical Resource Committee Meeting, September 24, 1986.
2. Kimbrough, R. D., H. Falk, P. Stehr, and G. Fires. "Health Implications of 2,3,7,8-Tetrachlorodibenzodioxin (TCDD) Contamination of Residential Soil," *J. Toxicol. Environ. Health* 14:47–93 (1984).
3. Porter, J. W. Memorandum to Regional Administrators, Region I–X, regarding interim final guidance on soil ingestion rates. U.S. EPA Office of Solid Waste and Emergency Response, January 27, 1989.
4. Calabrese, E. J., H. Pastides, R. Barnes, C. Edwards, P. Kostecki, E. Stanek, P. Veneman, and C. E. Gilbert. "How Much Soil Do Young Children Ingest: An Epidemiologic Study," *Reg. Toxicol. Pharm.* 10:1–15 (1989).
5. Binder, S., D. Sokal, and D. Maughan. "Estimating Soil Ingestion: the Use of Tracer Elements in Estimating the Amount of Soil Ingested by Young Children," *Arch. Environ. Health* 41:341–345 (1986).
6. Clausing, P., B. Brunekreef, and J. H. van Wijnen. "A Method for Estimating Soil Ingestion by Children," *Intern. Arch. Occup. Environ. Health* 59:73–82 (1987).

PART VI

Regulatory Considerations

CHAPTER 26

State of Florida Policy for Soil Treatment at Petroleum Contaminated Sites

Wm. Gordon Dean, Cherokee Groundwater Consultants, Inc., Tallahassee, Florida
Brian K. Cobb, Environmental Affairs Department, Tenneco, Inc., Marietta, Georgia

(NOTE: Work reported in this chapter was performed while the authors were employed by the Florida Department of Environmental Regulation, Bureau of Waste Cleanup, Tallahassee, Florida)

The State of Florida implemented the Early Detection Incentive (EDI) program in November 1987. This program provides incentives to responsible parties to report petroleum contamination sites by reimbursing responsible party cleanups or providing state cleanup of such sites. The application period ended December 31, 1988 with almost 10,000 applications received. A guidance policy has been established to deal with the contaminated soils at these EDI sites. While the policy is still evolving, the basic guidelines are well established and are codified in Chapter 17–70, Florida Administrative Code.

CLEANUP STANDARD FOR PETROLEUM CONTAMINATED SOILS

The cleanup of contaminated soils is an important part of the corrective process. The contaminated soils that are allowed to remain on site can significantly prolong the groundwater cleanup effort, resulting in greatly increased costs. Given Florida's unique geologic characteristics and the large number of petroleum contamination sites, establishment of a soil standard became a necessity.

Chapter 17-70.003 Florida Administrative Code (F.A.C.) is the state rule which contains all petroleum contamination site cleanup criteria. It defines excessive soil contamination as,

> "all soils saturated with petroleum or petroleum products and those soils that cause a total hydrocarbon reading of greater than 500 ppm on an organic vapor analysis instrument with a flame ionization detector . . . ."

This is the only definition related to contaminated soils contained in Chapter 17-70, F.A.C. The extent of contaminated soil may therefore be determined solely with a flame ionized organic vapor analyzer (OVA); no laboratory analyses are required. A number of choices and assumptions were evaluated to arrive at this policy. They are briefly summarized below.

Hazardous Waste Status

It was first necessary to determine whether excessively contaminated soils should be considered a hazardous waste. Petroleum dispensing facilities and petroleum products are specifically exempt from most RCRA hazardous waste regulations. Contaminated soils from these facilities, therefore, cannot be considered listed hazardous wastes. Five specific categories of petroleum refining wastes are listed as hazardous wastes in 40 CFR 261:

(a) dissolved air floatation float
(b) slop oil emulsion solids
(c) heat exchanger cleaning sludge
(d) API separator sludge
(e) leaded tank bottom sludge

None of these refining wastes should be encountered at EDI sites. If, however, soils were to become contaminated with these materials, those soils would then become hazardous wastes. The soils could also be considered hazardous if they fail one of the four hazardous characteristic tests: ignitability, corrosivity, reactivity, or toxicity (EPA Methods 1010 or 1020, 1110, 1210, and 1310, respectively). It was assumed that petroleum contaminated soils would not be corrosive or reactive. A solid would be considered ignitable if it met the following definition (40 CFR 261.21),

> ". . . is capable, under standard temperature and pressure, of causing fire through friction, absorption of moisture, or spontaneous chemical changes and when ignited, burns so vigorously and persistently that it creates a hazard."

Petroleum contaminated soils may be assumed to not meet this subjective definition. Finally, the soil would be considered toxic if its leachate failed the EP Toxicity test for one of eight metals or six pesticides. Of the 14 analytes, lead is the primary concern. Cadmium and chromium were considered minor concerns at these petroleum sites. Past field experience led to the conclusion that petroleum contaminated soils could not generally be considered characteristically toxic.

In summary, petroleum contaminated soils are not hazardous wastes under any of the current RCRA regulations. If and when the Toxicity Characteristic Leaching Procedure (TCLP) is officially adopted, it will replace the EP Toxicity test. A number of common organics will have maximum leachate concentration levels, in addition to the metals and pesticides. If the proposed levels are retained, it is very likely that most petroleum contaminated soils will be characteristically toxic. This will require major changes in the way assessments and remedial actions are conducted. These guidelines will be changed accordingly at that time.

Petroleum Contaminated Soil Standard

Given that these soils are not hazardous, it was then necessary to define a level for excessively contaminated soils. The U.S. Environmental Protection Agency has no set soil standards; levels are determined by site-specific risk assessments. The Department (Florida Department of Environmental Regulation) also determined soil cleanup standards with site-specific risk assessments. Florida adopted Maximum Contaminant Levels for eight organic chemicals in groundwater in 1984, but no soil standards were established. Assuming that the risks associated with contact from the soils were negligible, the goal of a soil standard is to protect the groundwater and to facilitate the site remediation. The number of sites in the EDI program made the continued use of site-specific standards logistically impossible. It was therefore decided that adoption of a quantifiable soil standard was essential.

The first possibility considered was some form of an odor and appearance standard. Petroleum products have distinctive and easily detectable odors and frequently stain affected soils. This method would have the advantage of zero cost and easy implementation. It was rejected because of potential health effects to workers and because it was not a quantifiable method.

Quantifiable detection methods which were identified were loosely grouped into two categories: field methods and laboratory methods. Laboratory methods offer the best quality assurance, but were rejected due to the required turnaround time, cost, and lack of existing numerical standards. Since much of the soil removal would be done in conjunction with a tank removal/replacement, the time constraints associated with laboratory analyses were a major concern. Both the liability associated with leaving an excavation open and the loss of income for extended periods of time were considered unacceptable.

It was therefore decided to choose a method of field detection which would optimize the following considerations: cost, reliability, ease of use, detection limit, detection range, response to hydrocarbons, and accuracy. It was also decided that the method should detect the total amount of hydrocarbons present, rather than identify particular compounds. A number of petroleum compounds (primarily the volatile and polynuclear aromatics) have been identified as harmful. However, they constitute only a small fraction of the number of compounds which comprise petroleum and petroleum products. The remaining compounds are not necessarily innocuous; they simply have not been classified due to a scarcity of toxicological data. Ignoring these compounds, given the toxicity of similar compounds, was unacceptable from a regulatory aspect.

Field Detection Methods

The following field instruments were identified as potentially acceptable: colorimetric tubes, photoionization detectors (PIDs), flame ionization detectors (FIDs), and portable gas chromatographs (GCs). All of the identified methods sample the air, rather than the soil matrix itself. This obviously requires the hydrocarbons to be volatile. This was decided to be an acceptable alternative to a complete bulk analysis since the volatile compounds are generally assumed to be both the most toxic and water soluble. Recently, new instrumentation has been reported to be capable of directly sampling the soil. These methods were not available in 1986 when Chapter 17-70 was being written. They will be closely evaluated as more data become available.

Colorimetric (Draeger, etc.) tubes were considered first since they are by far the least expensive option. The tubes are also the easiest to use, but are the least accurate and have limited detection ranges. In addition, the information from tubes can only be considered semiquantitative since the concentration is found by measuring the length of the tube which has changed color.

Portable gas chromatographs (GCs) were rejected for several reasons. GCs are the most expensive option, require an experienced operator, are relatively fragile, require long run times, and have difficulty in providing bulk analyses. They would, however, be the instrument of choice if identification of specific compounds was required.

The remaining two options, PIDs and FIDs, are essentially identical in cost, detection limit, ease of use, and accuracy. The FID has a higher response to the hydrocarbons of interest and is not adversely affected by high humidity. The FID will detect methane, whereas the PID will not. This can be a problem in areas with decaying vegetation, but can be overcome by comparing the results of the unfiltered sample with a sample passed through a carbon filter. Methane will pass through the filter, but hydrocarbons will not. A final decision was made to specify a FID organic vapor detector as the standard instrument for use in setting a soil standard.

All available literature and the data from several sites were evaluated to determine an appropriate standard. A draft EPA document on soil vapor monitoring at gasoline sites stated that FID readings in excess of 500 ppm were a clear indication of petroleum contamination, and that lower concentrations should cause no adverse effects on groundwater. Laboratory analyses of soils with vapor readings of 500 ppm presented no clear correlation; however, the results indicated that such a reading for fresh gasoline roughly corresponded to a total hydrocarbon concentration of 10 to 20 ppm. A total hydrocarbon concentration of 10 ppm had been recommended by California as a soil standard. A minor consideration was that 500 ppm was in the middle of the detection range for an FID and so should produce more accurate readings. Based on this information, a standard of 500 ppm vapor concentration as detected by a FID vapor detector was chosen as the definition of excessively contaminated soil.

A sampling methodology was also specified in Chapter 17-70 to ensure consistency. The method consists of ". . . sampling the headspace in a half-filled 16 ounce soil jar. The soil sample shall be brought to a temperature of 20°C (68°F) with a water bath and sampled five minutes thereafter."

A clarification was required due to confusion about whether cooling samples to 20° was required if the sampling was conducted during the summer. The intent of the method was to define a minimum acceptable temperature. Temperatures higher than this minimum would result in higher readings and would therefore represent a conservative error. The following guidance was issued to clarify the situation,

> "Although the method specifies a soil temperature of 20°C (68°F), it is not necessary to cool the sample if that temperature is slightly exceeded. Cooling is only required if the sample's temperature exceeds 29° to 32°C (85–90°F). However, an effort should be made to take all samples from a site at a constant temperature."

Chapter 17-70 is currently being revised. The proposed changes include eliminating the water bath requirement entirely and establishing an acceptable temperature range of 20° to 32°C.

The most common method of sampling in the field is to use a 16 ounce mason jar, with the solid top replaced with a layer of tin foil. The OVA probe is inserted through the foil after the sample has equilibrated.

The following language was also included in Chapter 17-70 to prevent the unfair exclusion of consultants who did not own a flame ionized detector,

> "Instruments with a photoionization detector may be used after a determination is made of that instrument's equivalent reading to 500 ppm with a flame ionization detector. Analytical instruments shall be calibrated in accordance with the manufacturer's instructions. Other analytical methods may be used subject to Bureau approval upon a demonstration that they provide accurate and verifiable results, and that the results may be calibrated to those achieved with an organic vapor analysis instrument."

Photovac, a major PID manufacturer, has issued a technical bulletin giving a procedure for calibrating their instrument to an equivalent FID reading. This procedure is currently being accepted, but Photovac is conducting tests to evaluate some Departmental concerns. HNU Corporation has recently completed comparison testing and will be issuing a technical bulletin shortly. Other PID instruments should be calibrated by measuring several gas concentrations with the PID and a FID, and plotting the relative response. The relative response will vary between instruments, but a PID reading of 40 to 130 ppm generally corresponds to a 500 ppm FID reading.

After adoption of the soil standard, several points of confusion were noted. First, the standard is a vapor concentration and not a laboratory analysis of total hydrocarbons. No laboratory analyses are required and laboratory data cannot be directly correlated to the vapor concentration. Second, the standard defines soils which are grossly contaminated and must be removed or treated if the cleanup is to be successful. The definition is *not* intended as an indicator that soils with concentrations below this level are clean or can be removed from the site and disposed in uncontaminated areas. Soils that are not "excessively contaminated" may still produce a harmful leachate. Third, soils with detectable vapor concentrations which are below 500 ppm are contaminated, although not "excessively contaminated." A lower bound of 10 ppm was recommended to prevent instrument interferences from giving false positive results. Soils with concentrations greater than 500 ppm *must* be remediated. Soils with concentrations between 10 and 500 ppm *may* be required to be remediated, depending upon the concentration and the apparent impact on the groundwater.

Diesel Contamination

One problem has been found with the soil standard. As noted in the above discussion of field instruments, only the volatile portion of the contamination is detected in the soil vapor. At some sites with only diesel or very weathered gasoline contamination, it was found that even soils approaching saturation had vapor readings below 500 ppm. Chapter 17-70 is currently undergoing minor revisions and a change in the soil standard has been proposed to address this problem. The proposed change will establish a separate soil standard of 50 ppm headspace for diesel contamination. As an interim measure, the guidelines for Initial Remedial Actions (IRAs) were changed as follows,

> ". . . To encourage removal of these soils, the following exceptions to the requirement that "excessively contaminated" soils be present will be allowed at sites where it is known that *only* diesel fuel contamination exists:
>
> a. As noted in guideline 1.a, a DER representative (a DER or contracted local government employee) may authorize removal of contaminated soil.
> b. An OVA reading higher than 50 ppm substantiated by one EPA 418.1 sample from the most apparently contaminated soils will be considered justification for removal and reimbursement.
> c. As with the gasoline contaminated sites, soils with OVA readings less than 10 ppm should not be removed as part of an IRA."

The intent of the EPA 418.1 sample is simply to verify that petroleum contamination is indeed present. A positive result is all that is required; no numerical standard must be met.

Used Oil Contamination

Chapter 17-70 excluded used oil from all standards. A ruling by the Department Secretary has resulted in a requirement for used oil to be included in Chapter 17-70. The proposed revisions include:

1. Limiting used oil Initial Remedial Actions to removal of free product and saturated soils.
2. Requiring sampling for priority pollutant organics, E.P. Toxicity, bulk metals, and total recoverable petroleum hydrocarbons (TRPH).
3. Establishing a numerical cleanup standard based on the TRPH concentration is under consideration.

Soil Standard Summary

In summary, Chapter 17-70.003(3) defines excessively contaminated soils as ". . . all soils saturated with petroleum or petroleum products and those soils that cause a total hydrocarbon reading of greater than 500 ppm on an organic vapor analysis instrument with a flame ionization detector . . . ." These excessively contaminated soils must be remediated. Soils with vapor readings between 10 and 500 ppm are contaminated, although not "excessively contaminated." These soils may require remediation, depending upon the concentration and the apparent effect on the groundwater. Revisions are proposed to establish separate standards for diesel (50 ppm) and used oil soil contamination.

ASSESSMENT METHODS

The horizontal and vertical extent of the contaminated soils must be defined during the contamination assessment as required by Chapter 17-70.008(3)(g). The extent of contamination is not limited to "excessively contaminated" soils; contaminated soils are also included. Although soils with vapor readings as low as 10 ppm could be considered contaminated, some judgment should be exercised when determining the assessment endpoint. The appropriate endpoint will be site-specific, but one between 10 and 100 ppm should generally be acceptable.

The appropriate testing method is specified in Chapter 17-70.003(3). It is, ". . . an organic vapor analysis instrument with a flame ionization detector in the survey mode upon sampling the headspace in a half-filled 16 ounce soil jar . . . ." As discussed in the Field Detection Methods section, the sample temperature should be between 68° and 90°F and equivalent instruments may be used as long as they have been calibrated to a flame ionization unit. A 5% duplicate sampling by either U.S. EPA Method 5030/8020 or U.S. EPA Method 418.1 may be conducted if desired. No other laboratory soil analyses are necessary, unless they are to comply with a disposal permit requirement.

The soil samples may be collected in a number of ways. The three most common methods are with hand augers, split spoon samplers, and with spoons from the side of an excavation. In all cases, correct decontamination procedures must be followed. The use of an in-situ soil gas procedure has been proposed by several consultants. This procedure is acceptable only if it is not feasible to collect an actual soil sample. The rationale for this decision must be discussed in the Contamination Assessment Report. The number of samples required must be determined on a site-specific basis in the field. A rough estimate would be one sample for every 10 to 30 feet around the horizontal perimeter and one sample for every 5 feet vertically.

Some confusion exists as to whether the soil assessment should be confined to the vadose (unsaturated) zone, or should continue below the water table. In the field it would be impossible to separate the waterborne contamination from that adsorbed to the soils. The soil assessment should therefore be confined to the vadose zone. The implicit assumption is made that the groundwater treatment will adequately address any residual soil contamination.

TREATMENT METHODS

Excavation followed by incineration, landfilling, or landfarming is the most commonly used treatment method. The use of in-situ vacuum extraction is increasing rapidly, while in-situ biodegradation remains in the experimental stage. Each method is discussed in detail below.

Incineration

Incineration is the preferred treatment method for excavated soils since it provides destruction of the contaminants. A number of asphalt plants, stationary incinerators, and mobile incinerators have been permitted by the Department's Bureau of Air Quality Management to incinerate petroleum contaminated soils. A generic permit with standard requirements is being prepared, but currently each permit contains slightly different requirements for influent/effluent sampling. The possible incorporation of handling and storage requirements in the air permit is also being evaluated. Current incineration costs average about $40 per ton of contaminated soils. These costs are dropping as the number of permitted facilities increases.

Landfilling

Landfilling of nonsaturated, nonhazardous soils in a lined landfill is an acceptable disposal option. The acceptance of these soils is entirely at the discretion of the landfill operator. The number of landfills accepting contaminated soils has been steadily declining due to concerns regarding potential future liability. The disposal costs have also increased for that reason. Currently, landfill costs range from $5 to $60 per ton of contaminated soil.

Landfarming

In some cases, treatment of contaminated soils may include landfarming. The process involves spreading the soils in a thin layer over an impermeable liner. The contaminant reduction is caused by a combination of volatilization, biodegradation, and photodegradation. The following conditions must be met:

1. An area large enough to allow spreading the soil 6 to 12 inches thick must be available for the duration of the treatment period.
2. The area must be covered by an impermeable liner, and provision must be made to prevent tearing the liner during the treatment operation.
3. Surface water controls to prevent water from entering or leaving the treatment area must be present.
4. The soil must not have a total recoverable petroleum hydrocarbon concentration greater than 500 ppm by U.S. EPA Method 418.1.
5. The soil must be tilled once every one to two weeks.

Any state-permitting requirements have been waived for landfarming operations conducted at a contamination site. However, a monitoring plan must be proposed to show that the treatment area has not been affected by the contaminated soil. The monitoring requirements are:

1. 100 cubic yards or less: visual inspection
2. 100 to 500 cubic yards: pre- and post-treatment soil sampling of the underlying soils with an organic vapor analyzer (OVA).
3. Over 500 yards: OVA sampling of the underlying soils, plus one or more monitoring wells sampled before and after the treatment period. The analyses should be the U.S. EPA methods specified in Chapter 17-70.008(9), FAC.

A permanent landfarming facility would require a solid waste permit, an air permit, and any applicable stormwater or NPDES permits. There are currently no permanent landfarming facilities in the state.

Vacuum Extraction

This relatively new technique is used to remove volatile organics from the soil and minor amounts of free product from the top of the water table. Generally used in-situ, it can remove the contaminant source and enhance further treatment without excavation. The extraction is accomplished by applying a vacuum to a well or series of wells constructed above the water table. Air will be drawn to the wells from the surrounding soils by the vacuum. As fresh air is brought into the formation, volatile contaminants will move from the soils into the air. They are then removed by the vacuum. The removed air can be exhausted directly to the atmosphere or treated to remove the contaminants. Vacuum extraction is of limited use in areas with very tight soil formations or thin unsaturated zones. It is not effective for oils and other nonvolatile contaminants.

A pilot project was undertaken by the Department in 1986 to remediate a petroleum site in Belleview, Florida. The results were generally excellent, with product recovery rates of up to 2000 pounds per day. Over 22,000 total pounds of hydrocarbons were extracted from the soil over the six-month test period. The relative reduction of hydrocarbons, based on wellhead concentrations, ranged from 95.9% to 99.7%.

The vacuum extraction technique is too new for standard design equations to have been developed. The radius of influence is the most important design factor. Determination of the radius requires a pilot study with a vacuum well and several closely spaced vacuum monitoring wells. The radius of influence may range from 10 to 100 feet, depending on the site lithology. The minimum effective vacuum is 0.25 to 0.5 inches of water. Recovery wellhead vacuums range up to 29 inches of mercury in several reported operating sites. Horsepower requirements ranged from 3 to 25 hp to achieve these vacuums. If an appreciable amount of free product is expected, an in-line air/fluid separator may be required. Costs range from $10 to $50 per gallon of product recovered, depending mostly upon whether exhaust gas treatment is required.

Bioremediation

The biodegradation of petroleum compounds by enhancing native bacteria or by introducing special microorganisms can theoretically be effective in-situ treatment. The environmental conditions which must be met include temperature, pH, salinity, oxygen, nitrogen, phosphorous, and trace minerals. The lack of oxygen is the factor that most limits degradation in subsurface soils. To date, no successful bioremediation projects have been completed in the state. The major problem appears to be transport of oxygen and nutrients.

Soil Fixation

Soil fixation is a process in which contaminated soils are excavated, treated, and returned to the excavation. Treatment involves mixing contaminated soils with a cementatious compound and/or chemical stabilizers. This method is not normally used at petroleum sites for economic reasons.

Soil Washing

Soil washing uses surfactants to remove contaminants from excavated soils. The leachate produced is contained, treated, and recycled to the washing operation. Reduction of contaminant levels to below detectable levels can be achieved with this technology. The cost ranges from $100 to $250 per ton of soil. This method is normally not used at petroleum sites for economic reasons.

DISPOSAL REQUIREMENTS

The treated soils may be returned to the excavation if the vapor concentration is less than 10 ppm. If the soils are to be disposed offsite, several options are available depending upon the residual contaminant concentration. The landfill requirements were discussed in the Treatment Methods section.

If the soils are to be used as clean fill, the following requirements must be met:

1. The sum of the benzene, toluene, xylenes, and ethylbenzene (BTEX) concentrations by U.S. EPA Method 5030/8020 is less than 100 ppb; and
2. The total recoverable petroleum hydrocarbon (TRPH) concentration by U.S. EPA Method 418.1 is less than 5 ppm.

The soils may be used in a roadbed if the following conditions are met:

1. The BTEX concentration is less than 200 ppb; and
2. The TRPH concentration is less than 100 ppm; and
3. The roadbed is at least one foot above the 100 year high water table (use of flood plane maps is acceptable).

The soils may be incorporated into an asphalt mix for road construction if:

1. The BTEX concentration is less than 500 ppb; and
2. The TRPH concentration is less than 500 ppm.
3. Specifically permitted plants may accept higher concentrations.

SUMMARY

The Florida minimum cleanup standard for petroleum contaminated soils is a 500 ppm vapor concentration, as measured by a flame ionized organic vapor detector. Separate standards are proposed for diesel and used oil contamination. These levels are most commonly achieved by excavation, followed by incineration, landfilling, or landfarming or by in-situ vacuum extraction. Several disposal options are available for the treated soils, depending upon the residual contaminant concentrations. The treated soil may be returned to the excavation; used as roadbed material, asphalt mix, or clean fill; or landfilled.

CHAPTER 27

Interpretation of State Guidelines and Requirements for Innovative Remedial Solutions: One Regulatory Agency's Experience

Anna Symington and **Audrey Eldridge,** Massachusetts Department of Environmental Protection, Western Region, Springfield, Massachusetts

This chapter is to address the widespread perception that the Massachusetts Department of Environmental Protection (the Department), along with other environmental regulators, are basically unyielding and narrow minded, and generally not open to innovative ideas or new technologies to assess and/or remediate sites. This perception enhances the ill-conceived image that bureaucracies are constantly requiring additional and unneeded studies and/or analyses, prior to (if ever) granting approval to perform remediation.

The Department has encountered many site-specific problems while assessing the feasibility of remedial technologies for petroleum contaminated soils. This chapter shall address the thought process and rationale utilized by some environmental regulators in determining the effectiveness and appropriateness of remedial actions, and will detail some of the solutions implemented. Examples will include the management and remediation of petroleum contaminated soils, and the regulatory insight and requirements pertaining to in-situ treatments, such as: soil flushing/groundwater recharge galleries; soil gas extraction with reinjection; vapor vacuum extraction combined with destruction via a catalytic afterburner; as well as the ultimate in regulatory response—the generation of new policies which have increased the efficiency of some site remediations and allowed for cost-effective petroleum contaminated soil removal and disposal.

HISTORICAL BACKGROUND

In 1986 the citizens of Massachusetts set a precedent, and voted for Referendum Question 4 (Q4), which mandated the Department to conduct an aggressive hazardous waste site identification, assessment, and remediation program. As part of Q4, the Massachusetts Oil and Hazardous Material Release Prevention and Response Act (M.G.L. ch. 21E) was revised to reflect the voters' demands. The revised M.G.L. ch. 21E included a provision that the Department produce and finalize a contingency plan which would address and detail the level of environmental assessment needed at locations to be investigated (LTBI), or sites, standardize the method of determining the need for site remediation, and outline the approach for deciding what remedial alternatives may be feasible for a specific site. As a result, in October 1988 the Massachusetts Contingency Plan (MCP) was finalized.

Requirements placed upon the Department's Bureau of Waste Site Cleanup (BWSC) resulting from the MCP and M.G.L. ch. 21E include: the listing and publishing of LTBIs and sites on a quarterly basis; conducting remedial actions at a site in order to achieve background levels, where feasible; listing 1,000 LTBIs and sites per year, by January, 1989; and producing a final remedial response plan for all confirmed sites within three to seven years of the date of the first listing of a site, depending upon the severity of the problem at the location. It is conceivable that by the year 1994, the Department may be overseeing the cleanup of over 1,000 sites. Approximately 60% to 80% of these sites are expected to be petroleum contaminated.

The Department is not only charged with protecting the environment, but also conducting site remediation if the Potentially Responsible Party (PRP) is not pursuing or willing to undertake the required work at a site within the established deadlines. It is in our best interest, as well as that of the PRP, to work with the consultant to allow progress to continue at a site. The Department therefore encourages the use of remedial technologies, including innovative technologies, to address contamination at sites in as cost-effective, productive, timely, and environmentally sound manner as is possible. Obviously, these technologies must be shown to be capable of addressing a specific situation by way of a well-designed plan. The issue of what is considered to be a well-designed plan often constitutes a source of conflict or tug-of-war between the Department and those who propose remediation plans.

COST-EFFECTIVE REMEDIAL METHODS

The Department has approved of several projects for in-situ remediation of petroleum contaminated soils, and in general, the Department will encourage in-situ remediation over excavation for offsite disposal. The two in-situ technologies most frequently proposed are soil flushing via a recharge gallery, and soil vacuum/venting methods. These methods have also been used successfully in conjunction with dual-phase recovery systems. Both soil remediation technologies have their advantages, and the use of such methods combined with a groundwater recovery and treatment system can often shorten the timeframe for all remedial actions to be completed at a site. However, there is an inherent need to understand certain aspects of specific sites, prior to the implementation of proposed remedial systems.

The soil flushing program can utilize water from the groundwater recovery and treatment program to "cleanse" or flush the contaminated soil. This can be seen as a plus, especially when compared to costly wastewater disposal and treatment options. As the introduction of a steady and long-term input of water to the subsurface will change the hydrological conditions at a site, a thorough understanding of the groundwater flow patterns prior to the recharge of such water is essential. Input from both the BWSC and Water Pollution Control regarding a Groundwater Discharge Permit will be required for such a project, and the concern regarding the creation of an artificial groundwater boundary, which may force migration to an "unaddressed" portion of the site, will have to be evaluated.

In addition, soil-flushing can only be effective if the petroleum contaminated soil is permeable enough to allow rapid percolation. Usually, the Department will require a "perc" test to determine the capability of the contaminated subsurface material to accept a certain flow rate of water. Conceptually, one would anticipate that the effectiveness of such a process would decrease with an increase in density, viscosity, and adsorption potential of the oil. Gasoline contaminated soils would have a greater potential for success utilizing this type of remediation, due in part to a higher solubility. This type of remediation may not be feasible for heavier oils, and a treatability study may be necessary prior to approving such an action.

Other factors that should be examined prior to approving such a recharge project would include: soil organic matter content; source and quality of water; saturated thickness of the aquifer; depth to groundwater; hydraulic conductivity; identification of the presence of any confining layers; mounding potential; and the location of any private or public water supply wells within a one-half mile radius of the disposal site.

A proposal for a soil vacuum remediation would require less site-specific information, but more data would be needed regarding the proposed instrumentation and equipment. Recently, consultants have been able to conduct relatively simple air "pump" tests, utilizing shallow vapor wells, a vacuum, and pressure sensitive probes to determine if applying a vacuum at a given location would have a measurable effect in the unsaturated material, thereby approximating the radius of such an effect. Such a simple test may give credence to the feasibility of a proposal to utilize soil vacuum extraction as a method of site remediation for volatile compounds.

Information required by the Department for soil-vacuum remediation is, in part, based upon the treatment method proposed for contaminant-laden vapors. The Division of Air Quality Control (DAQC) will issue a permit—if needed—after reviewing and approving the permit application and the engineering specifications and design. The Department has approved both carbon canister and catalytic combustion as acceptable technologies for the treatment of the collected vapors. However, data must be submitted to the Department which will document how the consultant will verify the efficiency and safety of the treatment system. As always, a strict air quality monitoring plan would be in force.

In the past, there has been much information gathered on the high costs associated with the use of carbon in the vapor-phase removal of volatile organic compounds. In part, this is due to the rapid utilization of available sites of the carbon during vapor-phase adsorption, as well as the costs of transportation and disposal of a hazardous waste. The level of treatment required to achieve a discharge concentration that will be satisfactory to the DAQC may require ultra-polishing. The Department has taken these costs into consideration when approving of two other proposed technologies, similar to the above-mentioned remedial actions: the treatment of organic vapors through self-contained, automatically regenerated carbon beds; and the reinjection of carbon canister-treated soil gas vapors to the excavated and encapsulated soil pile. As the latter does not result in an air discharge, the need to apply and obtain a permit from the DAQC is negated. The former option offers the advantages of onsite recycling of the spent carbon via regeneration, thus never encountering the high transporting and disposal fees that could arise if no onsite treatment methods were available.

Although biodegradation has great potential for remediating petroleum contaminated soil, and has had a substantial amount of research evaluating and documenting its feasibility, it is not a method that is often pursued by consultants. One of the Department's main concerns when evaluating soil bioremediation options would be the potential of uncontrolled volatilization becoming the main mechanism by which the more volatile organic compounds are removed from the soil. There must be care taken in designing a program that will be able to demonstrate that the reduction in total petroleum hydrocarbons or volatile compounds is directly a result of the *biodegradation* process, not aeration.

STATE GUIDELINES FOR PETROLEUM CONTAMINATED SOIL

Virgin petroleum oil products account for the majority of spills or releases responded to by the Department, pursuant to MGL ch. 21E. The cleanup of these releases, as well as the upgrading of underground storage tank (UST) facilities pursuant to the Public Safety Regulations (527 CMR 9.00) often involved the excavation of contaminated soils. Disposal options were not readily available in the Commonwealth, since virgin petroleum oils, when spilled, were considered unused waste oil, thereby subject to the provisions of the Hazardous Waste Regulations, MGL ch. 21C and 310 CMR 30.00, which list it as a hazardous waste. The few options that did exist proved costly, resulting in contaminated soils often being left as stockpiles at the sites of the releases. As the number of these oil-contaminated soil stockpiles increased, so did the potential threat to public health and safety, and to the environment. In response to this threat, the Department sought to provide for a cost-effective means for oil-contaminated soil removal and disposal, as well as increase the efficiency of some site remediations. Recycling Regulations—310 CMR 30.200—went into effect in 1987, which provided the means by which to accomplish this task. On December 22, 1987, the Department issued a new policy entitled, "Interim Policy—Management of Residuals Under MGL Chapter 21E From Spills/Releases of Virgin Petroleum Oils." This policy governed the storage, transport, treatment, and disposal of virgin oil-contaminated soils, and outlined the terms and conditions under which it could be used.

The intention of the Interim Policy was not to encourage soil excavation as a site remediation method, but to reduce the number and storage duration of virgin oil-contaminated soil stockpiles, thereby reducing the threat associated with them.

The Interim Policy defined "*virgin* petroleum oils" as *unused* distillate and residual petroleum oils which included gasoline, kerosene, diesel, and number 2, 4, and 6 heating oils. This policy provided for several disposal options, provided specific criteria were met. Excavated soils that exhibited volume/volume headspace concentrations of volatiles less than 1800 parts per million (ppm) for gasoline contaminated soils, or weight/weight concentrations of total petroleum hydrocarbons (TPH) less than 3000 ppm for soils contaminated with oil residuals could be managed under this policy. Once sampled and analyzed for the above parameters, this information was forwarded to the appropriate regional office for approval for disposal. If approved, soils could then be transported and disposed of in Massachusetts-approved, lined sanitary landfills, or transported to Massachusetts-approved asphalt batching plants. Soils that exhibited concentrations that exceeded these limits had to be managed under the Hazardous Waste Regulations.

For transport to an approved Massachusetts landfill, a standard Bill of Lading had to be obtained by the generator and authorized by the Department. Once the soil was disposed of, the completed Bill of Lading (signed by both the transporter and the receiving facility) was returned to the Department. For transport to an approved Massachusetts asphalt batching plant, a Bill of Lading, specifically designed and approved for that particular facility under its Recycling Permit, was utilized. Soils which exceeded the contaminant concentration limits allowed by the Interim Policy could be transported to an asphalt batching plant, provided the contaminant concentrations were within acceptable limits of the facility's Class A Recycling Permit, and the generator obtained a Recycling Permit as well.

In situ or onsite treatment of oil-contaminated soils was always encouraged, where practical, in lieu of excavating. The policy also permitted onsite disposal of contaminated soils, provided the soils were first treated to "allowable" contaminant levels, which was governed by the type of area where the release occurred. Onsite disposal was permitted at areas of moderate environmental impact if, after treatment, the soils exhibited contaminant levels of no more than 10 ppm volatiles, or 100 ppm TPH. For areas of low environmental impact, onsite disposal was permitted if contaminant levels, after treatment, did not exceed 100 ppm volatiles, or 300 ppm TPH.

The Interim Policy even had a provision for management of other nonhazardous material contaminated by virgin oil. Combustible debris, including absorbent pads, hay, etc., contaminated with virgin oil could be transported to solid waste incinerators or Resource Recovery Facilities. The same Bill of Lading used for landfill disposal was used to accommodate the approval and transport of these materials.

POLICY REVISION

As the Interim Policy was implemented through the year 1988, which actually became an invaluable trial period, the Department had an opportunity to examine how well this new tool was working. The policy had outlined quite specifically the protocol to be followed with regard to sampling, analysis, authorization, transport, storage, disposal, applicability, etc. The Bill of Lading also provided detailed instruction as to its handling. Despite these provisions, a simple tool became a complex one in the hands of some consultants, resulting in obstacles which tended to consume time and basically hold up progress. It was inevitable that the Department was seen as the culprit regarding any delays in processing Bills of Lading that were submitted for authorization.

Problems ranged from submission of incomplete Bills of Lading to actual misuse of the policy. Bills of Lading were being submitted without proper signatures, no indication of the disposal facility proposed for use, missing or insufficient sampling and analyses, and analyses without any indication of the method or instrument used. Some consultants tried an innovative, yet unsuccessful, approach by telephoning in requests for disposal authorization, assuring the Department that the supporting paperwork would be forwarded at a later date. This lack of information, or stalling, proved costly by way of delaying authorization anywhere from a couple of days to several weeks, whereas a properly completed Bill of Lading could be reviewed and authorized within the week of submission.

There were instances where some consultants grossly underestimated the volume of contaminated soil to be disposed of. The Department issues authorized Bill of Lading copies corresponding to the number of truckloads to be disposed of. Hence, if the Department issues 10 copies for the disposal of 200 cubic yards of soil, at 20 cubic yards per truck, and there are really 300 cubic yards of soil, that additional 100 cubic yards of soil cannot be disposed of without the submission of another Bill of Lading request and supporting analysis.

There were attempts to dispose of virgin oil-contaminated soils mixed with soils contaminated with used waste oil. The policy strictly stated that the only contaminated soils to be managed under such cases were those soils contaminated with virgin petroleum products.

The policy requires that the completed Bills of Lading be returned by the generator to the Department within five days of transport of the contaminated soils to the designated facility. Normally, the consultant takes on the task of acting as a representative for the generator; i.e., he/she will sample the soil, have the appropriate analysis performed, fill out the Bill of Lading, submit it to the Department for authorization, and then hand it off to the transporter. It would surely expedite matters if the consultant could carry this process one step further and ensure that the completed forms are returned to the Department. All too often, the completed forms are not returned, and the Department will have no official record that the soils have been removed.

The Department noted the problems encountered with the use of this policy, and realized by the year's end that some changes were warranted. On June 30, 1989, a revised version of this policy was introduced. The policy now in effect is Policy # WSC-89-001, entitled "Management Procedures for Excavated Soils Contaminated with Virgin Petroleum Oils." The changes offer additional flexibility as well as clarification, where needed, and include the following:

Authorization

The transport or disposal of soils to an approved facility requires prior authorization from the BWSC in the appropriate Departmental regional office. There was some confusion with regard to exactly what was considered the appropriate regional office. The appropriate regional office is determined by where the soil is coming from and where it is going; i.e., if the contaminated soil is generated in one region and disposed of at a facility located within the same region, that regional office would authorize/sign the Bill of Lading; if the contaminated soil is generated in one region and is then to be disposed of at a facility located in another region, both regional offices would then authorize/sign the Bill of Lading.

Storage

Storage of contaminated soil was only allowed at the site of generation for a period not to exceed four months. The revised policy allows for an extension of the storage period, if specifically authorized by the Department. The other change allows for storage offsite if the site of generation cannot accommodate the stockpile; the offsite property must be owned/operated by the same PRP. An authorized Bill of Lading would have to be used to transport the soil to the offsite storage location.

Bill of Lading Forms

A standard, all-inclusive Bill of Lading has been drafted by the Department to enable its use for all facilities, as opposed to each facility type having their own form. The revised Bill of Lading may now be used to transport contaminated soil/debris to sanitary landfills, asphalt batching/recycling plants, or incinerators/Resource Recovery Facilities.

Recycling Permit

Under the Interim Policy, those soils in excess of 1800 ppm volatiles, or 3000 ppm TPH, were considered to be outside the scope of the policy, and had to be managed in accordance with the requirements of MGL ch. 21C and 310 CMR 30.00. The generator can now have these soils transported to asphalt batching plants, provided the facility to be used has a Class A Recycling Permit, and the soil contaminant levels don't exceed the facility's allowed limits. Naturally, this is done only after authorization from the appropriate regional office.

Sampling

For those locations where sufficient storage space is unavailable, in-situ sampling may be permitted, instead of sampling excavated soils. Prior approval from the Department is required.

On-Site Reuse

There was some confusion regarding treatment of contaminated soils prior to onsite disposal/reuse. While the Interim Policy permitted onsite disposal, as discussed earlier, it was only permitted after the excavated soils were treated to attain the "allowable" contaminant levels. There were no provisions for those soils that initially exhibited those "allowable" contaminant levels prior to treatment. The revised policy provides for onsite reuse, provided the contaminated soils do not exceed the "allowable" levels, with or without treatment.

Analyses

The policy grants the Department the authority to request additional analysis if it deems it necessary. The Department has exercised this right in that it requires EPA Method 8240 performed on contaminated soils excavated from gasoline service/automotive stations and industrial sites. The Department requires this additional analysis per 100 cubic yards of contaminated soil to provide some assurance that chlorinated compounds are not present, given the nature of the site in question. While we have met with some resistance relative to requiring this additional analysis, we feel that it serves as a valuable tool in assuring soils are indeed contaminated with virgin petroleum products, and that there are no surprises.

It is hoped that these changes will facilitate and expedite completion and review of Bill of Lading requests and authorizations. Even so, there will be those who strive to test the system. An example of such is the gross misuse of the policy that occurs from time to time involving the submission of a Bill of Lading authorization request without any previous contact with the BWSC; i.e., no staff member has authorized soil excavation as a remedial action, or is familiar with the location/site in question. Such action could result in serious ramifications for both the consultant and the PRP.

CASE HISTORY (IT CAN WORK!)

During late December 1986, a consultant from XYZ, who was working at a tank pull at Anytown, Massachusetts gas station, was able to convince the Department official that the most environmentally sound manner to deal with the contaminated soil that had been excavated was to line the excavation hole with polyethylene, backfill the contaminated soil into the hole, and place the same liner material over the soil. As the PRP did not want to incur any additional liability or unnecessary expense by disposing of this material as a hazardous waste, and as the Soil Policy had not yet been drafted, an agreement was made to allow the backfilling, and a deadline for submittal of a proposal to address the soil issue was established.

Initially, XYZ had proposed an active soil-venting system for the entire site, utilizing vapor-phase carbon units for air controls. A vacuum would be applied to the nine manifolded vapor well points, and moisture traps and heaters would be used to prepare the air for the carbon treatment. The treated air effluent would then be reinjected to the subsurface. This process, along with the dual-phase recovery system that had been in operation for the groundwater contamination, would continue until the cleanup objectives for the site were obtained. Soil samples would be collected prior to the construction of the soil vapor system, and once construction was complete and remediation had commenced, samples would be collected and analyzed on a regular basis.

The Department was not adverse to this conceptual proposal, but did require some additional information. Although a review of the geology seemed to reinforce the hypothesis that this program would be feasible at this site, we required a pilot study to determine both the effectiveness of this technology at this site, and the appropriate placement for vapor wells. In addition, the Department required the proposed well screens depth and length be approved by the Department prior to the installations. As there had been a cleanup level established for the soil at the site, we required sampling ports at the vapor well points, so that when the time came to evaluate the remedial effectiveness, it would be possible to assess contamination levels at each individual station across the site, rather than sampling the influent port to the canisters, which would actually measure the cumulated effect and average concentration. The carbon canisters were required to be changed once the removal efficiency was less than 90%. Finally, the Department required engineering specifications for the project, and also informed the PRP that there might be a need to install additional well points. The need for such would be evaluated once the air "pump" test results were submitted.

An air "pump" test was performed, and the expected radius of influence for each well was calculated to be 35 feet at a given pressure. Based, to some degree, on the high costs of the carbon changeouts, and with an eye toward limiting PRP future liability, XYZ proposed to treat the collected vapors through a catalytic combustion furnace. Acceptable vapor well screens had been installed, and an engineering diagram for the proposed system had been submitted. A careful review of this diagram revealed that if the system's lower explosive level (LEL) meter sensed high explosive levels prior to vapors entering the catalytic combustion furnace, a vent would open, and some untreated gasoline vapors would be diverted to the ambient air.

As the release of untreated vapors from a remedial action to the ambient air would not be acceptable to DAQC, BWSC, or the manufacturer, a partial redesign of the proposed system was required. In addition, based upon the calculated radius of influence, the BWSC determined there would be the need to install offsite vapor wells, to totally address the contamination attributed to the gasoline release.

The manufacturer was able to satisfactorily redesign the system, and did away with the potential release by eliminating the vent, and by adding an automatic control which would shut down the system once a predetermined percentage of the LEL was reached. With this change, we all thought that the project was ready to be constructed, until the discovery of one final glitch. Evidently, the soil that started all of this in motion, the encapsulated gasoline contaminated soil still in the former tank pit, was inadvertently excluded in the redesign. As the soil liners were to be left intact, and there were no vapor wells proposed for the excavated portion of the site, the effectiveness of the proposed remediation would be minimal for that area of the site.

Because the Department felt it was much more important to get this remedial action in motion than it was to wait for the submission of another scope of work for an additional well vapor point, we gave approval to start up the system, with the understanding that an additional vapor well point would be installed in the encapsulated soil, and this point would be included in the manifold system. The proposed soil venting system, which was the second catalytic combustion furnace used for site remediation in the Northeast, was running 17 months after the first conceptual proposal was submitted.

The purpose of this account is to try to demonstrate that the Department is willing to be flexible, to work with PRPs and consultants, and that our main concerns include that the remedial system(s) be protective of public health, welfare, safety, and the environment, but also that the system be designed so that remedial actions conducted at a site are effective and, hopefully, final. XYZ has been able to design two very effective systems at the case study site. On an average, approximately 800,000 gallons of water per month are treated via air stripping, with a 90% to 95% removal efficiency. During the first two weeks of the soil vacuum system operation, over 290 pounds of BTEX were incinerated, with a destruction efficiency of 95% to 100%. The combination of the two systems will definitely shorten the time required to achieve an acceptable status for this site.

FINAL NOTE

It is recognized that the most frequent problem encountered by the Department when reviewing reports is that they are often submitted incomplete, indicating a lack of understanding of current environmental regulations. If this were somehow remedied, this would alleviate much frustration and delay in processing a determination on the part of the Department. The MCP now outlines the basic steps that must be taken during each phase of a site remediation. Insisting the Department is asking for unnecessary information can no longer be justified.

It is our hope that with understanding on the part of the PRPs and their consultants, regulators can be seen as part of the machinery that makes the whole process function, and not a wrench in the works. When consultants are able to understand the priorities of regulators, and state regulators can be forthcoming with their concerns regarding specific sites, reports can then be prepared that will address not only the needs of the PRP, but also those of the state, thereby conserving resources for both parties.

CHAPTER 28

An Analytical Manual for Petroleum and Gasoline Products for New Jersey's Environmental Program

Michael W. Miller and **Dennis M. Stainken,** New Jersey Department of Environmental Protection, Office of Quality Assurance, Trenton, New Jersey

During the 1960s and 1970s, major oil spills motivated many scientific studies on the fate and effect of petroleum and petroleum products. The development of analytical methods was sponsored by various agencies including the U.S. Environmental Protection Agency (EPA), United States Coast Guard, United States Department of Interior, and the U.S. Public Health Services. The methods utilized gas chromatography, infrared spectroscopy, and fluorescence spectroscopy.[1-7] The major effort of these methods was to identify the type of product, and establish the responsible party.

Many of these methods have been published in various formats by American Society for Testing and Materials (ASTM) and American Public Health Association (APHA).[1,2] These methods are either limited in scope and application or categorized as qualitative. Meanwhile, during the 1980s needs have increased for the quantitative analyses of petroleum and gasoline contamination of water and soil. National programs affected include Underground Storage Tanks, Superfund, Resource Conservation and Recovery Act, potable water, and groundwater monitoring.

Within New Jersey, the Department of Environmental Protection (DEP) administers several regulatory programs concerning petroleum products. Consequently, it has been unclear at times which methods were appropriate or required for a given program. In addition, many of the methods do not contain adequate quality control procedures. In the absence of regulatory/mandatory analytical procedures, the Department's Office of Quality Assurance has established a Department *Analytical Chemistry Manual for Petroleum Products in the Environment.*[8] The intent was to provide the Department with a manual to better achieve the regulatory goals of the Department and to be utilized by the Department's Laboratory Certification Program and Contract Laboratory Program. Development of the manual has been facilitated by the establishment of a Petroleum Products Methods Committee which evaluated the use of petroleum product analyses in DEP regulatory programs and the effectiveness of the individual methods.

The effort to establish the manual was divided into three phases. The regulatory objective and analytical needs of each program and definitions of petroleum products were determined in Phase I. In Phase II, an extensive review of currently existing or tentative methods was conducted. The methods/standards reviewed included those of federal and state agencies as well as ASTM, APHA, API (American Petroleum Institute), etc.[9] The results of Phases I and II were presented in the *Proceedings of Symposium on Waste Testing and Quality Assurance.*[10] Phase III consisted of the selection and editing of methods for New Jersey's Department of Environmental Protection Programs. These methods comprise the draft "State Manual."

The eight regulatory programs identified in DEP which may involve petroleum product analyses include Emergency Response, Environmental Claims Responsibility Act, New Jersey Pollution Discharge Elimination System, Resource Conservation and Recovery Act, Superfund, Underground Storage Tanks, New Jersey Spill Fund, and Residuals Management.

Petroleum products are defined in four New Jersey regulations:

1. N.J.A.C. 7:1E-1.3 Environmental Cleanup Responsibility Act Regulations. This act requires that an industrial property on which hazardous materials were stored, used, or manufactured be environmentally clean before the ownership of the property can change hands.
2. N.J.A.C. 7:20-1.4, Waste Oil. This act regulates the disposal of waste oil.
3. N.J.A.C. 7:26-8-13, Hazardous Waste. This act defines Hazardous Waste and regulates the transportation and disposal of the waste.
4. 58.10-23.116 Spill Compensation and Control Act. This act is similar to the federal Superfund act.

The definitions include all liquid petroleum hydrocarbons but do not mention petroleum products which contain nitrogen, sulfur, and oxygen in the molecule. The semisolid and solid asphalt materials are not included in the definitions.

Many of the programs have similar objectives. An itemization of the objectives is contained in Table 1. Determining the Total Petroleum Product Concentration in soil and water is a primary objective. The most common petroleum product contaminants in the environment are gasoline and #2 fuel oil. Obtaining a site survey for petroleum product contamination at moderate cost and in a timely manner is the second most stated objective.

Table 1. Program Objectives for Petroleum Product Analysis

Analytical Objectives	No. of Programs Affected	Emergency Response	ECRA	NJPDES	RCRA	CERCLA	UST	N. J. Spillfund	Residual Mngmt
Total Oil and Grease	3			X	X				X
Total Petroleum Products	6		X	X	X	X	X	X	
Low Cost Petroleum Analysis	2		X	X			X		
Site Survey Method	5	X	X		X	X	X		
Test for Gasoline (BETX)	5	X	X			X		X	X
Test for #2 Fuel Oil	6	X	X	X		X	X	X	
Determine Specific Priority Pollutants	4			X	X	X		X	
Responsible Party	4	X	X					X	
Low Detection Limits for Groundwater, Petroleum Products	2		X	X					
Test Kerosene—Jet Fuel	2		X			X			
Total Extraction from Soil of Petroleum Products	2		X				X		
Analysis for Petroleum Product Residues	3		X			X	X		
Biological Toxicity	2			X		X			
Composition of Waste Oil	1				X				
Marker Compounds Gasoline	2				X		X		
Hazardous Waste Characterization	1				X				
Field Instrument for Survey Work	4	X	X		X		X		
Define Contamination Plume	4		X		X	X	X		

Table 2 compares the current analytical methods available to achieve program objectives. To be considered, an analytical method has to be issued by a federal agency, state agency, or issued by ASTM or APHA. The methods currently in use do not totally satisfy the objectives of the New Jersey programs. For example, the GC and GC/MS EPA methods were designed to analyze priority pollutants for the NPDES program, not petroleum products. An example of this is the use of volatile organic compounds analyses, such as EPA 602, 624, 503.1, and 8240, to determine petroleum products.[5,6] The GC methods do not detect saturated organic compounds of petroleum. The higher boiling point volatile compounds will not clear the GC column during the analyses and will elute in the next sample analysis, contaminating the system. In addition to the GC problems, the mass spectrometer can be contaminated with the unresolved hydrocarbons. Also the nontarget-saturated organic compounds can be identified by class but not quantified by the mass spectrometer.

Another example of methods not fulfilling objectives is the Total Petroleum Hydrocarbon Methods which were not designed to fulfill the objectives of most of the programs in Table 1 and 2. If the sample contains gasoline, as much as 50% of the material is lost during sample preparation. Also, the method does not detect nonalkylated aromatics. Therefore, the method frequently yields low values. The method was designed to determine fuel oil #2 and heavier petroleum products. The need for an analytical method to determine total petroleum hydrocarbons in soil has resulted in the combination of the Soxhlet extraction from SW846-3540 and the analysis from EPA 418.1 by the various contract laboratories.[1,6] However, the soils method is not mandated, and each laboratory uses slightly different procedures. In addition, the quality control procedures are not specified, resulting in data of unknown quality.

METHODOLOGY

Methods are selected for the manual to satisfy the objectives of the eight mandated programs managed by the NJDEP. The manual has three major sections: Users Guide, Laboratory Methods, and Field Methods. Table 3 is an index of the draft laboratory methods. Qualitative, semiquantitative, and quantitative methods for aqueous and solid matrices are included. Each method is designed to be a complete standard operating procedure. The organization of the methods follows the format of the EPA 600 series methods.

A general outline for the analytical methods is presented in Table 4.[5] The introduction is designed to provide the site manager or technical advisor with specific information which will allow the choice of the appropriate analytical methods. Methods Identification and Scope sections define the analytes or parameters determined, the separation procedure, and the instrumentation.

New Jersey and federal regulations for certain programs cite specific methods which are listed in Table 5. All laboratories submitting data under these programs to the state must be New Jersey certified. Methods for the analysis of petroleum

Table 2. Current Analytical Methods Available To Achieve Program Objectives

Objectives/Methods	EPA 413.1	EPA 418.1	Soxhlet 418.1	EPA 601	602	624	625	503.1	EPA SW846 9071	Soil Gas PID	ASTM D3328	ASTM D3327	SW846 5030	Cal. WCB	EPA 239.2	Flor. 17-70.008	U.S. Coast Guard	ASTM D3650
Total Oil and Grease	X								X									
Total Petroleum Products		X	X											X				
Low Cost Analysis Petr.		X	X															
Site Survey Method		X	X			X												
Test for Gasoline (BETX)					X	X		X					X	X	X	X		
Test for #2 Fuel Oil		X	X			X	X											
Test for Kerosene—Jet Fuel					X	X	X						X	X		X		
Test for Petroleum Prod. Residues			X				X											
Responsible Party											X	X			X		X	X
Biological Toxicity						X	X											
Composition—Waste Oil						X	X											
Marker Compound—Gasoline				X	X	X		X						X	X	X		
Marker Compound—#2 Fuel Oil							X							X		X		
Hazardous Waste Characterization			X			X	X											
Determine Specific Priority Pollutants						X	X											
Define Contamination Plume		X			X	X	X			X								
Field Inst. for Survey work										X								

Table 3. Index of Petroleum Product Analytical Methods

NJDEP-OQA #	Ref. #	Title
QAM-001	413.1	Oil and Grease, Total Recoverable, Water, Gravimetric
QAM-002	413.2	Oil and Grease, Total Recoverable, Water, IR
QAM-003	413.3	Oil and Grease, Total Recoverable, Soxhlet, Soil
QAM-004	418.1	Total Recoverable, Petroleum Hydrocarbons, Water, IR
QAM-005	ERT-IR	Total Recoverable, Petroleum Hydrocarbons in Soil, IR
QAM-006	CAL-luft2	Quantitative Determination of Total Medium Weight Petroleum Products in Water
QAM-007	EPA602	Determination of Purgable Fuel Components in Water, GC
QAM-008	CAL602	Dissolved Total Petroleum Products in Water, GC
QAM-009	CAL-luft1	Total Volatile Petroleum Products in Soil, GC
QAM-010	CAL-luft2	Quantitative Determination of Total Medium Weight Petroleum Products in Soil
QAM-011	EPA624	Determination of Volatile Petroleum Products in Water, GC/MS
QAM-012	SW-8240	Determination of Volatile Petroleum Products in Soil, GC/MS
QAM-013	EPA625	Determination of Polynuclear Aromatic Hydrocarbons in Water, GC/MS
QAM-014	SW-8270	Determination of Polynuclear Aromatic Hydrocarbons in Soil, GC/MS
QAM-015	EPA610	Determination of PAH in Petroleum Contaminated Water, Soil and Sediment HPLC
QAM-016	EPA610	Determination of PAH in Petroleum Contaminated Water, Soil and Sediment GC
QAM-017	ASTM3328	Identification and Quantitation of Total Petroleum Products in Water, GC
QAM-018	ASTM3328	Identification and Quantitation of Total Petroleum Products in Soil, GC
QAM-019	SW-8015	Analysis of Oxygen Containing Petroleum Products, GC
QAM-020	SW-3810	Headspace Analysis of Petroleum Products in Water and Soils, GC

products are not covered in the present New Jersey laboratory certification regulations which apply to drinking water and wastewater. The EPA 600 series and 413.1 wastewater methods and EPA 500 series drinking water methods are used by laboratories for petroleum product analyses in water. When laboratories pass certification for the drinking water and wastewater, the assumption is made that the laboratory can apply the methods to petroleum contaminated matrices. The modifications of the cited methods in the Petroleum Methods Manual improve the applicability of the methods.

The sections listing the advantages and disadvantages of the method are a key to using the method in a specific site situation. These sections highlight the specific

Table 4. General Analytical Method Outline

Introduction

Method Identification
- Scope
- Regulatory Citation
- Objectives
- Type of Matrix
- Advantages of Method
- Limitations of Method

Analytical Method
- Reagents
- Apparatus
- Calibration
- Method Quality Control
- Sampling and Storage
- Procedure
- Calculations

Reporting Requirements

References

Table 5. Method Citations

1. ECRA Regulations N.J.A.C. 7:26B
   a. Total Petroleum Hydrocarbons, Aqueous—EPA 418.1.
   b. Gasoline Site—BETX by GC—EPA 602
   c. Heavy Industrial Site—Volatile Organics—GC/MS
   d. Heavy Industrial Site—Semi Volatile Organics—GC/MS
2. NJPDES Regulations N.J.A.C. 7:14A-147
   a. Non Petroleum Oil and Grease, Aqueous—EPA 413.1
   b. Petroleum—Oil and Grease—Aqueous—EPA 418.1
3. Water Pollution Control Act N.J.S.A. 58:10A-1
   a. U.S. EPA Methods Listed in Federal Register 40 CFR 136
   b. Total Petroleum Hydrocarbons—Aqueous 418.1
   c. Oil and Grease—Aqueous 413.1
4. Sludge Quality Assurance—N.J.A.C. 7:14-4, Oil and Grease—Standard Methods—503D, (1).

compounds and classes of compounds that can be determined by the method. The qualitative and quantitative advantages and limitations are also discussed.

The calibration section discusses calibration for specific petroleum compounds, reference petroleum products, and total petroleum products. Gas chromatographic methods are also calibrated with a homologous series of alkanes. This helps to classify the type of petroleum product present. Current methods do not use these compounds for calibration. The laboratory must generate a five point curve and check the curve daily or once in 20 samples. The continuing calibration check should be performed with a standard containing all the parameters of interest at the midpoint of the linear range.

Method quality control is the key to obtaining quality data. The elements of quality control are tabulated in Table 6. The "Initial Demonstration of Laboratory Capability" is the determination of the precision and accuracy of the method. This demonstration should be performed each time the method is modified or the analyst is changed. For aqueous methods, a QC (quality control) check sample is spiked into reagent water. In soil methods, a QC check sample is spiked into reagent silica. The multiple check samples are carried through the entire method.

Table 6. Elements of Quality Control

1. Initial Demonstration of Laboratory Capability
2. Daily Instrument Quality Control Check Standard
3. Daily Method Blank
4. Trip and Field Blanks for Aqueous Matrix
5. Determine Method Detection Limit
6. Weekly Analysis of a Low Level Sample to Confirm MDL
7. Matrix Spike and Matrix Spike Duplicate
8. Determine Surrogate Recovery
9. Maintain Control Charts

The daily analysis of an instrument QC check standard is a confirmation of the calibration standards. This analysis is very important in GC/MS and GC methods. The extra analysis takes 40 minutes to an hour. If the calibration sample is in error, all the sample analyses would be in error: a waste of 6 to 8 hours.

The blanks are items 3 and 4 of Table 6. The method blank checks the cleanliness of the laboratory. Contamination of sample with laboratory solvents is very common. A site cleanup could be triggered by contamination of the sample. Methylene chloride and tetrachloroethylene are industrial solvents that can be found with petroleum contamination and are common in the laboratory. Clean field blanks confirm the cleanliness of the sampling equipment. Clean trip blanks confirm that the sample containers and samples were not exposed to contamination. There have been cases of samples being transported with gasoline cans or gasoline generators and the sample analysis showed the presence of benzene and toluene.

The method detection limit (MDL) is dependent on the analyst and the instrumentation. The MDL is a qualitative number. At the MDL there is 95% confidence that the material is present. The actual MDL is important in a method, from a user's perspective. In many situations the presence or absence of a contaminant is important. A low level of an analyte is useful in tracking the leading edge of a contamination plume.

An important quality control check is the matrix spike and matrix spike duplicate analyses; they establish the accuracy and the precision of the method in a given sample matrix. The matrix spike also indicates the presence of interference.

Most methods involve a sample extraction or analyte separation. The surrogate is an independent means of evaluating the recovery. The surrogate compound is not found in the sample and performs similarly to the analytes in the sample.

Control charts for surrogate recovery, spike recovery, and duplicate precision help evaluate the laboratory performance of the method over a period of time.

The next item in the General Analytical Method Outline, Table 4, is sampling and sample storage. The quality of an analytical result is dependent upon the sampling procedure, sample preservation, and sample storage. The integrity of volatile components is the most difficult to maintain. Also, biological decomposition of aromatic and unsaturated petroleum hydrocarbons occurs in environmental samples. Therefore, sample preservation, container storage temperature, and holding time are specified in the method.

The Procedures and Calculation sections of the methods are written with a detailed description of each procedural step.

The Reporting Requirements and Deliverables section was not part of the original methods used to compile this draft manual. The data deliverables package from a sampling event is critical to the user. There are many laboratories performing the analysis of petroleum products. Data packages can vary from one sheet of numbers to a box full of data. The data user must define the data quality objectives. The data needs for a site survey, site monitoring, or site closure are different. If a case will end up in litigation, the data needs are extreme. The data user can request a partial report, but the amount of quality control and quality assurance is independent of the data deliverables. The laboratory must retain all QA/QC data.

The components of a complete data package are listed in Table 7. The components which should be part of every report are marked with an asterisk. The specific content depends on the type of analysis; i.e., GC, GC/MS, spectroscopic, or gravimetric. The key parts of the data report are the chain of custody, sample data, calibration, and quality control summary. The entire contents of the data report support the validity of the data.

The sample data cannot be used for enforcement or client defense without a complete Chain of Custody. The Chain of Custody traces the sample through the hand of every individual who touches the sample. The sample is tracked from the time it is taken in the field to the time the analysis is completed. The sample data package provides the information the site manager needs to evaluate the site. The sample data package contains the Data Summary, a list of the analytes found and the values, along with the method detection limits and method blank contamination. The supporting data, chromatograms, spectra, calculations, and GC/MS library proofs are included in this section.

The Methodology Review is a one paragraph description of the analytical method. This helps determine if the correct protocol was employed. The Nonconformance Summary alerts the user to problems which occurred during analysis which may qualify the data; i.e., holding times, interferences, recoveries.

The Quality Control Summary contains the information for evaluating the performance of the analysis. The quality control limits are specified in each method.

Table 7. Reporting Requirements and Deliverables

*1. Chain of Custody Documents

2. Sample Data Package
   *a. Sample Results Summary, Method Blank Results, Method Detection Limits
   *b. Sample Chromatograms or Spectra
   c. Quantitation Reports
   d. GC/MS Library Proofs

*3. Methodology Review

*4. Non-conformance Summary Report

5. Quality Control Summary
   *a. Surrogate Recovery Summary
   b. GC/MS Tuning Summary
   *c. Matrix Spike/Matrix Spike Duplicate
   d. Reagent Blank Summary

6. Standard Data Summary
   *a. Initial Calibration Data
   *b. Continuing Calibration Data
   c. Chromatograms or Spectra for Continuing Calibration

7. Raw QC Data Package
   a. QC Chromatograms or Spectra
   b. Quantitation Reports

The laboratory must experimentally determine the QC limits for every method the laboratory uses. For the sample data to be valid, the surrogate and matrix spike recoveries, relative percent difference between duplicates, method blanks, reagent blanks, and instrument QC must be analyzed at the interval specified, and meet method specifications.

The Department Petroleum Methods Manual currently contains 20 draft methods.[8] In this chapter, two methods will be discussed as examples: "Total Recoverable Petroleum Hydrocarbons in Soil" and "Dissolved Total Petroleum Fuel Hydrocarbons in Water."

"Total Recoverable Petroleum Hydrocarbons in Soil" is a 1,1,2 trichloro 1,2,2 trifluoro ethane (Freon) extraction and infrared spectroscopy analysis based on the EPA 418.1 method. Currently, there is no EPA method for soils. Commercial laboratories combine a Soxhlet extraction with EPA 418.1 and apply the method to #2 fuel oil and heavier molecular weight petroleum products.[4,6]

The NJDEP method is applicable to gasoline, kerosene, and fuel oils #2–6. This is a survey method which can limit the number of samples that must be analyzed by a quantitative method to evaluate a site. The major advantage is that the method is inexpensive and gives a good estimate of total petroleum hydrocarbons. The main disadvantage is that the sensitivity for aromatic and substituted aromatic compounds such as benzene, toluene, xylene, and naphthalene is poor. Light fuels can contain 12% to 20% of these aromatics.

Samples are taken in two-ounce jars with a minimum of air space, sealed, chilled to 4°C and analyzed within 7 *days*. The DEP method requires microextraction of 30 grams of soil and 30 ml of HCL/NaCl solution, with 3–20 mL portions of Freon in a sealed 125 mL flask on a shaker table. The nonpetroleum oils are removed by silica gel treatment. The extract is analyzed at 2930 cm-l by infrared spectroscopy.[11]

The method requires an initial five point calibration curve and a daily one point verification of the curve with reference petroleum samples. Table 8 lists the quality control required.

Table 8. Total Recoverable Petroleum Hydrocarbons in Soil, Quality Control

1. Initial determination of precision and accuracy using reagent silica.
2. Daily analyses of reagent water method blanks.
3. Monthly QC check sample (30g silica).
4. Matrix spike/matrix spike duplicate sample spiked with reference oil, #2 fuel oil.
5. Determination of method detection limit and monthly confirmation of method detection limit using a low level spike.
6. Control Charts.

This method has been used for the past 10 years by the U.S. EPA National Emergency Response Branch.[8] DeAngelis has published a similar method in the *Manual of Sampling and Analytical Methods for Petroleum Hydrocarbons in Ground Water and Soil.*[12] Recovery of #2 fuel oil spiked into soil is equivalent to the recovery from the Soxhlet extraction of soil. The recovery of gasoline is better by sealed microextraction (75% to 85%).

The NJDEP method, Dissolved Total Petroleum Fuel Hydrocarbons in Water, is adapted with modifications from the California *Leaking Underground Fuel Tanks Manual* and U.S. EPA method 602.[6,12] This is a purge and trap, gas chromatography method for the quantitative determination of dissolved petroleum fuel hydrocarbons in groundwater, wastewater, potable water, and surface water. The gas chromatograph is equipped with a column optimized for petroleum and a photoionization detector in series with a flame ionization detector. Quantitative values are obtained from the photoionization detector response for benzene, ethylbenzene, toluene, and xylenes (BETX). Methyl tert-butylether (MTBE) can also be determined. Total petroleum hydrocarbons are determined by integration of the total flame ionization detector response when the system is calibrated with reference standards.

The method gives a more accurate representation of dissolved petroleum products in water than USEPA 418.1 because aromatic compounds are detected. The disadvantage of the method is that the gas chromatograph conditions and columns are not designed to resolve the high molecular weight hydrocarbons present in fuel oil #2. An extraction method should be used for fuel oil such as NJDEP method, Quantitative Determination of Total Medium Weight Petroleum Product Concentration in Soil.[10]

The method requires an initial five-point calibration for BETX, MTBE, and total petroleum hydrocarbons. An internal standard method produces the most precise quantitative from the PID response; therefore, internal standardization is used to determine the BETX and MTBE concentrations. Because the FID response chromatogram is crowded with peaks, external standardization is used to determine total petroleum hydrocarbon concentration. A daily single-point confirmation of each type of calibration is also required.

Table 9 lists the quality control required. The requirements are similar to Table 8. The surrogate is added for the purge and trap method as a check of the operation of the purge and trap unit, independently of the internal standard.

Table 9. Dissolved Total Petroleum Fuel Hydrocarbons in Water, Quality Control

1. Initial determination of precision and accuracy using aqueous samples spiked separately with BETX-MTBE and synthetic fuel.
2. Daily analysis of reagent water method blanks.
3. Daily QC check samples for BETX-MTBE and synthetic fuel.
4. Matrix spike/matrix spike duplicate sample spiked with reference oil, #2 fuel oil.
5. Surrogate spike for all samples, blanks and standards.
6. Determination of the method detection limits and monthly analysis of a low level spike confirming the detections limits is required.
7. Control Charts.

A key section of the manual is a user's guide to help in the selection of the appropriate methods. The evaluation of a site for chemical contamination can be divided into five general areas: Site Survey, Site Contamination Confirmation, Remedial Investigation and Feasibility, Monitoring and Cleanup, and Closure. During the Site Survey, field analytical or laboratory survey methods are used. In the field, soil gas analysis is a quick inexpensive method of determining whether the site is contaminated by volatile petroleum components. The method is not effective for fuel oil #2. Soil gas can be determined using an organic vapor analyzer equipped with FID or PID detectors. A portable GC can also be used.[13] The soil gas methods will be added to the "Draft Manual" at a later date. Headspace analysis with a portable GC will detect gasoline and fuel oil #2, Manual Method QAM-020.[8] Samples sent to the laboratory for survey analyses can be analyzed by headspace GC or by extraction and infrared spectroscopy. Headspace methods are relatively inexpensive and are semiquantitative.

During the confirmation of site contamination and the Remedial Investigation/Feasibility Study, detailed information is required. Quantitation of petroleum products can be obtained by using the following methods: BETX by GC-PID, Total Petroleum Hydrocarbons by GC-FID, Total Extractable Polynuclear Aromatic Hydrocarbons (PAH) by HPLO, and specific priority pollutants by GC/MS. Petroleum product identification is obtained by using GC, IR, and GC/MS fingerprinting methods.

During the monitoring and cleanup stage qualitative and semiquantitative data are needed. The same analytical methods used in the survey stage can be used at sites with high levels of contamination. When low levels are required, the following methods can be used: BETX by GC-PID, PAH by HPLC, Total Petroleum Hydrocarbons by GC-FID.

When a site closure is being studied, low level quantitative results are required. The following methods can be used: BETX by GC-PID, PAH by HPLC, Total Petroleum Hydrocarbons by GC-FID, specific priority pollutant compounds by GC/MS.

In order to help field personnel or project managers select methods to meet the analytical objectives of their projects, "decision trees" were designed. Figure 1 and Figure 2 are two examples of the several "decision trees" used in the "Draft

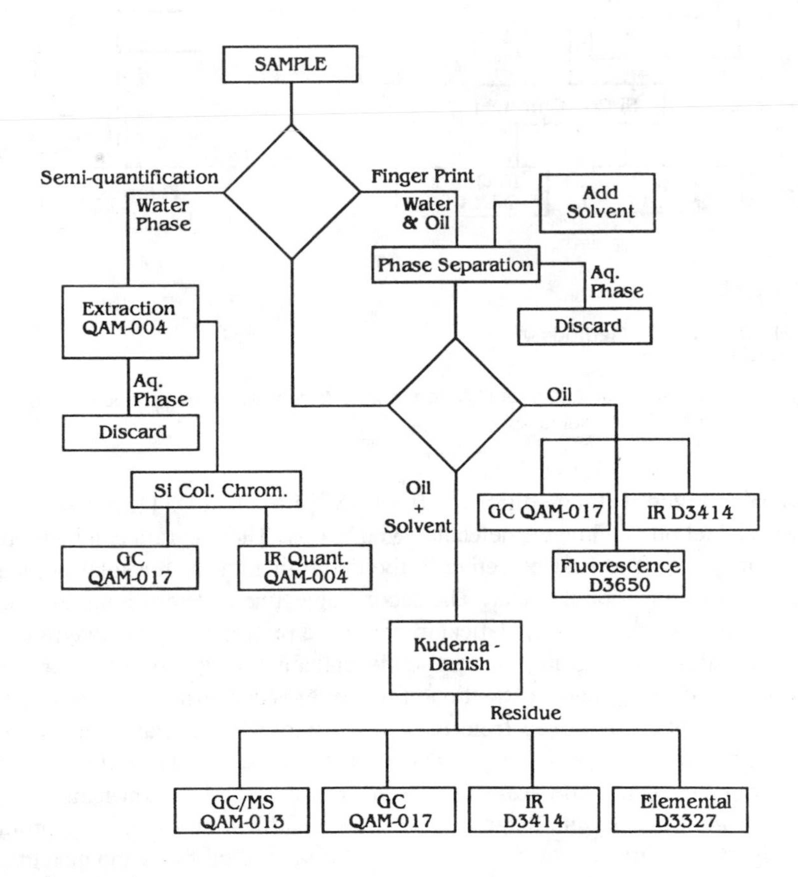

Figure 1. Identification and quantification of waterborne petroleum products; #2 fuel and heavier molecular weight oils.

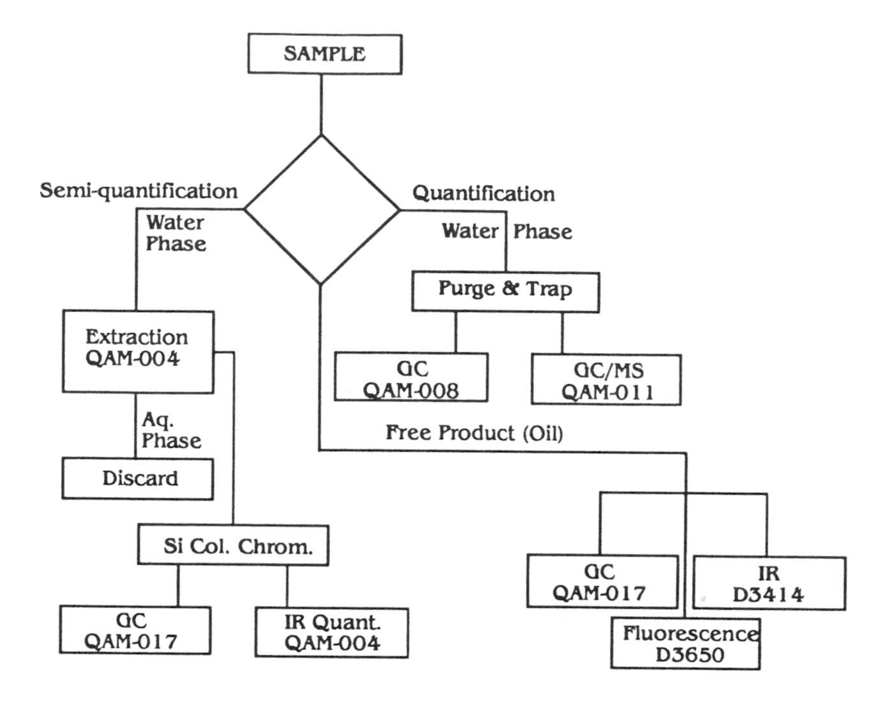

QAM= NJDEP-OQA Methods
D= ASTM Methods

Figure 2. Identification and quantification of waterborne petroleum products; gasoline, jet fuel, diesel, solvents.

Manual.'' Figure 1 is a modification from ASTM Method D3415.[2] The diagram is for #2 fuel oil and higher molecular weight oils. The diagram can be used for two objectives. The first objective is the determination of the total petroleum product contamination in water. The second objective is determining the source of the contamination. The analytical methods are referenced in the diagram. The left hand side of the diagram is for the semiquantitative analysis of the water soluble phase. A separatory funnel extraction of the water is performed with Freon. Non-petroleum oils are removed from the extract with silica gel and the extract is analyzed by infrared spectroscopy. The extract can also be analyzed by gas chromatography to determine aromatic content and an identification pattern.

The central path of the diagram is for the identification of floating product. The right side of the diagram is for the identification of oil films and quantitation of dissolved product. The identification can be achieved by gas chromatography, infrared spectroscopy, or fluorescence. Quantitation is accomplished by gas chromatography or gas chromatography/mass spectrometry. GC/MS is preferred for the positive identification of specific priority pollutant compounds.

A decision tree for the identification and quantification of light petroleum fuels in water is presented in Figure 2. The left hand side of the diagram is the survey method for total petroleum content. The semimicro sealed liquid-liquid extraction with Freon which retains volatile components is used. The nonpetroleum oils are removed by a silica gel column. The extract is analyzed by infrared spectroscopy. Gas chromatography can be used to obtain the aromatics and a pattern for identification.

The center path is for the identifications of the floating product. The right hand side is for the quantification and identification of specific compounds dissolved in water.

CONCLUSION

The analytical objectives of the DEP's programs that are concerned with petroleum product contamination are met by the manual. Problems with current methods are addressed and modifications provided. The methods specify quality control procedures and reporting requirements. This helps the manual user obtain analytical consistency between laboratories and improves data quality.

The NJDEP Office of Quality Assurance is responsible for New Jersey Laboratory Certification of drinking water and wastewater analytical methods. New regulations are being prepared which will require laboratory certification for sludge, solid waste, and petroleum product analyses. The Analytical Manual for Petroleum Products in the Environment will contain the regulatory methods which will be certified for petroleum product analyses.

Several steps need to be taken before the draft manual is issued. The methods in the manual should be formally validated. The methods and variations of these methods have been used by various laboratories in the past. Precision, accuracy, and detection limit data should be collected and evaluated for these methods using petroleum contaminated soil and water matrices. Also soil and water performance evaluation samples should be developed containing petroleum contamination and the field method section should be completed.

ACKNOWLEDGMENTS

Please note that the interpretations and opinions expressed in this chapter are those of the authors and should not be construed as official policy of the New Jersey Department of Environmental Protection.

The authors would like to thank the members of the Petroleum Products Analysis Committee for their contributions to the manual, and Robert Fischer, Anita Kopera, Floyd Genicola, Sharon Haas, Henry Hoffman, and Marc Ferko for the review of the manual and this chapter. The efforts of Karen Funari, Keshia Brown, Cecelia Durant and Eloise Wright in typing the manual and manuscript are appreciated.

REFERENCES

1. *Standard Methods for Examination of Water and Waste Water,* 16th ed. American Public Health Association, American Water Works Association, Water Pollution Control Federation, Washington DC, 1985.
2. "Water (II)," Volume 11.02. *Annual Book of ASTM Methods,* American Society for Testing and Materials, Philadelphia, PA, 1984.
3. *Oil Spill Identification System,* U.S. Coast Guard R & D Center, GC-D-52-77, 1977.
4. "Methods for Chemical Analyses of Water and Waste Water," Revised 1983, EPA 600-14-79-020. U.S. Environmental Protection Agency, 1979.
5. "Guidelines Establishing Test Procedures for the Analyses of Pollutants Under the Clean Water Act: Final Rule and Interim Final Rule and Proposed Rule" *Federal Register* 40 CFR Part 136. U.S. Environmental Protection Agency, 1984.
6. *Test Methods for Evaluating Solid Waste,* 3rd Edition, Office of Solid Waste Publication SW846. U.S. Environmental Protection Agency, 1986.
7. *Federal Register,* 40 CFR parts 141 and 142. U.S. Environmental Protection Agency, 1987.
8. Kane, M. *Manual of Sampling and Analytical Methods for Petroleum Hydrocarbons in Ground Water and Soil,* American Petroleum Institute, Washington, DC, Publication #4449, 1987.
9. Stainken, D., and M. Miller. "Establishing an Analytical Manual for Petroleum and Gasoline Products for New Jersey's Environmental Program," *Proceedings of Symposium on Waste Testing and Quality Assurance,* Vol. II, U.S.E.P.A. Office of Solid Waste, Washington DC, 1988.
10. Miller, M., and D. Stainken. *Analytical Chemistry Manual for the Petroleum Products in the Environment,* N.J. DEP, Document OQA-QAD-002, 1989 (unpublished).
11. Remeta, D. P., and M. Gruenfeld. "Emergency Response Analytical Methods for Use on Board Mobile Laboratories," Internal Communication, USEPA Hazardous Waste Engineering Research Laboratory, Release Control Branch, Edison, NJ, 1987.
12. DeAngelis, D. "Quantitative Determination of Hydrocarbons in Soil (Extraction Infrared Adsorption Method)," *Manual of Sampling and Analytical Method for Petroleum Hydrocarbons in Groundwater and Soil.* American Petroleum Institute, Washington, DC, Publication #4449, 1987, p.167.
13. Simmons, B., Ed. *Leaking Underground Fuel Tank Underground Storage Tank Closure,* California Leaking Underground Fuel Tank Task Force, State Water Resources Control Board, Sacramento, CA, 1989.

Glossary of Acronyms

ADEM	Alabama Department of Environmental Management
ADEQ	Arizona Department of Environmental Quality
ADI	acceptable daily intake
AERIS	Aid for Evaluating the Redevelopment of Industrial Sites
AFESC	Air Force Engineering and Services Center
ANOVA	analysis of variance
API	American Petroleum Institute
ATSDR	Agency for Toxic Substances and Disease Registry
BTEX	benzene, toluene, ethylbenzene, total xylenes
BWSC	Bureau of Waste Site Cleanup
CAL	corrective action limit
CDC	Center for Disease Control
CERCLA	Comprehensive Environmental Response, Compensation and Liability Act
CNS	central nervous system
COT	Cleanup Objectives Team
CPA	Canadian Petroleum Association
CPRC	Coordinated Permit Review Committee
DAQC	Division of Air Quality Control
DEC	Department of Environmental Conservation
DEP	Department of Environmental Protection
DERA	Defense Environmental Restoration Account
DOD	Department of Defense
DOH	Department of Health
DOHS	Department of Health Services
ECL	electrochemiluminescence
ECRA	Environmental Claims Responsibility Act (NJ)
EDI	Early Detection Incentive
EIL	environmental impairment liability
EOR	enhanced oil recovery
EPA	Environmental Protection Agency
FDER	Florida Department of Environmental Regulation
FID	flame ionization detector
GC	gas chromatograph
GC/MS	gas chromatography-mass spectrometry
HI	hazard index
HRGC/MS	high resolution gas chromatography-mass spectrometry
IARC	International Agency for Research on Cancer
IEPA	Illinois Environmental Protection Agency

IR	infrared
IRA	initial remedial action
IRP	Installation Restoration Program
IRTCC	Installation Restoration Technology Coordinating Committee
IS	indicator score
LED	light emitting diode
LEL	lower explosive level
LTBI	locations to be investigated
LUFT	leaking underground fuel tank
MCL	maximum contaminant level
MCLG	maximum contaminant level goal
MCP	Massachusetts Contingency Plan
MDL	method detection limit
MPCA	Minnesota Pollution Control Agency
MTBE	methyl tertiary butyl ether
MVSS	moderate vacuum steam stripper
NJDEP	New Jersey Department of Environmental Protection
NJPDES	New Jersey Pollution Discharge Elimination System
NRC	National Research Council
NTP	National Toxicology Program
ODASD(E)	Office of the Deputy Assistant Secretary of Defense (Environmental)
OM	organic material
OVA	organic vapor analyzer
PAA	polycyclic aromatic amine
PACE	Petroleum Association for Conservation of the Canadian Environment
PAH	polynuclear aromatic hydrocarbons
PCS	petroleum hydrocarbon contaminated soils
PID	photo ionization detector
PLIA	Pollution Liability Insurance Association
PPLV	preliminary pollutant limit value
PRDF	Petroleum Release Decision Framework
PRP	potentially responsible party
QC	quality control
RAFT	Risk Assessment/Fate and Transport (Modeling System)
RCRA	Resource Conservation and Recovery Act
R&D	research and development
RITZ	Regulatory and Investigative Treatment Zone
RP	responsible party
RSC	relative source contribution
RWQCB	Regional Water Quality Control Board (Los Angeles)
SARA	Superfund Amendments and Reauthorization Act of 1986
SBIR	Small Business Innovation Research
SD	standard deviation
SEM	standard error of the mean

SVE	soil vapor extraction
SVO	semivolatile
TCDD	2,3,7,8-tetra-chlorodibenzo-p-dioxin
TCE	trichloroethylene
TCLP	toxicity characteristic leaching procedure
TDH	Texas Department of Health
TPH	total petroleum hydrocarbons
TRPH	total recoverable petroleum hydrocarbon
UART	universal asynchronous receiver and transmitter
UCR	unit cancer risk
UCS	unconfined compressive strength
UST	underground storage tank
VO	volatile
VOC	volatile organic compound

List of Contributors

Abdel-Rahman, Mohamed S., Department of Pharmacology and Toxicology, New Jersey Medical School, University of Medicine and Dentistry of New Jersey, 185 South Orange Avenue, Newark, NJ 07103-2714

Barkach, John, The Dragun Corporation, 3240 Coolidge Highway, Berkley, MI 48072-1634

Barnes, Ramon M., Department of Chemistry, University of Massachusetts, Amherst, MA 01003

Baugh, Ann L., Unocal Corporation, P.O. Box 7600, Los Angeles, CA 90051

Bell, Charles E., Environmental Health Sciences Program, School of Public Health, University of Massachusetts, Amherst, MA 01003

Bishop, Mark, New England Testing Lab, North Providence, RI 02904

Block, Robert N., Remediation Technologies, Inc., 9 Pond Lane, Concord, MA 01742

Boyes, Stephen R., GeoSolutions, Inc., Gainesville, FL 32602

Bulman, Terri L., Campbell Environmental Ltd., Suite 93 Havelock Mall, West Perth, WA 6005 Australia

Byers, Dallas L., Shell Development, Westhollow Research Center, P.O. Box 1380, Houston, TX 77251-1380

Calabrese, Edward J., Environmental Health Sciences Program, School of Public Health, University of Massachusetts, Amherst, MA 01003

Carter, Chebryll J., Community Health Branch, Division of Health Assessment and Consultation, Agency for Toxic Substances and Disease Registry, Public Health Service, U.S. Department of Health and Human Services, Atlanta, GA 30333

Chen, David H., American Petroleum Institute, 1220 L Street, NW, Washington, DC 20005

Clark, Thomas P., Remediation Technologies, Inc., 9 Pond Lane, Concord, MA 01742

Clendending, L. Denise, Chevron Oil Field Research Company, La Habra, CA 90633

Cliff, Bruce L., VAPEX Environmental Technologies, 480 Neponset Street, Canton, MA 02021

Cobb, Brian K., Environmental Affairs Department, Tenneco, Inc., Building 400, Suite 106, 1395 S. Marietta Highway, Marietta, GA 30067 (NOTE: Work reported in Chapter 26 was performed while employed by Florida Department of Environmental Regulation, Bureau of Waste Cleanup.)

Dean, Wm. Gordon, Cherokee Groundwater Consultants, Inc., Tallahassee, FL 32301 (NOTE: Work reported in Chapter 26 was performed while employed by Florida Department of Environmental Regulation, Bureau of Waste Cleanup.)

Deans, John R., Box 600/95 Main St., Cape Cod Research, Inc., Buzzards Bay, MA 02532

Denahan, Barbara J., Environmental Science & Engineering, Inc., P.O. Box 1703, Gainesville, FL 32602

Denahan, Stephen A., Environmental Science & Engineering, Inc., P.O. Box 1703, Gainesville, FL 32602

Derammelaere, Ron, AWD Technologies, 10 W. Orange Avenue, South San Francisco, CA 94080

Dineen, Dennis, McLaren Environmental Engineering, 2855 Pullman Street, Santa Ana, CA 92705

Dixon, Brian G., Box 600/95 Main St., Cape Cod Research, Inc., Buzzards Bay, MA 02532

Dragun, James, The Dragun Corporation, 3240 Coolidge Highway, Berkley, MI 48072-1634

Draney, David, Chevron Corporation, San Francisco, CA 94120

Duffy, James, TreaTek, Inc., Technology Center, 2801 Long Road, Grand Island, New York 14072

Eldridge, Audrey, Massachusetts Department of Environmental Protection, Western Region, State House West, 436 Dwight Street, Springfield, MA 01103

Elliott, William G., Environmental Science & Engineering, Inc., P.O. Box 1703, Gainesville, FL 32602

Fang, H.Y., Department of Civil Engineering, Lehigh University, Bethlehem, PA 18015

Gilbert, Charles E., Environmental Health Sciences Program, School of Public Health, University of Massachusetts, Amherst, MA 01003

Gulledge, William P., Front Royal Group, Inc., 7900 Westpark Drive, Suite A300, McLean, VA 22102

Hartley, William R., Tulane University, School of Public Health & Tropical Medicine, Department of Environmental Health Sciences, 1501 Canal Street, New Orleans, LA 70112

Helgerson, Ron, Lockheed Aeronautical Systems Company, Burbank, CA

Hertz, Marvin B., Shell Development, Westhollow Research Center, P.O. Box 1380, Houston, TX 77251-1380

Hicks, Patrick, McLaren Environmental Engineering, 2855 Pullman Street, Santa Ana, CA 92705

Hijazi, Hazem M., Department of Civil Engineering, Lehigh University, Bethlehem, PA 18015

Hoag, George E., Environmental Research Institute and Department of Civil Engineering, University of Connecticut, Storrs, CT 06269

Holland, James, McLaren Environmental Engineering, 2855 Pullman Street, Santa Ana, CA 92705

Jank, Bruce E., Wastewater Technology Centre, Environment Canada, Burlington, Ontario, Canada, L7R 4A6

Janssen, Patricia L.D., ODASD (E), 206 N. Washington Street, Suite 100, Alexandria, VA 22314

Johnson, Paul C., Shell Development, Westhollow Research Center, P.O. Box 1380, Houston, TX 77251-1380

Kampbell, Don H., U.S. Environmental Protection Agency, P.O. Box 1196, Ada, OK 74820

Kostecki, Paul T., Environmental Health Sciences Program, School of Public Health, University of Massachusetts, Amherst, MA 01003

Lawes, Bernard C., E.I. Du Pont De Nemours & Company. (Inc.), Chemicals and Pigments Department, Chestnut Run Plaza, P.O. Box 80709, Wilmington, DE 19880-0709

Lovegreen, Jon R., Applied Geosciences Inc., 17321 Irvine Boulevard, Tustin, CA 92680

Ludvigsen, Phillip J., Automated Compliance Systems, Inc., 245 Highway 22 West, Bridgewater, NJ 08807

Marley, Michael C., VAPEX Environmental Technologies, 480 Neponset Street, Canton, MA 02021

Mason, Sharon A., The Dragun Corporation, 3240 Coolidge Highway, Berkley, MI 48072-1634

Miller, Michael W., New Jersey Department of Environmental Protection, Office of Quality Assurance, CN027, Trenton, NJ 08625

Nangeroni, Peter E., VAPEX Environmental Technologies, 480 Neponset Street, Canton, MA 02021

Nielsen, Bruce J., Engineering & Services Laboratory, Tyndall Air Force Base, FL 32403

Ohanian, Edward V., United States Environmental Protection Agency, Office of Drinking Water, Washington, DC 20460

Ostendorf, David W., Department of Civil Engineering, University of Massachusetts, Amherst, MA 01003

Pamukcu, Sibel, Department of Civil Engineering, Lehigh University, Bethlehem, PA 18015

Potter, Thomas L., Mass Spectrometry Facility, Chenoweth Laboratory, Massachusetts Agricultural Experiment Station, University of Massachusetts, Amherst, MA 01003

Sanford, John, Box 600/95 Main St., Cape Cod Research, Inc., Buzzards Bay, MA 02532

Scroggins, Rick P., Industrial Programs Branch, Environment Canada, Hull, Quebec

Shepherd, Greg, Southern Pacific Transportation Company, Southern Pacific Building, 1 Market Plaza, San Francisco, CA 94105

Skowronski, Gloria A., Department of Pharmacology and Toxicology, New Jersey Medical School, University of Medicine and Dentistry of New Jersey, 185 South Orange Avenue, Newark, NJ 07103-2714

Slater, Jill P., McLaren Environmental Engineering, 11101 White Rock Road, Rancho Cordova, CA 95670

Stainken, Dennis M., New Jersey Department of Environmental Protection, Office of Quality Assurance, CN027, Trenton, NJ 08625

Stanek, Edward J. III, Biostatistics and Epidemiology Program, School of Public Health, University of Massachusetts, Amherst, MA 01003

Stanley, Curtis C., Shell Oil Company, 910 Louisiana Street, Houston, TX 77002

Symington, Anna, Massachusetts Department of Environmental Protection, Western Region, State House West, 436 Dwight Street, Springfield, MA 01103

Tucker, William A., Environmental Engineering, Inc., P.O. Box 1703, Gainesville, FL 32602

Turkall, Rita M., Department of Pharmacology and Toxicology, New Jersey Medical School, and School of Health Related Professions, University of Medicine and Dentistry of New Jersey, 185 South Orange Avenue, Newark, NJ 07103-2714

Wilson, John T., U.S. Environmental Protection Agency, P.O. Box 1196, Ada, OK 74820

Winslow, Michael G., Environmental Science & Engineering, Inc., P.O. Box 1703, Gainesville, FL 32602

Wright, David, State of California Department of Health Services, Toxic Substances Control Division, Alternative Technology Section, 4000 P Street, Sacramento CA 94234

Ying, Anthony, TreaTek, Inc., 2801 Long Road, Grand Island, New York 14072

Index